ADVANCED CALCULUS

Avner Friedman
The Ohio State University

DOVER PUBLICATIONS, INC.
Mineola, New York

Copyright

Bibliographical Note

This Dover edition, first published in 2007, is an unabridged republication of the work published by Holt, Rinehart and Winston, Inc., New York, in 1971.

International Standard Book Number

ISBN-13: 978-0-486-45795-6
ISBN-10: 0-486-45795-8

www.doverpublications.com

PREFACE

This book is intended for students who have already completed a one year course in elementary calculus. These students are familiar with the powerful methods of calculus in solving a variety of problems arising in physics, engineering, and geometry. They are therefore sufficiently motivated towards a rigorous treatment of the subject.

Many of the ideas of advanced calculus are by no means simple or intuitive. (It took mathematicians two hundred years to develop them!) Furthermore, for most students this is the first substantial course in mathematics taught in a rigorous way. We have therefore attempted to keep this book on a comfortable level. For example, it has been our experience that the student does not feel at ease with calculus in two or more dimensions. We have therefore deferred the introduction of functions of several variables as long as possible. We feel that further clarity and simplicity is added by avoiding mixture of heuristic and rigorous arguments.

The book contains material for more than a year's course; some choice is thus offered. Our own inclination is to omit parts of several sections throughout Chapters 1–9, and present some of the topics of Chapter 10. Students may be assigned to do independent study based on some sections of Chapter 10; the sections in this chapter are independent of each other.

When we refer to a particular section, say to Theorem 1 of Section 1, it is understood that the reference is to Section 1 of the same chapter. When we wish to refer to Section 1 of a different chapter, say Chapter 2, then we write Section 2.1.

I would like to thank Andrew J. Callegari and Joseph E. D'Atri for reading the manuscript and making some useful comments, and Zeev Schuss for verifying the answers of many of the problems.

Tel Aviv, Israel   A.F.
May 1971

CONTENTS

part one

FUNCTIONS OF ONE VARIABLE

1

NUMBERS AND SEQUENCES

1. THE NATURAL NUMBERS

The positive integers 1, 2, 3, $\cdots$ are called *natural numbers*. Since we intend to do things rigorously, we cannot be satisfied with our everyday familiarity with these numbers, and we should try to axiomatize their properties. Let us first write down five statements concerning the natural numbers that we feel should be true:

(I)   1 is a natural number.

(II)   To every natural number n there is associated in a unique way another natural number n', called the *successor* of n.

(III)   1 is not a successor of any natural number.

(IV)   If two natural numbers have the same successor, then they are equal.

(V)   Let M be a subset of the natural numbers such that: (i) 1 is in M, and (ii) if a natural number is in M, then its successor also is in M. Then M coincides with the set of all the natural numbers.

From now on we consider the statements (I)–(V) to be axioms. They are called the *Peano axioms*. The natural numbers will be the objects occurring in the Peano axioms. Axiom (V) is called the *principle of mathematical induction*.

We denote the successor of 1 by 2, the successor of 2 by 3, and so on. Note that $2 \neq 1$. Indeed, if $2 = 1$, then 1 is the successor of 1, thus contradicting (III). Note next that $3 \neq 2$. Indeed, if $3 = 2$ then, by (IV), $2 = 1$, which is false. In general, one can show that all the numbers obtained by taking the successors of 1 any number of times are all different. The proof of this statement, which we shall not give here, is based on induction, that is, on Axiom (V).

We would like to state Axiom (V) in a form more suitable for application:

(V′) Let $P(n)$ be a property regarding the natural number n, for any n. Suppose that (i) $P(1)$ is true, and (ii) if $P(n)$ is true, $P(n')$ also is true. Then $P(n)$ is true for all n.

If we define M to be the set of all natural numbers for which $P(n)$ is true, then (V′) follows from (V). If, on the other hand, we define $P(n)$ to be the property that n belongs to M, then (V) follows from (V′). Thus (V) and (V′) are equivalent axioms.

The Peano axioms give us objects with which to work. We now proceed to define operations on these objects. There are two operations that we consider: addition $(+)$ and multiplication $(\cdot)$. To any given pair of natural numbers each of these operations corresponds another natural number. The precise definition of this correspondence is given in the following theorem.

THEOREM 1. *There exist unique operations "$+$" and "$\cdot$" with the following properties:*

$$n + 1 = n', \qquad n + m' = (n + m)', \tag{1}$$

$$n \cdot 1 = n, \qquad n \cdot m' = n \cdot m + n. \tag{2}$$

The proof will not be given here. We shall often write mn instead of $m \cdot n$.

THEOREM 2. *The following properties are true for all natural numbers* m, n, k:

$$m + n = n + m, \qquad mn = nm \quad \text{(the commutative laws)}, \tag{3}$$

$$(m + n) + k = m + (n + k), \qquad (mn)k = m(nk) \quad \text{(the associative laws)}, \tag{4}$$

$$m(n + k) = mn + mk, \qquad (n + k)m = nm + km \quad \text{(the distributive laws)}. \tag{5}$$

The proof of Theorem 2 can be given by induction; it is based on the properties (1) and (2).

We state, without proof, another theorem, known as the *trichotomy law:*

THEOREM 3. *Given any natural numbers m and n, one and only one of the following possibilities occurs:*

(i) $m = n$.
(ii) $m = n + x$ *for some natural number x.*
(iii) $n = m + y$ *for some natural number y.*

If (ii) holds, we write $m > n$ or $n < m$, and we say that m is *larger* or *greater* than n and that n is *smaller* or *less* than m. If either (i) or (ii) holds, we write $m \geq n$ or $n \leq m$, and say that m is *larger or equal* to n and that n is *less than or equal to m.*

PROBLEMS

1. If $n > m$, then $n + k > m + k$.

2. If $n > m$, $m > k$, then $n > k$.

3. If $n + k \geq m + k$, then $n \geq m$.

4. $(n + 1)^2 = n^2 + 2n + 1$, where $n^2 = n \cdot n$.

5. $(n + 1)^3 = n^3 + 3n^2 + 3n + 1$, where $n^3 = n^2 \cdot n$.

6. Prove the associative law $(m + n) + k = m + (n + k)$, by induction on k.

7. Prove the distributive law $(m + n)k = mk + nk$, by induction on n.

8. If $m > n$, then $mk > nk$. Conversely, if $mk > nk$, then $m > n$.

2. THE RATIONAL NUMBERS

To every natural number n we correspond a new symbol $(-n)$, or $-n$, called *minus n*. We also introduce the symbol 0 called *zero*. Next we define the operation of addition $(-n) + m$ as follows:

(i)   If $n = m$, then $(-n) + m = 0$.
(ii) If $n = m + x$, x a natural number, then $(-n) + m = -x$.
(iii) If $m = n + y$, y a natural number, then $(-n) + m = y$.

Note, by the trichotomy law, that one and only one of Cases (i), (ii), or (iii) occurs.

Next we define: $m + (-n) = (-n) + m$, $(-n) + (-m) = -(n + m)$, $0 + m = m + 0 = m$, $0 + (-n) = (-n) + 0 = -n$, and $0 + 0 = 0$.
Multiplication is defined as follows:

$$m \cdot (-n) = (-n) \cdot m = -(nm), \qquad (-m) \cdot (-n) = mn,$$

$$0 \cdot (-n) = (-n) \cdot 0 = 0 \cdot m = m \cdot 0 = 0 \cdot 0 = 0.$$

The symbols $-n$ are called the *negative integers*. The *integers* consist of the natural numbers (also called the *positive integers*), the negative integers, and 0. We state without proof:

THEOREM 1. *For any integers m, n, k, Rules (3), (4), and (5) of Section 1 are true.*

If a is a negative integer $-n$, then we define $-a$ as n. We also define -0 to be 0. Then, for any integer a, $a + (-a) = (-a) + a = 0$.

Given any two integers a, b, there is a solution x of the equation

$$a + x = b, \qquad x \text{ integer.} \tag{1}$$

Indeed, $x = (-a) + b$ is such a solution, since, by Theorem 1, $a + [(-a) + b] = [a + (-a)] + b = 0 + b = b$. If y is another solution, then $a + x = a + y$. Hence $(-a) + (a + x) = (-a) + (a + y)$. Using Theorem 1, we get $x = y$. Thus Equation (1) has a unique solution.

Notation. We write $b + (-a) = b - a = -a + b$.

If the solution x of (1) is a positive integer, we write $b > a$ or $a < b$, and we say that b is *larger* (or *greater*) than a and that a is *smaller* (or *less*) than b. If x is negative, then the equation $b + y = a$ has the positive solution $y = -x$. Hence $a > b$.

We now shall introduce *fractions*. These are symbols that we write in the form a/b or $\dfrac{a}{b}$, where a and b are any integers, and $b \neq 0$. These symbols are subject to the following definitions:

$$\frac{a}{b} = \frac{c}{d} \text{ if and only if } ad = bc \qquad (equality), \tag{2}$$

$$\frac{a}{b} + \frac{c}{d} = \frac{ad + bc}{bd} \qquad (addition), \tag{3}$$

$$\frac{a}{b} \cdot \frac{c}{d} = \frac{ac}{bd} \qquad (multiplication). \tag{4}$$

Note that if $a/b = c/d$ and $c/d = e/f$, then $a/b = e/f$.

The last two definitions are acceptable only if we can show that $b \neq 0$, $d \neq 0$ imply that $bd \neq 0$. This, however, can be checked by considering the four possibilities: b positive or negative, d positive or negative.

Definitions (3) and (4) would be most unnatural if it turned out that it is possible to have $a/b = a'/b'$, $c/d = c'/d'$ but $a/b + c/d$ is not equal to $a'/b' + c'/d'$ [or $(a/b) \cdot (c/d)$ is not equal to $(a'/b') \cdot (c'/d')$]. The following theorem shows that this cannot occur.

THEOREM 2.   If $a'/b' = a/b$ and $c'/d' = c/d$, then

$$\frac{a'}{b'} + \frac{c'}{d'} = \frac{a}{b} + \frac{c}{d}, \tag{5}$$

$$\frac{a'}{b'} \cdot \frac{c'}{d'} = \frac{a}{b} \cdot \frac{c}{d}. \tag{6}$$

Proof.   To prove (5) we have to show that

$$\frac{a'd' + b'c'}{b'd'} = \frac{ad + bc}{bd}$$

or

$$a'd'bd + b'c'bd = adb'd' + bcb'd'.$$

But this follows by multiplying the relation $a'b = ab'$ by dd', the relation $c'd = cd'$ by bb', and adding the resulting equalities. To prove (6) we have to show that

$$\frac{a'c'}{b'd'} = \frac{ac}{bd},$$

or

$$a'c'bd = acb'd'.$$

But this follows from

$$a'c'bd = (a'b)(c'd) = (ab')(cd') = acb'd'.$$

A fraction a/b is called *negative* if either $a > 0$, $b < 0$ or $a < 0$, $b > 0$. It is called *positive* if either $a > 0$, $b > 0$ or $a < 0$, $b < 0$. It is called *zero* if $a = 0$. It is easily seen that if a fraction c/d is equal to a fraction a/b, then they are either both positive, or both negative, or both zero.

THEOREM 3. *The following properties hold for any fractions a/b, c/d, e/f:*

$$\frac{a}{b} + \frac{c}{d} = \frac{c}{d} + \frac{a}{b}, \frac{a}{b} \cdot \frac{c}{d} = \frac{c}{d} \cdot \frac{a}{b} \quad (\text{the commutative laws}), \tag{7}$$

$$\left(\frac{a}{b} + \frac{c}{d}\right) + \frac{e}{f} = \frac{a}{b} + \left(\frac{c}{d} + \frac{e}{f}\right)$$

$$\qquad\qquad\qquad (\text{the associative laws}), \tag{8}$$

$$\left(\frac{a}{b} \cdot \frac{c}{d}\right) \cdot \frac{e}{f} = \frac{a}{b} \cdot \left(\frac{c}{d} \cdot \frac{e}{f}\right)$$

$$\frac{a}{b} \cdot \left(\frac{c}{d} + \frac{e}{f}\right) = \frac{a}{b} \cdot \frac{c}{d} + \frac{a}{b} \cdot \frac{e}{f}$$

$$\qquad\qquad\qquad (\text{the distributive laws}). \tag{9}$$

$$\left(\frac{c}{d} + \frac{e}{f}\right) \cdot \frac{a}{b} = \frac{c}{d} \cdot \frac{a}{b} + \frac{e}{f} \cdot \frac{a}{b}$$

Consider the equation

$$\frac{a}{b} + x = \frac{c}{d}, \qquad x \text{ fraction.} \tag{10}$$

It has a solution $x = (bc - ad)/bd$. If x is positive, then we write $a/b < c/d$ or $c/d > a/b$, and say that c/d is *larger* than a/b and that a/b is *less* than c/d. If x is negative, then the equation

$$\frac{c}{d} + y = \frac{a}{b}$$

has the positive solution $y = -x$, so that $a/b > c/d$. Note that c/d is positive (negative) if it is larger (smaller) than zero.

The definition of fractions is very intuitive and is, in fact, suggested by our experience with quotients of integers. There is, however, one disturbing feeling about the concept of fractions, due to the fact that fractions having *different* forms may be *equal* to each other. This makes it impossible to speak of *the* zero fraction (since there are many fractions $0/b$ taking the role of zero). We also cannot assert that Equation (10) has a *unique* solution. Similarly, the equation

$$\frac{a}{b} \cdot x = \frac{c}{d}, \qquad x \text{ fraction} \qquad (\text{where } a \neq 0) \tag{11}$$

does not have a *unique* solution.

To overcome this unpleasant situation, we introduce the concept of a rational number.

DEFINITION.   A *rational number* (a, b) (where a and b are integers, and $b \neq 0$) is the class of all the fractions e/f that are equal to a/b.

Addition and multiplication for rational numbers are given by

$$(a, b) + (c, d) = (ab + cd, bd), \tag{12}$$

$$(a, b) \cdot (c, d) = (ac, bd). \tag{13}$$

Theorem 2 shows that these definitions are meaningful. Theorem 3 implies:

THEOREM 4.   *Rational numbers satisfy the commutative laws for addition and multiplication, the associative laws for addition and multiplication, and the distributive laws.*

Let us write the analog of Equation (11) for rational numbers:

$$(a, b) \cdot x = (c, d), \qquad x \text{ rational (where } a \neq 0). \tag{14}$$

This equation has a unique solution $x = (bc, ad)$. We write this solution also in the form $(c, d)/(a, b)$ or $(c, d)(a, b)^{-1}$.

Let us write the analog of (10) for rational numbers:

$$(a, b) + x = (c, d), \qquad x \text{ rational.} \tag{15}$$

This equation also has a unique solution: $x = (bc - ad, bd)$. We write it also as $(c, d) - (a, b)$.

Note that there is a one-to-one correspondence between the integers a and the rational numbers $(a, 1)$. This correspondence $a \to (a, 1)$ is preserved under addition and multiplication. Indeed, this follows from the relations

$$(a, 1) + (b, 1) = (a + b, 1), \qquad (a, 1) \cdot (b, 1) = (ab, 1).$$

Hence, if we write an integer a in the form $(a, 1)$, we see that the integers can be identified with a subset of the rational numbers.

In what follows we shall adopt the definition of rational numbers as classes of fractions a/b. However, for brevity, we shall write the rational numbers (a, b) usually in the form a/b. When we write $a/b = c/d$, we mean that $(a, b) = (c, d)$, that is, $ad = bc$. The rational numbers $b/1$ will also be written, briefly, as b. In particular, the rational number zero will be denoted by 0.

PROBLEMS

1. Prove by induction on n that:
   (a) $1 + 2 + \cdots + n = n(n + 1)/2$.
   (b) $1 + 2^2 + \cdots + n^2 = n(n + 1)(2n + 1)/6$.
   (c) $1 + \alpha + \alpha^2 + \cdots + \alpha^n = (\alpha^{n+1} - 1)/(\alpha - 1)$ where α is rational, $\alpha \neq 1$.

2. If a, b are integers and $a \cdot b = 0$, then either $a = 0$ or $b = 0$.

3. If x and y are rational numbers and $x \neq 0$, $y \neq 0$, then $x \cdot y \neq 0$.

4. Prove Theorem 3, using Theorem 1.

5. Let r, s, t be rational numbers. If $r > s$, $s > t$, then $r > t$.

6. Let r, s, t be rational numbers. If $r > s$ and $t > 0$, then $rt > st$. If $r > s$ and $t < 0$, then $rt < st$.

3. THE REAL NUMBERS

As we already know, if a and b are rational numbers, then the equation $a + x = b$ has a unique rational solution. Similarly, the equation $ax = b$ has a unique rational solution, provided $a \neq 0$. However, quadratic equations

$$ax^2 + bx + c = 0$$

with rational a, b, c may not have rational solutions. Such an example is given in the following theorem.

THEOREM 1.   *The equation $x^2 = 2$ has no rational solution.*

Proof.   Suppose the assertion is false, that is, suppose there is a rational number x such that $x^2 = 2$. If $x > 0$, then we can write $x = m/n$ where m, n are natural numbers without a common factor. If $x < 0$, then we can write $x = -m/n$ with m, n natural numbers without a common factor. In both cases we get $2 = x^2 = m^2/n^2$. Hence

$$m^2 = 2n^2. \tag{1}$$

If m is odd then m^2 also is odd. Since the right-hand side of (1) is even, we would get a contradiction. Hence m is even, that is, $m = 2k$ where k is a natural number. Substituting this into (1) we find that $2n^2 = 4k^2$, or $n^2 = 2k^2$. We now can argue as before and conclude that n is an even number. Consequently, m and n have a common factor 2, which is a contradiction.

We shall extend the concept of a number and define real numbers. It will turn out that the equation $x^2 = 2$ (and many other quadratic equations) will have a solution that is a real number.

Notation. If a is one of the elements of a set A, then we write: $a \in A$; otherwise, we write: $a \notin A$.

Consider two sets A and B of rational numbers, having the following properties:

(I)   Every rational number belongs either to A or to B.
(II)   Each of the sets A, B is nonempty.
(III)   Each number in A is less than each number in B.

We then call the pair (A, B) a *Dedekind cut*, or, briefly, a *cut*.
There are four possibilities:

(i)   The set A contains a largest rational number α, that is, $\alpha \geq a$ for all $a \in A$. The set B does not contain a smallest rational number (that is, a number that is less than or equal to every number in B).

(ii)   The set B contains a smallest rational β, but the set A does not contain a largest rational number.

(iii)   The set A does not contain a largest rational number, and the set B does not contain a smallest rational number.

(iv)   The set A contains a largest rational α, and the set B contains a smallest rational β.

We can immediately exclude the last possibility. Indeed, if (iv) holds, then the number $(\alpha + \beta)/2$ is a rational number that does not belong either to A or to B; this contradicts (I).

We would like to exclude in the future all cuts (A, B) for which B has a smallest rational β. The reason is this. If we form a new cut $(\bar{A}, \bar{B})$ where $\bar{A}$ consists of β and of all the numbers in A, then the new cut will be "equivalent" to (A, B) in a way similar to that of a fraction a/b being "equivalent" to a fraction ac/bc. For the sake of convenient reference we state:

DEFINITION.   A *normalized cut* is a cut for which either (i) or (iii) holds. If a cut $(\bar{A}, B)$ satisfies (ii), then a cut (C, D) is called a *normalization* of (A, B) if C consists of the smallest rational number of B and of all the rationals in A.

We can now define the concept of a real number.

DEFINITION.   A normalized cut (A, B) is called a *real number*. If (i) holds, then we say that the real number (or the cut) is *rational*, and if (iii) holds, then we say that the real number (or the cut) is *irrational*.

We proceed to define addition and multiplication for real numbers.

DEFINITION OF ADDITION.   Let (A, B) and (C, D) be real numbers. Let $\bar{E}$ consist of all the rational numbers $a + c$ where $a \in A$, $c \in C$. Let $\bar{F}$ consist of all the rational numbers that do not belong to $\bar{E}$. Let (E, F) be a normalization of the cut $(\bar{E}, \bar{F})$. The real number (E, F) is called the *sum* of (A, B) and (C, D). We write

$$(E, F) = (A, B) + (C, D).$$

DEFINITION OF MULTIPLICATION.   Let (A, B) and (C, D) be real numbers. Consider four cases:

(a) A and C contain nonnegative rationals.
(b) A contains nonnegative rationals, but C does not contain nonnegative rationals.
(c) C contains nonnegative rationals, but A does not contain nonnegative rationals.
(d)   Neither A nor C contains nonnegative rationals.

We define a cut $(\bar{E}, \bar{F})$ as follows: If (a) holds, then $\bar{E}$ consists of all the rationals that are less than or equal to the numbers ac where $a \in A$, $c \in C$ and $a \geq 0$, $c \geq 0$. If (b) holds, then $\bar{E}$ consists of all numbers that are less than or equal to ac where $a \in A$, $c \in C$, and $a \geq 0$. If (c) holds, then $\bar{E}$ consists of all numbers that are less than or equal to ac where $a \in A$, $c \in C$, and $c \geq 0$. Finally, if (d) holds, then $\bar{F}$ consists of all the numbers that are larger than or equal to bd where $b \in B$, $d \in D$ and $b \leq 0$, $d \leq 0$. Denote by (E, F) the normalization of the cut $(\bar{E}, \bar{F})$. The real number (E, F) is called the *multiplication* or the *product* of (A, B) by (C, D). We write

$$(E, F) = (A, B) \cdot (C, D) \qquad \text{or} \qquad (E, F) = (A, B)(C, D).$$

We can correspond to every rational number α a real rational number (A, B), where A consists of all the rational numbers a that are $\leq \alpha$. This correspondence $\alpha \rightarrow (A, B)$ is preserved by addition and multiplication, that is, if

$$\alpha \rightarrow (A, B), \qquad \bar{\alpha} \rightarrow (\bar{A}, \bar{B}), \tag{2}$$

then

$$\alpha + \bar{\alpha} \rightarrow (A, B) + (\bar{A}, \bar{B}), \qquad \alpha\bar{\alpha} \rightarrow (A, B)(\bar{A}, \bar{B}). \tag{3}$$

It follows that the real rational numbers are the same, except for notation, as the rational numbers. From now on we shall denote real numbers also by Latin letters. When we speak of a rational number we may either mean a rational number as defined in Section 2 or a real rational number as defined in this section.

The student is certainly familiar with the intuitive concept of a real irrational number as an expanded decimal fraction

$$r = r_0 \cdot r_1 r_2 r_3 \cdots. \tag{4}$$

How does this concept relate to the above definition of a real number (A, B)? The answer is very simple. Take A to consist of all rational numbers $a = a_0 \cdot a_1 a_2 a_3 \cdots$ such that either $a_0 < r_0$ or $a_0 = r_0, \cdots, a_k = r_k, a_{k+1} < r_{k+1}$ for some k. In this way we correspond to every irrational r an irrational cut. This correspondence is preserved by addition and multiplication.

We state without proof:

THEOREM 2. *The following properties are true for all real numbers* r, s, t:

$$r + s = s + r, rs = sr \qquad (\text{the commutative laws}), \tag{5}$$

$$(r + s) + t = r + (s + t), (rs)t = r(st) \qquad (\text{the associative laws}), \tag{6}$$

$$r(s + t) = rs + rt, (s + t)r = sr + tr \qquad (\text{the distributive laws}). \tag{7}$$

A real number (A, B) is called *positive* if there exists a positive rational number a in A. It is called *negative* if it is neither positive nor zero. A number (A, B) is negative if, and only if, there exists a negative rational number in B.

More generally, we write $(C, D) > (A, B)$ if there exist numbers $\bar{c} \in C$ and $\bar{b} \in B$ such that $\bar{c} > \bar{b}$. We then also write $(A, B) < (C, D)$, and say that (C, D) is *larger* than (A, B) and that (A, B) is *smaller* than (C, D).

We write $(C, D) \geq (A, B)$ if either $(C, D) > (A, B)$ or $(C, D) = (A, B)$. It is easily seen that $(C, D) \geq (A, B)$ if, and only if, $d \geq a$ for all $d \in D, a \in A$.

We state without proof:

THEOREM 3. *Let r be a real number (A, B). Then $r \geq a$ for all $a \in A$, and $r < b$ for all $b \in B$. If r is irrational, then $r > a$ for all $a \in A$.*

Note that $r \geq a$ means: if (C, D) is the rational cut of a, then $b \geq c$ for any b in B and c in C.

THEOREM 4. *Let s and (A, B) be real numbers. If $s > (A, B)$, then there exists a number $\bar{b} \in B$ such that $s > \bar{b}$. If $s < (A, B)$, then there exists a number $\bar{a}$ in A such that $s < \bar{a}$.*

One can prove that there exists a unique real number x satisfying any equation of the form

$$(A, B) + x = (C, D).$$

We then write $x = (C, D) - (A, B)$. Similarly, it can be proved that if (A, B) $\neq 0$, then there exists a unique real solution x of the equation

$$(A, B)x = (C, D).$$

We denote it by $(C, D)/(A, B)$ or $(C, D)(A, B)^{-1}$.

The following theorem is called the *Archimedean law*.

THEOREM 5. *Let a and b be two positive real numbers. Then there exists a positive integer n such that $na > b$.*

Proof. Denote by A the set of all the rational numbers x satisfying $x < na$ for some positive integer n. Denote by B the set of all rational numbers not in A. Thus, $y \in B$ if, and only if, $y \geq na$ for all positive integers n. If the assertion of the theorem is false, then the set B is nonempty. A is also nonempty since $a < 2a$, so that $a \in A$. We claim that (A, B) is a cut. To prove it, it remains to show that if $x \in A$, $y \in B$, then $x < y$. But, clearly, for some positive integer n, $x < na \leq y$.

Denote by r the real number of the cut (A, B). For any positive integer m, $(m + 1)a$ is in A since $(m + 1)a < na$ if n is the positive integer $m + 2$. Hence, by Theorem 3, $(m + 1)a \leq r$, or $ma \leq r - a \leq \beta - a$ for any $\beta \in B$. Since m is arbitrary, $\beta - a$ must belong to B, so that $\alpha \leq \beta - a$ for any $\alpha \in A$, $\beta \in B$. Denote the cut of $r - a$ by (C, D). Then D consists of all the numbers $\beta - \alpha$, $\beta \in B$. From the last inequality and the statement preceding Theorem 3 we then deduce that $r - a \geq r$, a contradiction.

Applying Theorem 5 with $a = \varepsilon$, $b = 1$ we get:

COROLLARY. *For any positive real number ε there exists a positive integer n such that $1/n < \varepsilon$.*

THEOREM 6. *If α and β are positive rational numbers and $\alpha < \beta$, then there is a positive rational number x such that $\alpha < x^2 < \beta$.*

Proof. Suppose first that $\beta \leq 1$. By the corollary to Theorem 5, there is a positive integer n such that

$$\frac{1}{n} < \frac{1}{3}(\beta - \alpha). \tag{8}$$

Since $0 < \beta \leq 1$, there is a nonnegative integer k such that

$$\frac{k^2}{n^2} < \beta, \qquad \frac{(k+1)^2}{n^2} \geq \beta,$$

and $k < n$. We shall prove that $x = k/n$ satisfies the assertion of the theorem. Thus, what we have to show is that $k^2/n^2 > \alpha$. Since $(k+1)^2/n^2 \geq \beta$, we have

$$\beta - \frac{k^2}{n^2} \leq \frac{(k+1)^2}{n^2} - \frac{k^2}{n^2} = \frac{2k+1}{n^2} < \frac{3n}{n^2} = \frac{3}{n} < \beta - \alpha,$$

by (8). Hence $k^2/n^2 > \alpha$.

So far we have assumed that $\beta \leq 1$. If $\beta \geq 1$, then by what we have proved, there is a rational number y satisfying

$$\frac{\alpha}{\beta^2} < y^2 < \frac{\beta}{\beta^2} = \frac{1}{\beta}.$$

Now take $x = \beta y$.

We return to the equation $x^2 = 2$ and prove that it has a real solution.

THEOREM 7.   *The equation $x^2 = 2$ has a positive real irrational solution.*

Proof.   Define a cut (A, B) as follows: A consists of all the negative rationals and of all those positive rationals a satisfying $a^2 < 2$. Let $(E, F) = (A, B)(A, B)$. If $e \in E$ and e is not the largest rational in E, then either $e < 0$ or $e = a_1 a_2$ where $a_1^2 < 2$, $a_2^2 < 2$. We may assume that $a_2 \geq a_1 \geq 0$. Then $e \leq a_2 a_2 < 2$. Conversely, if $e < 2$ and e is rational, then, by Theorem 6, there is a positive rational x such that $e < x^2 < 2$. Since $x \in A$, e belongs to E. Since $x^2 \in E$ and $e < x^2$, e is not the largest number of E. We have thus proved that a rational number e belongs to E and is not the largest number of E, if, and only if, $e < 2$. This implies that the cut (E, F) coincides with the real number 2.

It remains to show that (A, B) is not a rational cut. If this is not true, then A has a largest rational number γ. From $(A, B)(A, B) = 2$ it follows [compare (2) and (3)] that $\gamma^2 = 2$. But this contradicts Theorem 2.

It is natural to ask: Suppose we form cuts (A, B) with the numbers in A and B being real numbers. Do we get a still more general concept of numbers than that of the real number? In other words, can we get a situation where A has no largest real number and B has no smallest real number? The answer is negative and is given in the following:

DEDEKIND'S THEOREM.   *If (A, B) is a cut of real numbers, then either A contains a largest real number or B contains a smallest real number.*

PROBLEMS

1. Prove Theorem 2 for positive real numbers.

2. Prove Theorem 3.

3. Prove Theorem 4.

4. The equation $x^2 = 3$ has no rational solutions.

5. The equation $x^2 = b$, b positive integer, has a real solution.

6. The equation $x^2 = r$, r a positive real number, has a real solution.

7. If a, b are rational numbers and $0 < a < b$, then there is a positive rational number x such that $a < x^3 < b$. [*Hint:* Choose x of the form k/n where $1/n < (b - a)/7$, and compare the proof of Theorem 6.]

8. The equation $x^3 = 2$ has a real solution.

9. If r and s are positive real numbers, then $r + s$ and rs also are positive real numbers.

10. If r,s,t are real numbers and if $r > s$, $s > t$, then $r > t$.

11. If r,s,t are real numbers and if $r > s$, then $r + t > s + t$.

12. Let r,s,t be real numbers and let $r > s$. Prove that $rt > st$ if $t > 0$, and $rt < st$ if $t < 0$.

13. Prove that there cannot exist two different positive real numbers x satisfying $x^2 = r$, where r is a given positive real number. (We denote the unique positive solution of $x^2 = r$ by $r^{1/2}$, or $\sqrt{r}$.)

14. Let r be a rational number and let s be an irrational number. Then $r + s$ is irrational.

15. Let r,s be real numbers and $r < s$. Then there is a rational number t such that $r < t < s$.

16. Let a,b be real numbers and $a < b$. Then there is an irrational number x such that $a < x < b$. [*Hint:* Take $x = m/n\sqrt{2}$, m integer, n positive integer.]

17. Prove Dedekind's theorem as follows: Define a cut $(\bar{C}, \bar{D})$ of rational numbers: $c \in \bar{C}$ if, and only if, $c \le \bar{a}$ for some $\bar{a}$ in A. Let (C, D) be the normalization of $(\bar{C}, \bar{D})$. Prove that (C, D) is either the largest real number in A or the smallest real number in B.

4. BOUNDED SETS

From now on, whenever we refer to a number without further specifications, we always mean a real number.

In order to have a better intuitive conception of the real numbers, we introduce a geometric language for them. We call a real number a *point* and the set of all real numbers the *real line*. The point 0 is called the *origin* of the real line. Given any points a,b, with $a < b$, we call the set of points x satisfying $a < x < b$ the *open interval* with *endpoints* a and b. The point a is also called the *initial point* or the *left endpoint* and point b is also called the *terminal point* or the *right endpoint*. We denote this open interval by (a,b), or by $\{x; a < x < b\}$, or by $a < x < b$. Similarly, we define the *closed interval* $[a,b]$ ($\{x; a \leq x \leq b\}$, or $a \leq x \leq b$) as the set of all points satisfying $a \leq x \leq b$. The *half-open intervals* $[a,b)$ and $(a,b]$ are defined by $a \leq x < b$ and $a < x \leq b$, respectively. If $a > c$, then we say that point a lies *to the right* of (or *above*) c and the point c lies *to the left* of (or *below*) a.

In order to use the above geometric language more effectively, we draw a straight line and call it the *real line*. We mark on it points that are to signify the real numbers. Since the student is surely familiar with the manner by which we mark these points, in Figure 1.1 we shall only give the picture of the real line with the specific designation of a few real numbers (or points). Notice that our definitions of intervals, a to the right of b, and so on, fit very well with what we see on the real line. Thus, for instance, if the real number a is to the right of c, then in Figure 1.1 the point designating a will appear further to the right than the point designating c. It actually can be proved that there is a one-to-one correspondence (which fits in with Figure 1.1) of the real numbers and of the points of a line, whereby by "the points" we mean those determined by the usual geometric axioms.

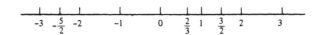

Figure 1.1

Let I be a given set on the real line. Let f be a rule that corresponds to every point x in I a real number $f(x)$. Then we call f a *function* and I the *domain* of f. When we speak of *a function* $f(x)$ *defined on a set* I we mean that f is a function with domain I. To get a better idea of the nature of a given function, we often resort to a picture. We introduce a coordinate system (x,y) in the plane. The x-axis is given the role of the real line as before. The y-axis is also given the role of the real line, but now the line is drawn vertically, with $a > b$ if a lies above b. We take the domain I of f to be a subset of the x-axis, and the points $f(x)$ to belong to the y-axis. We then draw points $(x,f(x))$ in the plane, and thus get a picture of the function. The set of all points $(x,f(x))$, when x varies in I, is called the *graph* of f.

An important function is the function $f(x) = |x|$ defined by

$$|x| = \begin{cases} x & \text{if } x \geq 0, \\ -x & \text{if } x < 0. \end{cases}$$

It is called the *absolute value function*. The number $|x|$ is called the *absolute value* of x.

THEOREM 1.   *The following relations hold:*

$$|x + y| \leq |x| + |y|, \tag{1}$$

$$|xy| = |x|\,|y|. \tag{2}$$

The proof is obtained by treating separately for cases: $x \geq 0$, $y \geq 0$; $x \geq 0$, $y < 0$; $x < 0$, $y \geq 0$; and $x < 0$, $y < 0$.

A set G of real numbers is said to be *bounded above* if there is a number K such that $x \leq K$ for all $x \in G$. Such a number K is called an *upper bound* of G. The set G is *bounded below* if there is a number L such that $x \geq L$ for all $x \in G$. Such a number L is called a *lower bound* of G. If G is bounded above and below, then we say that G is *bounded*. This is clearly the case if, and only if, there is a positive number M such that $|x| \leq M$ for all $x \in G$. Such a number M is called a *bound* of G. A set G is said to be *unbounded* if it is not bounded.

Let G be a set bounded above. A number M is called the *least upper bound* (*l.u.b.*) or the *supremum* (*sup*) of G if the following two conditions hold:

$$x \leq M \qquad \text{for all } x \in G. \tag{3}$$

For any $\varepsilon > 0$, there is an $\bar{x}$ in G satisfying $\bar{x} > M - \varepsilon$. (4)

Condition (3) means that M is an upper bound of G. Condition (4) means that for any $\varepsilon > 0$, $M - \varepsilon$ is not an upper bound of G. Since any number smaller than M has the form $M - \varepsilon$ for some $\varepsilon > 0$, there cannot be an upper bound smaller than M, that is, M is the smallest upper bound. We write the supremum M in the form

$$M = \sup_{x \in G} x,$$

or

$$M = \text{l.u.b.}_{x \in G} x.$$

THEOREM 2.   *Every nonempty set G bounded above has a supremum.*

Proof.   Denote by B the set of all the rational numbers that are upper bounds of G. Denote by A the set of all rational numbers not in B. We shall prove that (A,B) is a cut, that is, that the conditions (I)–(III) of Section 3 are satisfied. (I) holds by the definition of A. As for (II), B is nonempty since G is bounded above. Next, since G is nonempty, there is a point x in G. But then any rational number less than x cannot be an upper bound of G. Hence it belongs to A. This proves that also A is nonempty. To prove (III) we must show that $a < b$ for any $a \in A$, $b \in B$. If this is false, then there exist points $\bar{a} \in A$, $\bar{b} \in B$ such that $\bar{a} \geq \bar{b}$. Since $\bar{b}$ is an upper bound, $\bar{a} \geq \bar{b} \geq x$ for all $x \in G$. Hence $\bar{a}$ is also an upper bound and therefore it must belong to B. This is impossible, however, since $\bar{a} \in A$.

Having proved that (A,B) is a Dedekind cut, we designate by M the real number associated with it. We shall prove that M satisfies (3) and (4). Suppose (3) is false. Then there exists a point x in G such that $x > M$. Then, by Theorem 4 of Section 3, there is a number $\bar{b} \in B$ such that $x > \bar{b}$. Since $\bar{b}$ is an upper bound of G, $\bar{b} \geq x$, which is impossible.

Next we prove that (4) holds. Suppose that (4) is false. Then there exists an $\varepsilon > 0$ such that $x \leq M - \varepsilon$ for all $x \in G$. Hence $M - \varepsilon$ is an upper bound. Let N be a rational number satisfying: $M - \varepsilon < N < M$. Then N is also an upper bound of G. It therefore belongs to B. But, by Theorem 3 of Section 3, $M \leq b$ for all $b \in B$; in particular, $M \leq N$. This is impossible.

Let G be a set bounded below. A number m is called the *greatest lower bound* (*g.l.b.*) or the *infimum* (*inf*) of G if the following two conditions hold:

$$x \geq m \qquad \text{for all } x \in G. \tag{5}$$

$$\text{For any } \varepsilon > 0, \text{ there is an } \bar{x} \text{ in } G \text{ satisfying } \bar{x} < m + \varepsilon. \tag{6}$$

Condition (5) means that m is a lower bound of G, and Condition (6) means that m is the largest of all the lower bounds.

We write the infimum in the form

$$m = \inf_{x \in G} x,$$

or

$$m = \underset{x \in G}{\text{g.l.b.}} x.$$

Analogously to Theorem 2 we have:

THEOREM 3.   *Every nonempty set G bounded below has an infimum.*
Suppose we are given a rule that assigns to each positive integer n an object a_n. Then we write this rule explicitly in the form

$$\{a_1, a_2, \ldots, a_n \ldots\}$$

or, more briefly, in the form $\{a_n\}$. Instead of saying: the rule $\{a_n\}$, we shall say: the *sequence* $\{a_n\}$.

Let G be a set of objects. Suppose we can find a one-to-one correspondence between the objects of G and a subset of the positive integers Then we say that G is *countable*, or *denumerable*. When G is countable as well as infinite (that is, for any positive integer m, G contains more than m objects), we can form a sequence $\{g_n\}$ where g_n varies over the elements of G as n varies over the natural numbers. (The proof is omitted.)

We mention without proof several facts. (a) The rational numbers form a countable set. (b) The real numbers do not form a countable set. (c) Let H be a countable set of objects, say G_α, where each G_α is a countable set of objects, say $g_{\alpha\beta}$. Then the set of all the objects $g_{\alpha\beta}$ (called the *union* of the sets G_α) is again a countable set.

PROBLEMS

1. A set G of real numbers is bounded if, and only if, there exists an interval (a,b) such that G is contained in it.

2. Prove Theorem 1.

3. Prove that $|x| - |y| \le |x - y|$.

4. Prove by induction that

    $$|x_1 + x_2 + \cdots + x_n| \le |x_1| + |x_2| + \cdots + |x_n|.$$

5. If $y \ne 0$, then $|x/y| = |x|/|y|$.

6. Prove, by induction, Bernoulli's inequality:

    $$(1 + x)^n \ge 1 + nx \qquad (n \text{ positive integer}, x \ge -1).$$

7. For any positive integers n,k with $n > k$, we define

    $$\binom{n}{k} = \frac{n(n-1)\cdots(n-k+1)}{k!}.$$

    For any number a we define $a^1 = a$, $a^{n+1} = a^n a$. Prove, by induction, the *binomial theorem*:

$$(1 + x)^n = 1 + nx + \binom{n}{2}x^2 + \cdots + \binom{n}{k}x^k + \cdots + nx^{n-1} + x^n,$$

where n is a positive integer and x a real number.

8. Find the infimum and supremum of the following sequences:

(a) $\{(-1^n)\}$; (b) $\left\{\dfrac{1}{n}\right\}$; (c) $\left\{\dfrac{n}{n+1}\right\}$;

(d) $\left\{\dfrac{(-1)^n}{n}\right\}$; (e) $\left\{\dfrac{1}{n^2+1}\right\}$; (f) $\left\{2 + \dfrac{1}{n!}\right\}$.

5. LIMITS OF SEQUENCES

Let $\{a_n\}$ be a sequence of real numbers. Suppose there exists a number L having the following property:

For any $\varepsilon > 0$, all the numbers a_n, with the possible exception of a finite number of them, lie in the interval $(L - \varepsilon, L + \varepsilon)$. In other words, for any $\varepsilon > 0$ there exists a positive integer n_0 such that

$$|a_n - L| < \varepsilon \qquad \text{if } n \geq n_0. \tag{1}$$

Then we say that the sequence $\{a_n\}$ *converges* to L (or is *convergent* to L, or *has limit* L), and we call L the *limit* of $\{a_n\}$. We express this property in any one of the following notations:

$$\lim_{n \to \infty} a_n = L, \qquad \lim_{n} a_n = L, \qquad a_n \to L \qquad \text{if } n \to \infty.$$

If $\{a_n\}$ is not convergent (to any number), then we say that it *diverges*, or that it is *divergent*.

A sequence $\{b_n\}$ is called a *subsequence* of $\{a_n\}$ if

$$b_1 = a_{n_1}, \qquad b_2 = a_{n_2}, \ldots, \qquad b_k = a_{n_k}, \ldots$$

where $n_1 < n_2 < \cdots < n_k < \cdots$. We write the subsequence also in the form $\{a_{n_k}\}$.

The proofs of the following theorems are left to the student.

THEOREM 1. *If* $\lim a_n = L$, *then*

$$\lim_{k \to \infty} a_{n_k} = L, \tag{2}$$

for any subsequence $\{a_{n_k}\}$.

THEOREM 2.   *If* $\lim a_n = L$ *and* $\lim a_n = M$, *then* $L = M$.

DEFINITION.   Let $\{a_n\}$ be a sequence of real numbers. A number N is called a *limit point* (a *cluster point*, or a *point of accumulation*) of the sequence if there exists a subsequence $\{a_{n_k}\}$ such that $\lim_k a_{n_k} = N$.

Assume that $\{a_n\}$ is bounded above, say $a_n \leq C$ for all n. Then each limit point N is also $\leq H$. Indeed, let $\{a_{n_k}\}$ be a subsequence of $\{a_n\}$ converging to N. Then, for any $\varepsilon > 0$,

$$N - \varepsilon < a_{n_k} < N + \varepsilon$$

if k is sufficiently large. Hence $N \leq a_{n_k} + \varepsilon \leq C + \varepsilon$. We thus have proved that $N \leq C + \varepsilon$ for any $\varepsilon > 0$. This implies that $N \leq C$. Indeed, if $N > C$, then $N = C + \delta$ for some $\delta > 0$. But then the inequality $N \leq C + \varepsilon$ cannot be satisfied for $\varepsilon = \delta/2$.

We have proved that if a sequence $\{a_n\}$ is bounded above (by a number C), then the set of all its limit points is also bounded above (by C). Denote by M' the supremum of the set of all the limit points of the sequence $\{a_n\}$. We call M' the *limit superior* of $\{a_n\}$ and write it in any one of the forms:

$$M' = \limsup_{n \to \infty} a_n, \qquad M' = \limsup_n a_n,$$

$$M' = \overline{\lim_{n \to \infty}} a_n, \qquad M' = \overline{\lim_n} a_n.$$

THEOREM 3. M' *is a limit point of* $\{a_n\}$. *In other words, the limit superior is the largest limit point of* $\{a_n\}$.

Proof.   From the definition of M' it follows that for any $\varepsilon > 0$ there exists a limit point ξ_ε of $\{a_n\}$ such that

$$M' \geq \xi_\varepsilon > M' - \frac{\varepsilon}{2}.$$

We may assume that $\xi_\varepsilon < M'$, for if $\xi_\varepsilon = M'$, then the proof is already complete. From the definition of a limit point it follows that there exists a number $a_{n(\varepsilon)}$ such that

$$|a_{n(\varepsilon)} - \xi_\varepsilon| < \frac{\varepsilon}{2}, \ |a_{n(\varepsilon)} - \xi_\varepsilon| < M' - \xi_\varepsilon.$$

Hence, for any $\varepsilon > 0$ there is a number $a_{n(\varepsilon)}$ satisfying

$$M' - \varepsilon < a_{n(\varepsilon)} < M'.$$

Now take $\varepsilon = 1$ and denote the corresponding $a_{n(\varepsilon)}$ by a_{n_1}. Next take $\varepsilon = \varepsilon_2$ such that

$$\varepsilon_2 < \tfrac{1}{2}, \varepsilon_2 < M' - a_{n_1}.$$

Denote the corresponding point $a_{n(\varepsilon)}$ by a_{n_2}. Since

$$M' - a_{n_2} < \varepsilon_2 < M' - a_{n_1},$$

$a_{n_2} \neq a_{n_1}$. Next take $\varepsilon = \varepsilon_3$ such that

$$\varepsilon_3 < \tfrac{1}{3}, \qquad \varepsilon_3 < M' - a_{n_2},$$

and denote the corresponding point $a_{n(\varepsilon)}$ by a_{n_3}. Then

$$M' - a_{n_3} < \varepsilon_3 < M' - a_{n_2}.$$

It follows that $a_{n_3} \neq a_{n_2}$ and $a_{n_3} \neq a_{n_1}$. In fact, $a_{n_1} < a_{n_2} < a_{n_3} < M'$. Continuing in this way step by step, we construct a sequence $\{a_{n_k}\}$ of mutually distinct numbers, such that

$$0 < M' - a_{n_k} < \frac{1}{k}.$$

We shall prove that $\lim_k a_{n_k} = M'$. For any $\varepsilon > 0$, there exists a natural number k_0 such that $1/k_0 < \varepsilon$ (by the corollary to Theorem 5 of Section 3). Hence if $k \geq k_0$, then

$$|M' - a_{n_k}| < \frac{1}{k} \leq \frac{1}{k_0} < \varepsilon.$$

This completes the proof that $\lim a_{n_k} = M'$.

The sequence $\{a_{n_k}\}$ may not be a subsequence of $\{a_n\}$ since we have not proved that $n_1 < n_2 < \cdots$. If we can construct a subsequence $\{b_m\}$ of $\{a_{n_k}\}$ that is at the same time a subsequence of $\{a_n\}$, then, by Theorem 1, $\lim b_m = M'$, so that M' is a limit point of $\{a_n\}$. To construct $\{b_m\}$, set $h_1 = 1$. Note that since $a_{n_i} \neq a_{n_j}$ if $n_i \neq n_j$, the indices n_i are distinct from one another. Hence there are indices n_k larger than n_{h_1}. Let h_2 be the smallest positive integer such that $n_{h_2} > n_{h_1}$. Similarly, let h_3 be the smallest positive integer such that $n_{h_3} > n_{h_2}$, and so on. Now take $b_m = a_{n_{h_m}}$.

Let $\{a_n\}$ be a sequence bounded below. Denote by m' the infimum of the set of the limit points of $\{a_n\}$. We call m' the *limit inferior* of $\{a_n\}$ and write it in any one of the forms

$$m' = \liminf_{n \to \infty} a_n, \qquad m' = \liminf_n a_n,$$

$$m' = \varliminf_{n \to \infty} a_n, \qquad m' = \varliminf_n a_n.$$

The following analog of Theorem 3 is valid:

THEOREM 4. *m′ is a limit point of* $\{a_n\}$. *In other words, the limit inferior is the smallest limit point of* $\{a_n\}$.

Note that $\overline{\lim}\, a_n = \underline{\lim}\, a_n = L$ if, and only if, $\lim a_n = L$.

EXAMPLE 1. Let $a_n = n/(n + 1)$. Then $a_n \to 1$ as $n \to \infty$. Indeed, for any $\varepsilon > 0$,

$$\left|\frac{n}{n + 1} - 1\right| = \frac{1}{n + 1} < \varepsilon \qquad \text{if } n \geq n_0,$$

where n_0 is any positive integer such that $1/n_0 < \varepsilon$.

EXAMPLE 2. Let $a_n = (n^2 - n)/(n^2 + 1)$. Then $\lim_n a_n = 1$. Indeed,

$$\left|\frac{n^2 - n}{n^2 + 1} - 1\right| = \frac{n + 1}{n^2 + 1} < \frac{2n}{n^2} = \frac{2}{n} < \varepsilon \qquad \text{if } n \geq n_0,$$

where n_0 is any positive integer such that $1/n_0 < \varepsilon/2$.

EXAMPLE 3. $\lim_n \sqrt{n}(\sqrt{n + 1} - \sqrt{n}) = 1/2$. We first write

$$\sqrt{n}(\sqrt{n + 1} - \sqrt{n}) = \sqrt{n}\,\frac{(n + 1) - n}{\sqrt{n + 1} + \sqrt{n}} = \frac{\sqrt{n}}{\sqrt{n + 1} + \sqrt{n}}$$

$$= \frac{1}{1 + \sqrt{1 + 1/n}}.$$

Then

$$\left|\sqrt{n}(\sqrt{n + 1} - \sqrt{n}) - \tfrac{1}{2}\right| = \frac{1}{2} - \frac{1}{1 + \sqrt{1 + 1/n}} = \frac{\sqrt{1 + 1/n} - 1}{2(\sqrt{1 + 1/n} + 1)}$$

$$< \frac{(1 + 1/n) - 1}{2(1 + 1)} = \frac{1}{4n} < \varepsilon$$

if $n \geq n_0$, where $n_0 > 1/4\varepsilon$.

EXAMPLE 4. Let $0 < \alpha < 1$. Then $\lim_{n \to \infty} \alpha^n = 0$. To prove it, we write $\alpha = 1/(1 + h)$ where $h > 0$. By Bernoulli's inequality

$$(1 + h)^n \geq 1 + nh.$$

Hence

$$0 < \alpha^n \le \frac{1}{1 + nh} < \frac{1}{nh} < \varepsilon \qquad \text{if } n \ge n_0,$$

where $n_0 > \dfrac{1}{\varepsilon h}$.

EXAMPLE 5.   Let $0 < \alpha < 1$. Then $\lim_n n\alpha^n = 0$. To prove it, we again write $\alpha = 1/(1 + h)$, $h > 0$. By the binomial theorem,

$$(1 + h)^n = 1 + nh + \frac{n(n - 1)}{2} h^2 + \cdots > \frac{1}{2} n(n - 1)h^2.$$

Hence

$$n\alpha^n < \frac{2n}{n(n - 1)h^2} = \frac{2}{(n - 1)h^2} < \varepsilon \qquad \text{if } n \ge n_0 + 1,$$

where $1/n_0 < \varepsilon h^2/2$.

EXAMPLE 6.   Let n be a positive integer. It will be shown in Section 6 of Chapter 2 that for any $a > 0$ there is a unique $x > 0$ satisfying: $x^n = a$. We write $x = a^{1/n}$, or $x = \sqrt[n]{a}$. Let $0 < \alpha < 1$. Prove that $\lim_n \alpha^{1/n} = 1$. We begin by writing $\alpha^{1/n} = 1/(1 + h_n)$ for some $h_n > 0$. Then

$$\alpha = \frac{1}{(1 + h_n)^n} \le \frac{1}{1 + nh_n}.$$

Hence, for any $\varepsilon > 0$, $h_n \le (1/\alpha - 1)/n < \varepsilon$ if $n \ge n_0$ for a suitable n_0. Next,

$$|\alpha^{1/n} - 1| = 1 - \alpha^{1/n} = 1 - \frac{1}{1 + h_n} = \frac{h_n}{1 + h_n} < h_n < \varepsilon$$

if $n \ge n_0$.

EXAMPLE 7.   We shall prove that $\lim_n \sqrt[n]{n} = 1$. We begin by writing $\sqrt[n]{n} = 1 + h_n$, $h_n > 0$. Then

$$n = (1 + h_n)^n > \tfrac{1}{2}n(n - 1)h_n^2,$$

by the binomial theorem. It follows that

$$h_n < \sqrt{\frac{2}{n - 1}} < \varepsilon \qquad \text{if } n > n_0, \, 1/n_0 < \tfrac{1}{2}\varepsilon^2.$$

Hence $|\sqrt[n]{n} - 1| < \varepsilon$ if $n > n_0$.

EXAMPLE 8.   If $a_n = (-1)^n$, then lim sup $a_n = 1$, lim inf $a_n = -1$.

EXAMPLE 9.   Let $a_n = 1/n$ for n even and $a_n = 1 + 2/n$ for n odd. Then lim sup $a_n = 1$, lim inf $a_n = 0$.

PROBLEMS

1.   Prove Theorem 4.

2.   If $a_n \to L$, then $|a_n| \to |L|$.

3.   Find the limits of the sequences:

(a) $\dfrac{n^2 - 1}{2n^2 + 3}$;   (b) $\dfrac{(-1)^n}{\sqrt{n}}$;   (c) $\dfrac{\sqrt{n+1}}{n+7}$;

(d) $\sqrt{n+1} - \sqrt{n}$;   (e) $\dfrac{3n^2 + 2n - 1}{5n^2 + 8}$;   (f) $\dfrac{n^4 + 3n^2 - n}{2n^4 + n^3 + 1}$.

4.   Find the limit points of the sequences:

(a) $2 + (-1)^n$;   (b) $(2^n + (-1)^n)/2^n$;

(c) $(-1)^n \dfrac{\sqrt{n} - 2}{\sqrt{n} + 1}$;   (d) $n(\sqrt{n+1} - \sqrt{n})$.

5.   Let $0 < \alpha < 1$. Prove that $\lim\limits_{n} n^2 \alpha^n = 0$.

6.   Let $0 < \alpha < 1$. Prove that $\lim\limits_{n} n^k \alpha^n = 0$, where k is a fixed positive integer.

7.   Let $\alpha > 1$. Prove that $\sqrt[n]{\alpha} \to 1$ if $n \to \infty$.

8.   Find the limit points of the sequence $\{a_n\}$ where

$$a_n = \frac{\sin \tfrac{1}{2} n^2 \pi}{1 + \cos \tfrac{1}{2} n^2 \pi}.$$

9.   Prove that $\sqrt[n]{n + \sqrt{n}} \to 1$ if $n \to \infty$.

6.   OPERATIONS WITH LIMITS

THEOREM 1.   *Let $\{a_n\}$, $\{b_n\}$ be two sequences of real numbers. If* $\lim a_n = A$ *and* $\lim b_n = B$, *then* $\lim(a_n + b_n)$ *exists and equals* $A + B$.

In short,

$$\lim(a_n + b_n) = \lim a_n + \lim b_n. \tag{1}$$

Proof. By assumption, for any $\varepsilon > 0$ there exist positive integers n_0 and n_1 such that

$$|a_n - A| < \frac{\varepsilon}{2} \qquad \text{if } n \geq n_0,$$

$$|b_n - B| < \frac{\varepsilon}{2} \qquad \text{if } n \geq n_1.$$

Let $n_2 = \max(n_0, n_1)$. If $n \geq n_2$, then

$$|(a_n + b_n) - (A + B)| = |(a_n - A) + (b_n - B)|$$

$$\leq |a_n - A| + |b_n - B| < \frac{\varepsilon}{2} + \frac{\varepsilon}{2} = \varepsilon.$$

Hence $\{a_n + b_n\}$ converges to $A + B$.

THEOREM 2.   *A convergent sequence is bounded.*

Proof. Let $\{a_n\}$ be a convergent sequence with limit A. From the definition of convergence it follows that $|a_n - A| < 1$ for all $n \geq n_0$, where n_0 is some natural number. Hence

$$|a_n| = |a_n - A + A| \leq |a_n - A| + |A| < 1 + |A| \qquad \text{if } n \geq n_0.$$

Let

$$K = |a_1| + |a_2| + \cdots + |a_{n_0 - 1}| + 1 + |A|.$$

Then $|a_n| \leq K$ for all $n \geq 1$. This completes the proof.

THEOREM 3.   *Let $\{a_n\},\{b_n\}$ be sequences of real numbers. If* $\lim a_n = A$ *and* $\lim b_n = B$, *then* $\lim_n(a_n b_n)$ *exists and equals* AB.

In short,

$$\lim(a_n b_n) = (\lim a_n) \cdot (\lim b_n). \tag{2}$$

Proof. By Theorem 2,

$$|b_n| < K \qquad \text{for all } n \geq 1.$$

Let ε be any positive number. Then, by our assumptions, there exist positive integers n_0 and n_1 such that

$$|a_n - A| < \frac{\varepsilon}{2K} \qquad \text{if } n \geq n_0,$$

$$|b_n - B| < \frac{\varepsilon}{2(|A| + 1)} \qquad \text{if } n \geq n_1.$$

Let $n_2 = \max(n_0, n_1)$. Then, if $n \geq n_2$,

$$|a_n b_n - AB| = |(a_n b_n - A b_n) + (A b_n - AB)|$$

$$\leq |b_n| |a_n - A| + |A| |b_n - B| < K \frac{\varepsilon}{2K} + |A| \frac{\varepsilon}{2(|A| + 1)} \leq \varepsilon.$$

This completes the proof.

From Theorem 3 it follows that for any real number λ,

$$\lim(\lambda a_n) = \lambda \lim a_n. \tag{3}$$

THEOREM 4. *Let $\{b_n\}$ be a sequence of real numbers. If $\lim b_n = B$ and $B \neq 0$, then there exists a positive integer n^* such that $b_n \neq 0$ for all $n \geq n^*$; in fact, $|b_n| \geq |B|/2$ for all $n \geq n^*$.*

Proof. For $\varepsilon = |B|/2$ there exists n^* such that

$$|b_n - B| < \frac{|B|}{2} \qquad \text{if } n \geq n^*.$$

But then

$$|b_n| = |(b_n - B) + B| \geq |B| - |b_n - B| > |B| - \frac{|B|}{2} = \frac{|B|}{2} > 0.$$

THEOREM 5. *Let $\{a_n\}, \{b_n\}$ be sequences of real numbers. If $\lim a_n = A$, $\lim b_n = B$, and if $B \neq 0$, then the sequence of numbers a_n/b_n for $n \geq n^*$ (n^* as in Theorem 4) is convergent to A/B.*

In short,

$$\lim \frac{a_n}{b_n} = \frac{\lim a_n}{\lim b_n} \qquad \text{if } \lim b_n \neq 0. \tag{4}$$

Proof. From Theorem 4 we have

$$|b_n| > \frac{|B|}{2} \quad \text{if } n \geq n^*.$$

Now let ε be any positive number and choose positive integers n_0, n_1 such that

$$|a_n - A| < |B| \frac{\varepsilon}{4} \quad \text{if } n \geq n_0, \tag{5}$$

$$|b_n - B| < |B|^2 \frac{\varepsilon}{4(|A| + 1)} \quad \text{if } n \geq n_1. \tag{6}$$

Let $n_2 = \max(n^*, n_0, n_1)$. Then, for any $n \geq n_2$,

$$\left| \frac{a_n}{b_n} - \frac{A}{B} \right| = \frac{|a_n B - b_n A|}{|b_n B|} = \frac{|(a_n B - AB) + (AB - b_n A)|}{|b_n B|}$$

$$\leq \frac{|a_n - A|}{|b_n|} + \frac{|B - b_n| |A|}{|b_n B|} < 2 \frac{|a_n - A|}{|B|} + 2 \frac{|B - b_n| |A|}{|B|^2} < \varepsilon$$

by (5) and (6). This completes the proof.

THEOREM 6. *Let $\{a_n\}, \{b_n\}$ be convergent sequences of real numbers. If $a_n \leq b_n$ for all n sufficiently large, then* $\lim a_n \leq \lim b_n$.

Proof. Let $A = \lim a_n$, $B = \lim b_n$. By assumption, $a_n \leq b_n$ for all $n \geq n_0$. For any $\varepsilon > 0$ there exist positive integers n_1, n_2 such that

$$|a_n - A| < \frac{\varepsilon}{2} \quad \text{if } n \geq n_1$$

$$|b_n - B| < \frac{\varepsilon}{2} \quad \text{if } n \geq n_2.$$

Let $n_2 = \max(n_0, n_1, n_2)$. Then, if $n \geq n_2$,

$$A < a_n + \frac{\varepsilon}{2} \leq b_n + \frac{\varepsilon}{2} \leq B + \frac{\varepsilon}{2} + \frac{\varepsilon}{2} = B + \varepsilon.$$

Thus, for every $\varepsilon > 0$, $A \leq B + \varepsilon$. This implies that $A \leq B$. Indeed, if $A > B$, then we get $A > B + \varepsilon$ with $\varepsilon = (A - B)/2 > 0$, a contradiction.

PROBLEMS

1. Prove that $\lim(a_n - b_n) = \lim a_n - \lim b_n$, provided $\lim a_n$ and $\lim b_n$ exist.

2. Give an example where $\lim a_n$ and $\lim b_n$ do not exist, but $\lim(a_n + b_n)$ exists.

3. If $\{a_n\}$ is a bounded sequence and if $\{b_n\}$ is a sequence converging to 0, then $\{a_n b_n\}$ converges to 0.

4. Let $\{a_n\}$ be a sequence of nonnegative numbers. Then $\varliminf a_n \geq 0$.

5. Let $\{a_n\}$ be a sequence of positive numbers. If $\lim a_n = A$, then $\lim \sqrt{a_n} = \sqrt{A}$.

6. Let $\{a_n\}$ be a sequence of positive numbers satisfying $a_{n+1} < ra_n$, where r is a fixed positive number smaller than 1. Prove that $a_n \to 0$ if $n \to \infty$.

7. If $a_n > 0$ for all n and $\varlimsup(a_{n+1}/a_n) = \theta < 1$, then $a_n \to 0$ if $n \to \infty$.

8. Use the previous problem to prove that
    (a) For any positive α, $\alpha^n/n! \to 0$ if $n \to \infty$;
    (b) $n!/n^n \to 0$ as $n \to \infty$;
    (c) For any $0 < \alpha < 1$, and for any positive integer k, $n^k \alpha^n \to 0$ if $n \to \infty$.

9. If $a_n \to 0$ as $n \to \infty$, then
    $$(a_1 + a_2 + \cdots + a_n)/n \to 0 \text{ as } n \to \infty.$$

10. If $a_n \to A$ as $n \to \infty$, then
    $$(a_1 + a_2 + \cdots + a_n)/n \to A \text{ as } n \to \infty.$$

11. Calculate the limit of $\{a_n\}$ where
    (a) $a_n = n(1 - \sqrt[3]{1 - 3/n})$;

    (b) $a_n = \dfrac{1}{n^2 + 1} + \dfrac{1}{n^2 + 2} + \cdots + \dfrac{1}{n^2 + n}$;

    (c) $a_n = \dfrac{1}{\sqrt{n^2 + 1}} + \dfrac{1}{\sqrt{n^2 + 2}} + \cdots + \dfrac{1}{\sqrt{n^2 + n}}$.

12. Let $\alpha \geq \beta > 0$, $a_n = (\alpha^n + \beta^n)^{1/n}$. Prove that $\lim a_n = \alpha$.

13. Let $\{a_n\},\{b_n\}$ be sequences bounded above. Prove

$$\overline{\lim}(a_n + b_n) \leq \overline{\lim}\, a_n + \overline{\lim}\, b_n. \tag{7}$$

[*Hint:* Set $\overline{\lim}(a_n + b_n) = C$, $\overline{\lim}\, a_n = A$, $\overline{\lim}\, b_n = B$. For any $\varepsilon > 0$, $a_n \leq A + \varepsilon$, $b_n \leq B + \varepsilon$ for all $n \geq n_0$. Show that $C < A + B + 3\varepsilon$.]

14. Let $\{a_n\},\{b_n\}$ be sequences bounded from above. Prove

$$\underline{\lim}(a_n + b_n) \geq \underline{\lim}\, a_n + \underline{\lim}\, b_n. \tag{8}$$

15. Let $a_n \leq b_n$ for all n sufficiently large. Prove

$$\overline{\lim}\, a_n \leq \overline{\lim}\, b_n, \qquad \underline{\lim}\, a_n \leq \underline{\lim}\, b_n. \tag{9}$$

16. If $a_n \geq 0$, $b_n \geq 0$, and if $\lim a_n$ exists and is positive, then

$$\overline{\lim}(a_n b_n) = (\lim a_n)(\overline{\lim}\, b_n).$$

17. Use Theorems 1, 3, and 5 in order to prove that

(a) $\displaystyle\lim_{n \to \infty} \frac{\sqrt{n} - n + 1}{2\sqrt{n} + n + 2} = -1.$

(b) $\displaystyle\lim_{n \to \infty} \frac{n^3 - n^2 + n - 8}{2n^3 + 1} = \frac{1}{2}.$

(c) $\displaystyle\lim_{n \to \infty} \frac{\sqrt{n^2 + 1} + \sqrt{n} - \sqrt{n^2 - 1}}{n + 1} = 0.$

7. MONOTONE SEQUENCES

A sequence $\{a_n\}$ of real numbers is called *monotone increasing* if $a_n \leq a_{n+1}$ for all n. It is *strictly monotone increasing* if $a_n < a_{n+1}$ for all n.

A sequence $\{a_n\}$ is called *monotone decreasing* if $a_n \geq a_{n+1}$ for all n. It is *strictly monotone decreasing* if $a_n > a_{n+1}$ for all n.

A *(strictly) monotone* sequence is a sequence that is either (strictly) monotone increasing or (strictly) monotone decreasing

THEOREM 1. *If a monotone increasing sequence is bounded above, then it is convergent.*

Proof. Let $\{a_n\}$ be a monotone increasing sequence, bounded above. Denote by A the l.u.b. of the set $\{a_n\}$ Then $a_n \leq A$ for all n, and for any $\varepsilon > 0$ there exists a number a_{n_0} such that

$$a_{n_0} > A - \varepsilon.$$

It follows that

$$a_n \geq a_{n_0} > A - \varepsilon \qquad \text{if } n \geq n_0.$$

Thus $|a_n - A| < \varepsilon$ if $n \geq n_0$. This shows that $a_n \to A$ if $n \to \infty$.
The previous proof shows that $a_n \leq \lim_m a_m$ for any n.

Similarly one can prove that:

THEOREM 2.   *If a monotone decreasing sequence is bounded below, then it is convergent.*
If we denote the decreasing sequence by $\{b_n\}$, then we have: $b_n \geq \lim_m b_m$ for all n.

Application.   We use Theorem 2 to prove that if $0 < \alpha < 1$, then $\lim \alpha^n = 0$. Since $\alpha^n > \alpha^{n+1}$, Theorem 2 shows that $A = \lim \alpha^n$ exists. Also, $\lim_n \alpha^{n+1}$ exists and is equal to A. From

$$\alpha^{n+1} = \alpha \cdot \alpha^n$$

and from (3) of Section 6, we get

$$A = \lim \alpha^{n+1} = \alpha \lim \alpha^n = \alpha A.$$

Thus $A(1 - \alpha) = 0$. Since $\alpha \neq 1$, it follows that $A = 0$.
An important example of a monotone sequence is given in the following theorem.

THEOREM 3.   *The sequence*

$$b_n = \left(1 + \frac{1}{n}\right)^n \tag{1}$$

is monotone increasing, and $2 \leq b_n < 3$.

Proof.   From the binomial theorem we get

$$b_n = 1 + n\left(\frac{1}{n}\right) + \frac{n(n-1)}{2!}\left(\frac{1}{n}\right)^2 + \cdots + \frac{n(n-1)\cdots(n-k+1)}{k!}\left(\frac{1}{n}\right)^k$$

$$+ \cdots + \frac{n!}{n!}\left(\frac{1}{n}\right)^n$$

$$= 1 + 1 + \frac{1}{2!}\left(1 - \frac{1}{n}\right) + \cdots + \frac{1}{k!}\left(1 - \frac{1}{n}\right)\left(1 - \frac{2}{n}\right)\cdots\left(1 - \frac{k-1}{n}\right) + \cdots$$

$$+ \frac{1}{n!}\left(1 - \frac{1}{n}\right)\left(1 - \frac{2}{n}\right)\cdots\left(1 - \frac{n-1}{n}\right).$$

We can write a similar expression for b_{n+1}. Since

$$\frac{1}{k!}\left(1 - \frac{1}{n}\right)\left(1 - \frac{2}{n}\right)\cdots\left(1 - \frac{k-1}{n}\right)$$

$$< \frac{1}{k!}\left(1 - \frac{1}{n+1}\right)\left(1 - \frac{2}{n+1}\right)\cdots\left(1 - \frac{k-1}{n+1}\right),$$

it follows that $b_n < b_{n+1}$.

From the last expression for b_n we also see that

$$b_n < 1 + 1 + \frac{1}{2!} + \frac{1}{3!} + \cdots + \frac{1}{n!}.$$

Since $n! \geq 2^{n-1}$, we get

$$b_n < 1 + 1 + \frac{1}{2} + \left(\frac{1}{2}\right)^2 + \cdots + \left(\frac{1}{2}\right)^{n-1} = 1 + \frac{1 - 1/2^n}{1 - 1/2} < 1 + 2 = 3.$$

Finally, $b_n \geq b_1 = (1 + 1)^1 = 2$.

Applying Theorem 1 to the sequence (1), we conclude that $\lim b_n$ exists. This limit is denoted by e. Thus

$$e = \lim_{n \to \infty}\left(1 + \frac{1}{n}\right)^n. \tag{2}$$

The student is surely familiar with the importance of this number.

One often encounters sequences $\{a_n\}$ having the following property: For any positive number N there exists an n_0 such that

$$a_n > N \qquad \text{if } n \geq n_0.$$

When this occurs we shall say that $\{a_n\}$ *converges to infinity* and write: $\{a_n\}$ converges to ∞, or

$$\lim_{n \to \infty} a_n = \infty, \qquad \text{or} \qquad a_n \to \infty \qquad \text{if } n \to \infty. \tag{3}$$

We wish to stress that we have not defined here a number or a symbol ∞. We have merely expressed the property described at the beginning of this paragraph by the short notation (3). Moreover, if $\lim a_n = \infty$, then the sequence $\{a_n\}$ is certainly not a convergent sequence.

If for any positive N there is an n_0 such that

$$a_n < -N \qquad \text{if } n \geq n_0,$$

then we say that $\{a_n\}$ *converges to minus infinity* and write: $\{a_n\}$ converges to $-\infty$, or

$$\lim_{n \to \infty} a_n = -\infty, \qquad \text{or} \qquad a_n \to -\infty \qquad \text{if } n \to \infty. \tag{4}$$

THEOREM 4. *A monotone increasing sequence is either convergent (to a real number) or is convergent to ∞.*

Proof. Let $\{a_n\}$ be a monotone increasing sequence. If it is bounded from above then, by Theorem 1, it is convergent to a real number. Suppose then that $\{a_n\}$ is not bounded from above. Then, for any positive N there is there is an a_{n_0} such that

$$a_{n_0} > N.$$

It follows that $a_n > N$ if $n \geq n_0$ Hence $\lim a_n = \infty$.

PROBLEMS

1. Show that $\lim(1 + 1/2n)^n = \sqrt{e}$.

2. Prove that $\lim(1 + 1/(n + 2))^{3n} = e^3$.

3. Use the relation $1 + 2/n = (1 + 1/n)(1 + 1/(n + 1))$ to show that $(1 + 2/n)^n \to e^2$ if $n \to \infty$.

4. Show that $(1 + 3/n)^n \to e^3$ if $n \to \infty$.

5. Define a sequence by

$$b_1 = \sqrt{2}, \qquad b_2 = \sqrt{2 + \sqrt{2}},$$

and, in general,

$$b_{n+1} = \sqrt{2 + b_n}.$$

Prove by induction that $\{b_n\}$ is monotone increasing, bounded, and $b_n < 2$. Compute $\lim b_n$.

6. Define a sequence $\{b_n\}$ by $b_1 = \sqrt{c}, b_{n+1} = \sqrt{c + b_n}$ where $c > 0$. Prove by induction that $b_n < b_{n+1}, b_n < 1 + \sqrt{c}$, and find $\lim b_n$.

7. Let $c_n = \sqrt{n^2 + n} - n$. Prove that $\{a_n\}$ is monotone and find its limit.

8. Let $b_1 > 0, b_{n+1} = 3(1 + b_n)/(3 + b_n)$. Show that $\{b_n\}$ is monotone, $0 < b_n < 3$, and deduce that $\lim b_n = \sqrt{3}$.

9. Prove that $\lim(n^n/n!e^n)$ exists.

10. Let $x_1 > y_1 > 0, x_{n+1} = (x_n + y_n)/2, y_{n+1} = \sqrt{x_n y_n}$. Show that $y_n < y_{n+1} < x_{n+1} < x_n, y_1 < x_{n+1} < x_n, 0 < x_{n+1} - y_{n+1} < (x_1 - y_1)/2^n$. Deduce that $\lim x_n = \lim y_n$.

11. Define $x_1 = \sqrt{2}$, $x_n = \sqrt{2x_{n-1}}$. Prove that $\{x_n\}$ is monotone and find $\lim x_n$.

12. If $\lim a_n = \infty$ and $\lim b_n = B$, then $\lim(a_n + b_n) = \infty$.

13. If $\lim a_n = -\infty$ and $\lim b_n = B$, then $\lim(a_n + b_n) = -\infty$.

14. If $\lim a_n = \infty$ and $\lim b_n = B$, and if $B \neq 0$, then $\lim a_n b_n = \infty$ when $B > 0$, and $\lim a_n b_n = -\infty$ when $B < 0$.

15. If $\lim a_n = \infty$ and $\lim b_n = \infty$, then $\lim(a_n b_n) = \infty$.

16. A monotone decreasing sequence is either convergent to some real number or is convergent to $-\infty$.

17. For any real number r there exists a strictly monotone decreasing sequence of rational numbers s_n such that $s_n \to r$ if $n \to \infty$.

8. BOLZANO–WEIERSTRASS THEOREM

Let G be a set of numbers. Suppose there is a number N with the following property: For any $\varepsilon > 0$, there exists a number $a \neq N$, $a \in G$, satisfying

$$|a - N| < \varepsilon. \tag{1}$$

We then say that N is a *limit point* (also a *cluster point*, or a *point of accumulation*) of G.

THEOREM 1. *If N is a limit point of a set G, then there exists a sequence $\{a_n\}$ of mutually distinct points that belong to G such that $\{a_n\}$ converges to N.*

Proof. Taking $\varepsilon = 1$ in (1), we see that there is a point a_1 in N such that $|a_1 - N| < 1$ and $a_1 \neq N$. Let $\varepsilon_2 > 0$ be such that $\varepsilon_2 < 1/2$ and $\varepsilon_2 < |a_1 - N|$. Applying (1) with $\varepsilon = \varepsilon_2$, we conclude that there exists a point a_2 in G such that

$$|a_2 - N| < \tfrac{1}{2}, \qquad |a_2 - N| < |a_1 - N|. \tag{2}$$

The second inequality shows that $a_2 \neq a_1$.

Next we apply (1) with $\varepsilon = \varepsilon_3$ where $\varepsilon_3 < 1/3$ and $\varepsilon_3 < |a_2 - N|$. We conclude that there exists a point a_3 in G such that

$$|a_3 - N| < \tfrac{1}{3}, \qquad |a_3 - N| < |a_2 - N|. \tag{3}$$

The second inequalities in (2) and (3) show that $a_3 \neq a_2$, $a_3 \neq a_1$.

Proceeding in this way step by step, we construct a sequence $\{a_n\}$ of mutually distinct points, satisfying

$$|a_n - N| < \frac{1}{n}. \tag{4}$$

From this it follows that $a_n \to N$ if $n \to \infty$.

Denote by G' the set of all limit points of a set G. We call G' the *derived set* of G. The union of the two sets G and G' is called the *closure* of G. If $G' \subset G$, then we say that G is a *closed set*.

Let $\{I_n\}$ be a sequence of closed intervals such that $I_1 \supset I_2 \supset \cdots \supset I_n \supset I_{n+1} \supset \cdots$, and such that the lengths λ_n of the intervals I_n satisfy: $\lambda_n \to 0$ if $n \to \infty$. Then we call $\{I_n\}$ a sequence of closed *nested* intervals, and say that the sequence forms a *nest*.

THEOREM 2. *If $\{I_n\}$ is a sequence of closed nested intervals, then there exists a unique real number ξ that belongs to all the intervals I_n.*

Proof. Write I_n in the form

$$\{a_n \leq x \leq b_n\}.$$

Since $I_n \supset I_{n+1}, a_n \leq a_{n+1} \leq b_{n+1} \leq b_n$ for all n. From Theorem 1 of Section 7, it follows that

$$\lim a_n = \xi$$

and

$$\lim b_n = \eta$$

exist, and from Theorem 6 of Section 6, it follows that $\xi \leq \eta$. Since $a_n \leq \xi \leq \eta \leq b_n$ for all n, the point ξ belongs to each interval I_n.

Now let ξ' be any real point contained in each interval I_n. We shall prove that $\xi' = \xi$. Since

$$a_n \leq \xi \leq b_n, \qquad a_n \leq \xi' \leq b_n,$$

it follows that $|\xi - \xi'| \leq b_n - a_n$. By assumption, $b_n - a_n \to 0$ if $n \to \infty$. This implies that $\xi' = \xi$.

BOLZANO–WEIERSTRASS THEOREM (FOR SETS). *Every bounded, infinite set of real numbers has at least one limit point.*

Proof. Let G be a bounded, infinite set. Then there exist numbers α, β such that

$$\alpha \leq x \leq \beta \qquad \text{for all } x \text{ in } G.$$

We shall designate by I_1 the closed interval $[\alpha,\beta]$. Denote by γ_1 the midpoint $(\alpha + \beta)/2$ of I_1 and consider the two closed intervals $[\alpha,\gamma_1]$ and $[\gamma_1,\beta]$. At least one of them must contain an infinite number of points of G. Denote that interval by I_2. We divide I_2 into two closed intervals, by introducing the midpoint γ_2 of I_2. One of these intervals must contain an infinite number of points of G. We designate that closed interval by I_3.

Proceeding in this manner step by step, we get a sequence $\{I_n\}$ of closed intervals satisfying $I_1 \supset I_2 \supset I_3 \supset \cdots$. The length of I_n is $(\beta - \alpha)/2^n$. Therefore, the sequence $\{I_n\}$ forms a nest. By Theorem 2, there exists a unique point ξ that is contained in all the intervals I_n. We claim that ξ is a limit point of G.

To prove it, let ε by any positive number. Choose n such that $(\beta - \alpha)/2^n < \varepsilon$. Since $\xi \in I_n$ and since the length of I_n is $(\beta - \alpha)/2^n$,

$$|x - \xi| < \varepsilon \qquad \text{if } x \in I_n.$$

Since I_n contains an infinite number of points of G, there is certainly a point $\bar{x}$ in G such that $\bar{x} \neq \xi$ and such that $|\bar{x} - \xi| < \varepsilon$. Thus, ξ is a limit point of G.

BOLZANO–WEIERSTRASS THEOREM (FOR SEQUENCES).   *Every bounded sequence of real numbers has at least one limit point.*

Proof.   Let $\{a_n\}$ be a bounded sequence of real numbers. If one of the numbers occurs an infinite number of times in the sequence, say $a_{n_k} = \xi$ for $k = 1,2,\ldots$, where $n_1 < n_2, < \cdots$, then ξ is a limit point of $\{a_n\}$. We therefore may assume from now on that this situation does not occur. Set $m_1 = 1$. Then there is a positive integer m_2, such that $a_n \neq a_1$ if $n \geq m_2$. Next, there is a positive integer $m_3 (m_3 > m_2)$ such that $a_n \neq a_{m_2}$ if $n \geq m_3$. Proceeding in this way step by step, we construct a sequence $\{a_{m_1}, a_{m_2}, a_{m_3}, \cdots\}$ of mutually distinct numbers. By the Bolzano–Weierstrass theorem for sets, the set of points $\{a_{m_1}, a_{m_2}, a_{m_3}, \cdots\}$ has a limit point ξ. Hence, by Theorem 1, ξ is a limit point of the set $\{a_{m_k}\}$. Thus there exists a sequence $\{a_{\bar{m}_k}\}$ whose elements occur in the set $\{a_{m_k}\}$, such that $a_{\bar{m}_k} \to \xi$ if $k \to \infty$. We do not know that $\{a_{\bar{m}_k}\}$ is a subsequence of $\{a_{m_k}\}$, since we have not shown that $\bar{m}_1 < \bar{m}_2 < \cdots$. If we construct a subsequence $\{a_{\tilde{m}_k}\}$ of $\{a_{\bar{m}_k}\}$ such that $\tilde{m}_1 < \tilde{m}_2 < \cdots$, then $\{a_{\tilde{m}_k}\}$ is a subsequence of $\{a_{m_k}\}$ (and hence a subsequence of $\{a_n\}$), and from $\lim_k a_{\tilde{m}_k} = \xi$ it follows that ξ is a limit point $\{a_n\}$. Let $\tilde{m}_1 = \bar{m}_1$. Let $\tilde{m}_2 = \bar{m}_{k_2}$ where k_2 is such that $\bar{m}_{k_2} > \tilde{m}_1$. Let $\tilde{m}_3 = \bar{m}_{k_3}$ where k_3 is such that $\bar{m}_{k_3} > \tilde{m}_2$, and so on. Then $\{a_{\tilde{m}_k}\}$ is a subsequence of $\{a_{\bar{m}_k}\}$ with $\tilde{m}_1 < \tilde{m}_2 < \cdots$.

PROBLEMS

1. Find the closure of an interval (a,b).

2. Find the closure of the set of points $(-1)^n n/(n + 1)$, $n = 1,2,\ldots$.

3. Show that a closed interval is a closed set.

4. Is the set of all integers a closed set?

5. Denote by R the set of all rational numbers. Prove that the derived set R' is the set of all the real numbers.

6. Let $x_{n+1} = a/(1 + x_n)$ where $x_1 > 0$, $a > x_1^2 + x_1$. Show that the intervals $[x_1,x_2],[x_3,x_4],\ldots$ form a nest, and that the point ξ common to all these intervals is a root of $x^2 + x = a$.
    [*Hint*: Show that $x_{n+1} - x_n$ and $x_n - x_{n-1}$ have different signs, and $|x_{n+1} - x_n| \le \theta |x_n - x_{n-1}|$ for some $\theta < 1$.]

7. Let $\{I_n\}$ be a sequence of closed intervals, and denote by λ_n the length of I_n. Assume that $I_1 \supset I_2 \supset \cdots \supset I_n \supset I_{n+1} \supset \cdots$. Show that (i) $\lim \lambda_n$ exists; (ii) if $\lim \lambda_n > 0$, then the set of points that are contained in all the closed intervals I_n form a closed interval of length $\lim \lambda_n$.

8. Give an example of an unbounded set that has no limit points.

9. If G is a bounded set, then its derived set G' is also a bounded set. Is the converse also true?

10. Prove the following theorem: A number M' is the limit superior of a sequence $\{a_n\}$ bounded above if, and only if, for any $\varepsilon > 0$, the following two properties hold:
    (a) There is an infinite number of n's for which $a_n > M' - \varepsilon$;
    (b) There is only a finite number of n's for which $a_n > M' + \varepsilon$.
    [*Hint*: Use Theorem 2, Section 5 and the Bolzano–Weierstrass theorem.]

11. Prove the following theorem: A number m' is the limit inferior of a sequence $\{a_n\}$ bounded below if, and only if, for any $\varepsilon > 0$, the following properties hold:
    (a) There is an infinite number of n's for which $a_n > m' + \varepsilon$;
    (b) There is only a finite number of n's for which $a_n < m' - \varepsilon$.
    [*Hint:* Use Theorem 3, Section 5 and the Bolzano–Weierstrass theorem.]

2

CONTINUOUS FUNCTIONS

1. DEFINITION OF CONTINUOUS FUNCTION

When we speak of an interval (on the real line) we mean either a closed interval, an open interval, or a half-open interval. A point x_0 is called an *interior* point of an interval I if x_0 is contained in I but is not an endpoint of I. An open interval (a,b) containing a point x_0 is called a *neighborhood* of x_0. The interval $(x_0 - \delta, x_0 + \delta)$ is called a δ-neighborhood of x_0.

DEFINITION.   Let $f(x)$ be a function defined on an interval I (that is, its domain is I) and let c be any point of I. Suppose that for every $\varepsilon > 0$ there is a $\delta > 0$ such that

$$|f(x) - f(c)| < \varepsilon \qquad \text{if } |x - c| < \delta \text{ and } x \in I. \tag{1}$$

Then we say that $f(x)$ is *continuous* at c.

Condition (1) can be reworded: If x lies in a δ-neighborhood of c as well as in I, then $f(x)$ lies in an ε-neighborhood of $f(c)$. Note that δ will depend, in general, on ε.

Condition (1) also can be written as follows:

$$|f(c + h) - f(c)| < \varepsilon \qquad \text{if } |h| < \delta \text{ and } c + h \in I. \tag{2}$$

Roughly speaking, f is continuous at c if the values $f(x)$ of f are arbitrarily close to $f(c)$ when the values of x ($x \in I$) are sufficiently close to c.

Note that if c is the left endpoint of I, then condition (1) becomes

$$|f(x) - f(c)| < \varepsilon \qquad \text{if } c < x < c + \delta, x \in I. \tag{3}$$

In the following three examples, the domain I of each of the functions considered can be any interval.

EXAMPLE 1. $f(x) = x^2$ is continuous at every point c. To prove it, take any $\varepsilon > 0$. We must find a number δ such that if $|x - c| < \delta$, then $|x^2 - c^2| < \varepsilon$. Now, if $\delta < 1$, then $|x + c| \le |x - c| + 2|c| < 1 + 2|c|$. We therefore have

$$|x^2 - c^2| = |x + c| \, |x - c| < (1 + 2|c|)|x - c| < (1 + 2|c|)\delta \le \varepsilon$$

if $\delta \le \varepsilon/(1 + 2|c|)$. Thus the continuity condition (1) holds with $\delta = \min\{1, \varepsilon/(1 + 2|c|)\}$.

EXAMPLE 2. $f(x) = x^n$ (n positive integer) is continuous at every point c. To prove it, write

$$x^n - c^n = (x - c)(x^{n-1} + x^{n-2}c + \cdots + xc^{n-2} + c^{n-1}).$$

If $|x - c| < \delta$ and $\delta < 1$, then $|x| < 1 + |c|$. Hence

$$|x^{n-1-i}c^i| < (1 + |c|)^{n-1}.$$

It follows that

$$|x^n - c^n| < n(1 + |c|)^{n-1}|x - c| < n(1 + |c|)^{n-1}\delta \le \varepsilon$$

if $\delta \le \varepsilon/n(1 + |c|)^{n-1}$. We take $\delta = \min\{1, \varepsilon/n(1 + |c|)^{n-1}\}$.

EXAMPLE 3. $f(x) = \sin x$ is continuous at each point c. To prove it, we use the trigonometric formula

$$\sin x - \sin c = 2 \sin \frac{x - c}{2} \cos \frac{x + c}{2}$$

and the inequality $|\sin y| \le |y|$. We find that

$$|\sin x - \sin c| \le 2 \left| \sin \frac{x - c}{2} \right| \le |x - c|.$$

Hence, if $\delta = \varepsilon$, then $|x - c| < \delta$ implies that $|\sin x - \sin c| < \varepsilon$.

THEOREM 1.   *Let $f(x)$ be defined on an interval I and let it be continuous at a point c of I. Then for any sequence $\{x_n\}$ of points of I, if $\lim_n x_n = c$, $\lim_n f(x_n)$ exists and equals $f(c)$.*

Proof.   Take a sequence $\{x_n\}$ where $x_n \in I$ for each n and $\lim x_n = c$. We have to show that for any $\varepsilon > 0$ there is a positive integer n_0 such that

$$|f(x_n) - f(c)| < \varepsilon \qquad \text{if } n \geq n_0. \tag{4}$$

Since $f(x)$ is continuous at c, condition (1) holds. Since also $\lim x_n = c$, we can find a positive integer n_0 such that

$$|x_n - c| < \delta \qquad \text{if } n \geq n_0.$$

But then we can apply (1) to $x = x_n$, $n \geq n_0$. This yields (4).

We shall now prove a converse of Theorem 1.

THEOREM 2.   *Let $f(x)$ be defined on an interval I and let c be a point of I. Assume that for any sequence $\{x_n\}$ of points of I, if $\lim_n x_n = c$, then $\lim f(x_n)$ exists and equals $f(c)$. Then f is continuous at c.*

Proof.   Suppose $f(x)$ is not continuous at c. Then there exists a positive number ε such that property (1) is not satisfied no matter what δ is. Thus, if $\delta = 1/n$ where n is any positive integer, then there is a point x_n such that $x_n \in I$, $|x_n - c| < 1/n$, but

$$|f(x_n) - f(c)| \geq \varepsilon. \tag{5}$$

It is clear that $\lim x_n = c$. On the other hand, from (5) we see that it is impossible that $\lim f(x_n)$ will both exist and be equal to $f(c)$. We have thus derived a contradiction to the assumptions of the theorem.

Let I be an interval and let ξ be an interior point of I. Denote by I' the set of all points x of I that are different from ξ. Let $f(x)$ be a function defined on either I or I'. Suppose that there is a number A such that the following holds: For every $\varepsilon > 0$ there is a $\delta > 0$ such that

$$|f(x) - A| < \varepsilon \qquad \text{if } 0 < |x - \xi| < \delta \text{ and } x \in I. \tag{6}$$

Then we say that $f(x)$ *converges* (or *tends*) to A as x converges (or *tends*) to ξ, and write

$$\lim_{x \to \xi} f(x) = A, \qquad \text{or } f(x) \to A \text{ as } x \to \xi.$$

We call A the *limit of $f(x)$* as $x \to \xi$.

EXAMPLE 4.   If $f(x)$ is defined on an interval I and ξ is an interior point of I, then $f(x)$ is continuous at ξ if, and only if, $\lim\limits_{x \to \xi} f(x) = f(\xi)$.

Suppose next that the condition (6) is replaced by

$$|f(x) - A| < \varepsilon \qquad \text{if } \xi < x < \xi + \delta, x \in I;$$

here ξ may be either an interior point of I or the left endpoint of I. We then say that $f(x)$ converges (or tends) to A as x converges (or tends) to ξ *from the right*, and write

$$\lim_{x \to \xi+} f(x) = A, \qquad \lim_{x \to \xi+0} f(x) = A,$$

$$\lim_{x \searrow \xi} f(x) = A, \qquad \text{or } f(x) \to A \text{ as } x \searrow \xi.$$

We call A the *right limit* of $f(x)$ at ξ. We also denote A by $f(\xi +)$ or by $f(\xi + 0)$. Thus

$$\lim_{x \to \xi+} f(x) = \lim_{x \to \xi+0} f(x) = \lim_{x \searrow \xi} f(x) = f(\xi+) = f(\xi + 0).$$

Suppose now that the condition (6) is replaced by

$$|f(x) - A| < \varepsilon \qquad \text{if } \xi - \delta < x < \xi, x \in I;$$

here ξ may be either an interior point of I or the right endpoint of I. Then we say that $f(x)$ converges to A *from the left*, and write

$$\lim_{x \to \xi-} f(x) = \lim_{x \to \xi-0} f(x) = \lim_{x \nearrow \xi} f(x) = f(\xi-) = f(\xi - 0) = A.$$

We call A the *left limit* of $f(x)$ at ξ.

DEFINITION.   Let $f(x)$ be a function defined on an interval I and let ξ be a point of I. If ξ is not the right endpoint of I and $f(\xi + 0)$ exists, then we say that $f(x)$ is *right continuous* at ξ. If ξ is not the left endpoint of I and $f(\xi - 0)$ exists, then we say that $f(x)$ is *left continuous* at ξ.

Note that if f is both right and left continuous at an interior point ξ of I, and if $f(\xi + 0) = f(\xi - 0) = f(\xi)$, then f is continuous at ξ. Conversely, if f is continuous at ξ then it is both right continuous and left continuous, and $f(\xi + 0) = f(\xi - 0) = f(\xi)$.

If ξ is the left endpoint of the interval on which f is defined, then $f(x)$ is continuous at ξ if, and only if, f is right continuous at ξ and $f(\xi+) = f(\xi)$.

If a function f is defined in an interval I containing a point ξ and is not continuous at ξ, then we say that f is *discontinuous* at ξ. If $f(x)$ is not defined at ξ (but is defined for all $x \in I$, $x \neq \xi$), then we still say that f is *discontinuous* at ξ provided, for any definition of $f(\xi)$, f is discontinuous at ξ.

EXAMPLE 5.   The function $f(x) = x \sin 1/x$, defined for $x > 0$, has right limit 0 at $x = 0$. Indeed, for any $\varepsilon > 0$,

$$\left| x \sin \frac{1}{x} - 0 \right| = \left| x \sin \frac{1}{x} \right| \le |x| < \varepsilon \qquad \text{if } |x - 0| < \delta,$$

provided $\delta = \varepsilon$.

EXAMPLE 6.   The function $f(x) = \sin 1/x$, $x > 0$, does not have right limit at 0. To prove this we remark that if a right limit A exists then for any sequence of points $\{x_n\}$

$$\lim f(x_n) = A \qquad \text{if } x_n > 0, x_n \to 0 \text{ as } n \to \infty. \tag{7}$$

Indeed, this follows from the proof of Theorem 1. Take two sequences. The first one $\{x_n'\}$ is given by $x_n' = 1/2n\pi$ and the second one $\{x_n''\}$ is given by $x_n'' = 1/(2n\pi + \pi/2)$. Then $x_n' > 0$, $\lim x_n' = 0$, $x_n'' > 0$, $\lim x_n'' = 0$, but

$$\sin \frac{1}{x_n'} = 0, \qquad \sin \frac{1}{x_n''} = 1.$$

This shows that (7) cannot hold. Thus, the function $\sin 1/x$ has no right limit at 0.

DEFINITION.   If a function f, defined on an interval I, is continuous at each point of I, then we say that f is *continuous on I* (or *in I*).

PROBLEMS

1.   The function $|x|$ is continuous for all x.

2.   The function $\operatorname{sgn} x$ is defined as follows:

$$\operatorname{sgn} x = \begin{cases} 1 & \text{if } x > 0, \\ -1 & \text{if } x < 0, \\ 0 & \text{if } x = 0. \end{cases}$$

Prove that it is right continuous and left continuous at 0.

3. $\sqrt{x}$ is continuous for all $x \ge 0$.

4. $\cos x$ is continuous for all x.

5. $1/x$ is continuous for all $x > 0$. Does it have a right limit at 0?

6. At which point is the following function discontinuous?

$$f(x) = \begin{cases} x^2 - 5 & \text{if } x < -2, \\ 1 + 2x^3 & \text{if } -2 \leq x < 1, \\ 2 + \sqrt{x} & \text{if } x \geq 1. \end{cases}$$

7. Draw graphs of sin $1/x$ and of x sin $1/x$. These graphs will illustrate very clearly the behavior of these functions at $x = 0$.

8. Denote by G the set of all limit points of sequences $\{\sin 1/x_n\}$ where $x_n > 0$, $\lim x_n = 0$. Prove that G coincides with the interval $[-1,1]$.

9. Consider the function

$$f(x) = \begin{cases} x & \text{if } x \text{ is rational}, \\ 1 - x & \text{if } x \text{ is irrational}. \end{cases}$$

Prove that f is continuous only at $x = 1/2$.

DEFINITION.   We shall denote by (a,∞) the set of all points x satisfying $x > a$. Similarly, we denote by $[a,\infty)$ the set of all points x satisfying $x \geq a$. We call (a,∞), $[a,\infty)$ *unbounded intervals* and we say that the intervals (a,∞), $[a,\infty)$ are *unbounded from the right*. We write (a,∞) and $[a,\infty)$ also in the form $\{x; a < x < \infty\}$ (or $a < x < \infty$) and $\{x; a \leq x < \infty\}$ (or $a \leq x < \infty$), respectively.

The intervals $(-\infty,b)$ and $(-\infty,b]$ are defined by the relations $x < b$ and $x \leq b$, respectively. We say that these intervals are *unbounded from the left*. We write

$$(-\infty,b) = \{x; -\infty < x < b\}, \qquad (-\infty,b] = \{x; -\infty < x \leq b\}.$$

By the unbounded interval $(-\infty,\infty)$, we mean the set of all the real numbers. An interval that is not unbounded is also called a *bounded interval*.

The definitions of continuity, right and left continuity, and so on, as well as Theorems 1 and 2, immediately extend to the case where I is an unbounded interval.

DEFINITION.   Let $f(x)$ be defined in an interval (a,∞). Suppose that there is a number A such that the following holds. For any $\varepsilon > 0$ there is a number N in (a,∞) such that

$$|f(x) - A| < \varepsilon \qquad \text{if } x > N. \tag{8}$$

We then say that $f(x)$ *converges* (or *tends*) *to A as x converges* (or *tends*) *to ∞*, and write

$$\lim_{x \to \infty} f(x) = A, \qquad \text{or } f(x) \to A \text{ as } x \to \infty.$$

We also say that $f(x)$ is *continuous at* ∞ and write $f(\infty) = A$.

Similarly, one defines the concept of $f(x)$ *converges to A as x converges to* $-\infty$. Instead of (8), we have

$$|f(x) - A| < \varepsilon \qquad \text{if } x < -N.$$

We now write

$$\lim_{x \to -\infty} f(x) = A, \qquad f(x) \to A \text{ as } x \to -\infty, \qquad \text{or } f(-\infty) = A.$$

There is one more concept we would like to define. Let I be an interval and let $\xi \in I$. Let $f(x)$ be defined for all $x \neq \xi$, $x \in I$. Suppose that for every positive number Ω there exists a $\delta > 0$ such that

$$f(x) > \Omega \qquad \text{if } 0 < |x - \xi| < \delta, x \in I.$$

Then we say that $f(x)$ *converges to* ∞ *as* $x \to \xi$. We write

$$\lim_{x \to \xi} f(x) = \infty. \tag{9}$$

Note that f is not continuous at ξ; in fact, it is not even defined at ξ.

Similarly, one defines the concepts

$$\lim_{x \to \xi} f(x) = -\infty, \qquad f(\xi + 0) = \infty, \qquad f(\xi - 0) = \infty, \text{ and so on.}$$

One also defines the analogous concepts with ξ replaced by ∞. For instance,

$$\lim_{\xi \to \infty} f(\xi) = -\infty$$

if for any $\Omega > 0$ there is an $N > 0$ such that

$$f(x) < -\Omega \qquad \text{if } x > N.$$

PROBLEMS

10. Let $f(x) = 1/(x^2 - 1)$ if $x \neq 1$. Find $f(1 + 0)$, $f(1 - 0)$, $f(-1 + 0)$, $f(-1 - 0)$, $f(\infty)$, $f(-\infty)$.

11. Find $f(\infty)$ for each of the following functions:

(a) $\sqrt{x + 1} - \sqrt{x}$;

(b) $\sqrt{x}(\sqrt{x + 1} - \sqrt{x})$;

(c) $\dfrac{x^2 - 1}{2x^2 + 3}$;

(d) $\dfrac{x}{\sqrt{1 + x^2}}$.

12. Prove that (9) holds if, and only if, for any sequence $\{x_n\}$ with $x_n \in I$, $x_n \neq \xi$, $x_n \to \xi$ as $n \to \infty$, the sequence $\{f(x_n)\}$ converges to ∞.

2. OPERATIONS WITH CONTINUOUS FUNCTIONS

Let f and g be two functions defined on a set I. Then the functions $f + g, f - g, fg$, and f/g (if $g \neq 0$) are defined by

$$(f + g)(x) = f(x) + g(x),$$
$$(f - g)(x) = f(x) - g(x),$$
$$(fg)(x) = f(x)g(x),$$
$$(f/g)(x) = f(x)/g(x),$$

for any x in I.

THEOREM 1. *Let f and g be functions defined on an interval I and continuous at a point ξ of I. Then $f + g$ is also continuous at ξ.*

First proof. We have to prove that for any $\varepsilon > 0$ there exists a $\delta > 0$ such that

$$|(f(x) + g(x)) - (f(\xi) + g(\xi))| < \varepsilon \qquad \text{if } |x - \xi| < \delta, x \in I. \qquad \textbf{(1)}$$

Since f and g are continuous at ξ, there exist positive numbers δ_1 and δ_2 such that

$$|f(x) - f(\xi)| < \frac{\varepsilon}{2} \qquad \text{if } |x - \xi| < \delta_1, x \in I,$$

$$|g(x) - g(\xi)| < \frac{\varepsilon}{2} \qquad \text{if } |x - \xi| < \delta_2, x \in I.$$

It follows that if $x \in I$ and $|x - \xi| < \delta$, where $\delta = \min(\delta_1, \delta_2)$, then

$$|(f(x) + g(x)) - (f(\xi) + g(\xi))| = |(f(x) - f(\xi)) + (g(x) - g(\xi))|$$

$$\leq |f(x) - f(\xi)| + |g(x) - g(\xi)| < \frac{\varepsilon}{2} + \frac{\varepsilon}{2} = \varepsilon.$$

This completes the proof.

Second proof. In view of Theorem 2 of Section 1, it suffices to show the following: If $x_n \in I$, $x_n \to \xi$ as $n \to \infty$, then

$$f(x_n) + g(x_n) \to f(\xi) + g(\xi) \qquad \text{as } n \to \infty. \qquad \textbf{(2)}$$

Since f and g are continuous at ξ, Theorem 1 of Section 1 implies that

$$f(x_n) \to f(\xi) \qquad \text{and} \qquad g(x_n) \to g(\xi) \qquad \text{as } n \to \infty. \qquad \textbf{(3)}$$

Now use Theorem 1 of Section 1.6 to deduce (2) from (3).

THEOREM 2. *Let f and g be functions defined on an interval I and continuous at a point ξ of I. Then fg is also continuous at ξ.*

Also, this theorem can be given two proofs; the first one is a direct proof analogous to the first proof of Theorem 1, and the second proof is indirect and is analogous to the second proof of Theorem 1. We shall give here only the latter proof.

Proof. In view of Theorem 2 of Section 1, it suffices to prove the following: If $x_n \in I$, $x_n \to \xi$ as $n \to \infty$, then

$$f(x_n)g(x_n) \to f(\xi)g(\xi) \qquad \text{as } n \to \infty. \tag{4}$$

Since f and g are continuous at ξ, Theorem 1 of Section 1 implies that (3) holds. Now use Theorem 3 of Section 1.6 to deduce (4) from (3).

THEOREM 3. *Let f and g be continuous functions defined on an interval I and continuous at a point ξ of I, and assume that g(x) ≠ 0 for all x in I. Then f/g is also continuous at ξ.*

We shall give a direct proof, analogous to the first proof of Theorem 1. First we need a lemma.

LEMMA. *Let g be as in Theorem 3. Then there is an η-neighborhood $(\xi - \eta, \xi + \eta)$ of ξ such that $|g(x)| \geq |g(\xi)|/2 > 0$ for all x in this neighborhood and in I.*

Proof. Since g is continuous at ξ and since $g(\xi) \neq 0$, for the positive number $\varepsilon = |g(\xi)|/2$ there is associated a positive number η such that

$$|g(x) - g(\xi)| < \frac{|g(\xi)|}{2} \qquad \text{if } |x - \xi| < \eta \quad \text{and} \quad x \in I.$$

It follows that

$$|g(x)| = |(g(x) - g(\xi)) + g(\xi)| \geq |g(\xi)| - |g(x) - g(\xi)|$$
$$> |g(\xi)| - \frac{|g(\xi)|}{2} = \frac{|g(\xi)|}{2}$$

if $|x - \xi| < \eta, x \in I$.

Proof of Theorem 3. Since f and g are continuous at ξ, for any $\varepsilon > 0$ there exist positive numbers δ_1 and δ_2 such that

$$|f(x) - f(\xi)| < |g(\xi)| \frac{\varepsilon}{4} \qquad \text{if } |x - \xi| < \delta_1, x \in I, \tag{5}$$

$$|g(x) - g(\xi)| < |g(\xi)|^2 \frac{\varepsilon}{4(|f(\xi)| + 1)} \qquad \text{if } |x - \xi| < \delta_2, x \in I. \tag{6}$$

We can write

$$\left|\frac{f(x)}{g(x)} - \frac{f(\xi)}{g(\xi)}\right| = \frac{|f(x)g(\xi) - f(\xi)g(x)|}{|g(x)g(\xi)|}$$

$$\leq \frac{|f(x)g(\xi) - f(\xi)g(\xi)|}{|g(x)g(\xi)|} + \frac{|f(\xi)g(\xi) - f(\xi)g(x)|}{|g(x)g(\xi)|}$$

$$\leq \frac{|f(x) - f(\xi)|}{|g(x)|} + |f(\xi)| \frac{|g(\xi) - g(x)|}{|g(x)g(\xi)|}.$$

If we take $\delta = \min(\eta, \delta_1, \delta_2)$, then we can use the lemma and (5) and (6). We find that

$$\left|\frac{f(x)}{g(x)} - \frac{f(\xi)}{g(\xi)}\right| < \varepsilon \qquad \text{if } |x - \xi| < \delta, x \in I.$$

This completes the proof.

DEFINITION.   Let $g(x)$ be a function defined on an interval I. Denote by J_0 the set of all numbers $g(x)$ when x varies in I. Let $f(x)$ be a function defined on an interval J containing the set J_0. Then, for each $x \in I$ there is defined the number $f(g(x))$. We now define a function h by $h(x) = f(g(x))$, and call it the *composite function*. It is also written in the form $f \circ g$.

THEOREM 4.   *Let f and g be as in the previous definition. Let $g(x)$ be continuous at a point ξ of I and let $f(x)$ be continuous at the point $g(\xi)$. Then the composite function $h = f \circ g$ is continuous at ξ.*

Proof.   Since f is continuous at $g(\xi)$, for any $\varepsilon > 0$ there is a positive number η such that

$$|f(y) - f(g(\xi))| < \varepsilon \qquad \text{if } |y - g(\xi)| < \eta, y \in J. \tag{7}$$

Since $g(x)$ is continuous at ξ, given the positive number η there exists a positive number δ such that

$$|g(x) - g(\xi)| < \eta \qquad \text{if } |x - \xi| < \delta, x \in I. \tag{8}$$

From (8) we see that we can apply (7) with $y = g(x)$ provided $|x - \xi| < \delta$, $x \in I$. Hence

$$|f(g(x)) - f(g(\xi))| < \varepsilon \qquad \text{if } |x - \xi| < \delta, x \in I.$$

This proves the continuity of $h(x)$ at ξ.

EXAMPLE 1.   Let $f_1, f_2, ..., f_n$ be continuous functions. By Theorem 1, $f_1 + f_2$ is continuous. By the same theorem, $f_1 + f_2 + f_3 = (f_1 + f_2) + f_3$ is also a continuous function. Proceeding in this way step by step, we conclude that $f_1 + f_2 + \cdots + f_n$ is a continuous function. Similarly, by using Theorem 2 we conclude that the product $f_1 f_2 ... f_n$ is continuous.

Suppose we just know that the function $f(x) = x$ is continuous. The function $f_1(x) = c$ (c a fixed number) is clearly also continuous. Taking $f_1 = c, f_2 = f_3 = \cdots = f_{m+1} = x$, it follows that $cx^m = f_1 f_2 ... f_{m+1}$ is continuous. Taking $f_1 = a_0 x^n, f_2 = a_1 x^{n-1}, ..., f_n = a_{n-1} x_1, f_{n+1} = a_n$ where the a_i are numbers, we further have that $f_1 + f_2 + \cdots + f_{n+1}$ is continuous, that is,

$$a_0 x^n + a_1 x^{n-1} + a_2 x^{n-1} + \cdots + a_{n-1} x + a_n \tag{9}$$

is continuous. Any function of the form (9) is called a *polynomial* (in the variable x).

EXAMPLE 2.   A quotient of two polynomials

$$\frac{a_0 x^n + a_1 x^{n-1} + \cdots + a_{n-1} x + a_n}{b_0 x^m + b_1 x^{m-1} + \cdots + a_{m-1} x + b_m} \tag{10}$$

is called a *rational function*. By Theorem 3, a rational function is continuous on any interval where the denominator is never 0.

EXAMPLE 3.   If $g(x)$ is continuous on an interval I and if $g(x) \geq 0$ on I, then $\sqrt{g(x)}$ is continuous. Indeed, this follows from Theorem 4 with $f(x) = \sqrt{x}$.

EXAMPLE 4.   If $g(x)$ is continuous, then $|g(x)|$ is continuous. Indeed, this follows from Theorem 4 with $f(x) = |x|$.

PROBLEMS

1. Consider the function $[x]$ = largest integer $\leq x$. Investigate the continuity of the functions: $[x]$; $x - [x]$; $[x] + [-x]$. Draw the graphs of these functions.

2. Investigate the continuity of the function $f(x) = \sin(\pi[x]/2)$. Draw the graph of this function.

3. Find $\lim\limits_{x \to 0+} (\sqrt{1 + x^2} - 1)/x$.

4. Find the points ξ where the function $f(x) = \tan x$ is not continuous. Compute $f(\xi+)$ and $f(\xi-)$.

5. Find the points of discontinuity of the following functions and compute the right and left limits (if they exist):

(a) $\dfrac{1}{\sin x - \cos x}$; (b) $\dfrac{x}{\sin x + \cos x}$; (c) $(x^2 + 1)\sin 1/x$;

(d) $\dfrac{1}{\tan x + \sqrt{3}}$; (e) $\cos 1/x$; (f) $x \sin 1/x^2$.

6. Give an example of a function $f(x)$ such that $|f(x)|$ is continuous at some point ξ but $f(x)$ is not continuous at ξ.

7. Let $f(x) = A \cos x + B \sin x$ where A and B are numbers. Prove that if $\lim\limits_{x \to \infty} f(x)$ exists, then $A = B = 0$.

8. If $\lim\limits_{x \to \infty} f(x) = A$ and $\lim\limits_{x \to \infty} g(x) = B$, then $\lim\limits_{x \to \infty} f(x)g(x) = AB$.

9. If $\lim\limits_{x \to \xi} f(x) = \infty$ and $\lim\limits_{x \to \xi} g(x) = B$, then $\lim\limits_{x \to \xi} f(x)g(x)$ is equal to ∞ if $B > 0$ and to $-\infty$ if $B < \infty$.

3. MAXIMUM AND MINIMUM

We sometimes use the word *constant* instead of the word *number*. A function f satisfying $f(x) = c$ for all x in its domain of definition, where c is a constant, is called a *constant function*.

DEFINITION.   Let $f(x)$ be a function defined on a set I. Consider the set G of its values $f(x)$, where x varies in I. If G is bounded above, then we say that f is *bounded above*. This is the case if, and only if, there exists a constant N such that

$$f(x) \le N \qquad \text{for all } x \in I. \tag{1}$$

The supremum of G is called the *supremum* (or the *least upper bound*) of f on I. We denote it by

$$\sup_{x \in I} f(x), \qquad \text{or l.u.b. } f(x).$$

If G is bounded below, then we say that f is *bounded below*. The infimum of G is called the *infimum* (or the *greatest lower bound*) of f on I. It is denoted by

$$\inf_{x \in I} f(x), \qquad \text{or g.l.b. } f(x).$$

If G is a bounded set, then we say that f is a *bounded* function on I. This is the case if, and only if, there exists a positive constant M such that

$$|f(x)| \le M \qquad \text{for all } x \in I. \tag{2}$$

If f is not bounded, then we say that it is *unbounded*.

THEOREM 1. *Let f be a continuous function on a closed, bounded interval $[a,b]$. Then f is bounded.*

Proof. Suppose f is unbounded. That means that there does not exist a number M for which (2) holds. Thus, for any positive integer n there is a point x_n in $[a,b]$ for which

$$|f(x_n)| > n. \tag{3}$$

The sequence $\{x_n\}$ is a bounded sequence. Hence, by the Bolzano–Weierstrass theorem, there exists a subsequence $\{x_{n_k}\}$ that is convergent to some point ξ. Since this point ξ belongs to $[a,b]$, f is defined at ξ and $f(\xi)$ is some real number.

Now, by Theorem 1 of Section 1,

$$\lim_{k \to \infty} f(x_{n_k}) = f(\xi).$$

Since $\{f(x_{n_k})\}$ is a convergent sequence, it must also be a bounded sequence (by Theorem 2 of Section 1.6). Thus,

$$|f(x_{n_k})| \le N \qquad (k = 1,2,\ldots) \tag{4}$$

for some constant N. However, from (3) we infer that

$$|f(x_{n_k})| > n_k \qquad (k = 1,2,\ldots).$$

Since $n_k \to \infty$ as $k \to \infty$, this contradicts (4).

DEFINITION. Let $f(x)$ be a function defined on a set I and bounded above. Suppose there exists a point ξ in I such that

$$f(x) \le f(\xi) \qquad \text{for all } x \in I, \tag{5}$$

that is,

$$f(\xi) = \sup_{x \in I} f(x).$$

Then we call ξ a *maximum point* of f and we call $f(\xi)$ the *maximum* of f on I. We also say that f has a maximum, and that it assumes (or attains) its maximum on I at the point ξ.

Similarly, we define a *minimum point* η of f on I by

$$f(x) \geq f(\eta) \qquad \text{for all } x \in I. \tag{6}$$

If (6) holds, then we say that $f(\eta)$ is the *minimum* of f on I. We also say that f has a minimum, and that it assumes its minimum on I at η.

THEOREM 2. *Let $f(x)$ be a continuous function on a closed, bounded interval $[a,b]$. Then f has maximum and minimum on I.*

Proof. By Theorem 1, f is bounded on $[a,b]$. That is, the set G of all points $f(x)$ where x varies in $[a,b]$ is bounded. Denote by M its supremum. Then

$$f(x) \leq M \qquad \text{for all } a \leq x \leq b, \tag{7}$$

and for any $\varepsilon > 0$ there is a point x_ε in $a \leq x \leq b$ such that

$$f(x_\varepsilon) > M - \varepsilon. \tag{8}$$

Take $\varepsilon = 1/n$, $n = 1,2,\ldots$. Denote the corresponding point x_ε by z_n. From (7) and (8) we have

$$|f(z_n) - M| < \frac{1}{n}. \tag{9}$$

By the Bolzano–Weierstrass theorem there is a subsequence $\{z_{n_k}\}$ that converges to some point ξ. Since ξ belongs to $[a,b]$, and since $f(x)$ is continuous at ξ,

$$\lim_k f(z_{n_k}) = f(\xi).$$

Since, by (9),

$$\lim_k f(z_{n_k}) = M,$$

we conclude that $f(\xi) = M$. From (7) we then get

$$f(x) \leq f(\xi) \qquad \text{for all } a \leq x \leq b.$$

This shows that f has maximum, attained at point ξ.

The proof that f has minimum is similar.

PROBLEMS

1. Find which of the following functions is bounded:

(a) $2x \sin 1/x + \cos 1/x$ for $x \neq 0$;

(b) $\dfrac{1-x}{1+x}$ for $-1 < x < 1$;

(c) $\dfrac{\sin x}{x}$ for $0 < x \leq 1$;

(d) $\dfrac{x^2 + 1}{x^3 + 1}$ for $0 < x < \infty$;

(e) $\tan x$ for $0 < x < \pi/2$;

(f) $\dfrac{1 - \cos x}{x^2}$ for $x \neq 0$.

2. Find the minimum and maximum, if they exist, of

   (a) $x - [x]$ on the interval $[0,3]$;

   (b) $\sqrt{5 - x^2}$ on the interval $[1,2]$;

   (c) $x^2 + 3x - 5$ on the interval $[-1,1]$.

3. Let f be continuous on the whole real line and assume that for some $\lambda > 0, f(x + \lambda) = f(x)$ for all x. Prove that f is bounded on $(-\infty,\infty)$ and that it has maximum and minimum.

4. Let f be a continuous function on an unbounded interval $[b,\infty)$. Assume that $f(x) \to A$ as $x \to \infty$, where A is some real number. Prove that f is bounded on $[b,\infty)$.

5. Let f be as in the previous problem. Does f have a maximum in $[b,\infty)$?

4. INTERMEDIATE VALUES

The following theorem is sometimes called the *intermediate value theorem*.

THEOREM 1.   *Let $f(x)$ be a continuous function on a closed, bounded interval $[a,b]$. Assume that $f(a) \neq f(b)$ and let C be any number lying between $f(a)$ and $f(b)$. Then there exists at least one point c in the interval (a,b) such that $f(c) = C$.*

In other words, a continuous function taking values A and B at the endpoints of an interval also must take any intermediate value at some intermediate point of the interval.

Proof.   For the sake of definiteness assume that $f(a) < f(b)$. The proof in the case where $f(a) > f(b)$ is similar. Denote by c_1 the midpoint of $[a,b]$. If $f(c_1) = C$, then the proof of the theorem is complete with $c = c_1$. If $f(c_1) \neq C$, then there are two cases: (i) $f(c_1) < C$, and (ii) $f(c_1) > C$. If (i) holds, then we denote by I_1 the interval $[c_1, b]$ whereas if (ii) holds, then we denote by I_1 the interval $[a, c_1]$. In either case we have the situation that the value of f at the left endpoint of I_1 is less than C, whereas the value of f at the right endpoint of I_1 is larger than C.

Next we divide I_1 into two subintervals by introducing its midpoint c_2. We may assume that $f(c_2) \neq C$. We then define I_2 to be that subinterval with the property that the value of f at the left endpoint is less than C whereas the value of f at the right endpoint is larger than C.

Proceeding in this way step by step, we construct a sequence of closed intervals $I_1, I_2, I_3, \ldots$ such that $I_1 \supset I_2 \supset I_3 \supset \ldots$, and such that the length of each interval I_n is $(b - a)/2^n$. $f(x)$ is less than C at the left endpoint of each I_n, and $f(x)$ is larger than C at the right endpoint of each I_n. Write the interval I_n in the form

$$I_n = \{x ; a_n \leq x \leq b_n\}.$$

Then $\{a_n\}$ is monotone increasing, $\{b_n\}$ is monotone decreasing, and $b_n - a_n = (b - a)/2^n \to 0$ as $n \to \infty$. It follows that $\lim_n a_n$ and $\lim_n b_n$ exist and are equal. Denote the common limit by ξ. Since

$$f(a_n) < C, \qquad f(b_n) > C$$

and since f is continuous, we have

$$f(\xi) = \lim_n f(a_n) \leq C, \qquad f(\xi) = \lim_n f(b_n) \geq C.$$

Hence $f(\xi) = C$. This completes the proof.

From Theorem 2 of Section 3 it follows that if f is a continuous function on a closed, bounded interval, then it assumes its maximum, say M, and its minimum, say m. From Theorem 1 it also follows that f assumes any intermediate value. Hence:

COROLLARY. *Let $f(x)$ be a continuous function on a closed, bounded interval $[a,b]$. Denote by G the set of all numbers $f(x)$, where x varies in $[a,b]$. Then G coincides with a closed interval $[m,M]$.*

PROBLEMS

1. Show that for any $\delta > 0$ the function

$$f(x) = \begin{cases} \sin \dfrac{1}{x} & \text{if } 0 < x \leq \delta, \\ 0 & \text{if } x = 0 \end{cases}$$

takes intermediate values (as in the assertion of Theorem 1). Note that this function is discontinuous at 0.

2. If $f(x)$ is continuous in a closed, bounded interval $[a,b]$ and if $f(a) = f(b)$, then f assumes any value different from and between the maximum and the minimum at least twice (that is, at least at two different points of $[a,b]$).

3. If there exists a continuous function $f(x)$ in a closed bounded interval, which takes every value exactly twice, then f must assume its maximum at both endpoints. Similarly, f must assume its minimum at both endpoints. Since this yields a contradiction, such a continuous function f cannot exist.

4. If f is right continuous and left continuous on a closed, bounded interval $[c,d]$, and if f satisfies the intermediate value property in every subinterval $[a,b]$ of $[c,d]$, then f is continuous on $[c,d]$.

5. If $f(x)$ is continuous in $[0,2a]$ and if $f(0) = f(2a)$, then $f(x) = f(x + a)$ for some x in the interval $[0,a]$. [Hint: Apply Theorem 1 to $f(x) - f(x + a)$.]

5. MONOTONE FUNCTIONS AND INVERSE FUNCTIONS

Let $f(x)$ be a function defined on an interval I. Suppose that for every two points x_1 and x_2 of I,

$$\text{if } x_1 < x_2, \qquad \text{then } f(x_1) \leq f(x_2). \tag{1}$$

Then we say that $f(x)$ is *monotone increasing* in I. If instead of (1) we assume that

$$\text{if } x_1 < x_2, \qquad \text{then } f(x_1) < f(x_2), \tag{2}$$

then we say that $f(x)$ is *strictly monotone increasing* in I.

Now suppose that instead of (1) we have

$$\text{if } x_1 < x_2, \qquad \text{then } f(x_1) \geq f(x_2). \tag{3}$$

Then we say that $f(x)$ is *monotone decreasing* in I. Finally, if instead of (3) we have

$$\text{if } x_1 < x_2, \qquad \text{then } f(x_1) > f(x_2), \tag{4}$$

then we say that $f(x)$ is *strictly monotone decreasing* in I.

A (*strictly*) *monotone* function is a function that is either (strictly) monotone increasing or (strictly) monotone decreasing.

THEOREM 1. *If $f(x)$ is a continuous function on a closed, bounded interval $[a,b]$, and if it assumes every value between $f(a)$ and $f(b)$ exactly once, then $f(x)$ is strictly monotone in this interval.*

Proof. It suffices to consider the case where $f(a) < f(b)$. We shall prove that condition (2) holds whenever $a \le x_1 < x_2 \le b$. We first show that

$$\text{if } a < x_1 < b, \qquad \text{then } f(a) < f(x_1) < f(b). \tag{5}$$

To show this we have to rule out the following possibilities:

    (i) $f(x_1) < f(a)$;
    (ii) $f(x_1) > f(b)$.

If (i) holds, then take any number C between $f(x_1)$ and $f(a)$. Then C lies also between $f(x_1)$ and $f(b)$. By Theorem 1 of Section 4, f takes this value C at least once in each interval (a,x_1) and (x_1,b). This contradicts an assumption of the theorem.

Case (ii) can be ruled out in a similar way. In fact, if (ii) holds, then any number C between $f(b)$ and $f(x_1)$ is assumed at least once in each interval (a,x_1) and (x_1,b).

From (5) it follows that if $a < x_2 \le b$, then $f(a) < f(x_2)$. We therefore can apply (5) with $b = x_2$. We conclude that if $a < x_1 < x_2 \le b$, then

$$f(a) < f(x_1) < f(x_2).$$

This proves (2) whenever $a \le x_1 < x_2 \le b$.

THEOREM 2.   *A bounded monotone function f in an interval I is right and left continuous.*

Proof. It will be enough to consider the case where f is monotone increasing. We shall prove that f is left continuous; the proof that f is right continuous is similar.

To prove left continuity at a point ξ, denote by M the supremum of the set of points $f(x)$ where x varies in the interval $[a,\xi)$, a being any point i.. I that is less than ξ. Then

$$f(x) \le M \qquad \text{if } a \le x < \xi, \tag{6}$$

and for any $\varepsilon > 0$ there exists a point $\bar{x}$ such that

$$a \le \bar{x} < \xi$$

and

$$f(\bar{x}) \ge M - \varepsilon.$$

Since f is monotone increasing, we conclude that $f(x) \ge M - \varepsilon$ if $\bar{x} < x < \xi$. Combining this with (6) we get

$$|f(x) - M| < \varepsilon \qquad \text{if } \xi - \delta < x < \xi$$

where $\delta = \xi - \bar{x}$. This shows that $\lim_{x \to \xi-} f(x)$ exists. Thus $f(x)$ is left continuous at ξ.

DEFINITION.   Let $f(x)$ be a function defined on a set I. Denote by J the set of all points $f(x)$ where x varies in I. Suppose that f is one to one, that is, $f(x_1) \neq f(x_2)$ if $x_1 \neq x_2$. Then we define a function g with domain J as follows: Let $y \in J$ and let x be the point in I such that $f(x) = y$. Then we set $g(y) = x$. The function g thus defined is called the *inverse function* of f. It is sometimes denoted by f^{-1}.

Since $g(y) = x$ if $y = f(x)$, it follows that

$$g(f(x)) = x, \qquad f(g(y)) = y. \tag{7}$$

If f is a strictly monotone function, then $f(x_1) \neq f(x_2)$ if $x_1 \neq x_2$. Thus a strictly monotone function has an inverse function g. If f is strictly monotone increasing, then g is also strictly monotone increasing. Indeed, otherwise there are points y_1, y_2 such that

$$y_1 > y_2$$

and

$$g(y_1) \leq g(y_2).$$

Setting $y_1 = f(x_1)$, $y_2 = f(x_2)$, we then have, by (7), $g(y_1) = x_1$, $g(y_2) = x_2$. Hence

$$x_1 \leq x_2$$

and

$$f(x_1) > f(x_2).$$

This contradicts the assumption that f is strictly monotone increasing.

THEOREM 3.   *Let $f(x)$ be a strictly monotone, continuous function in a closed, bounded interval $[a,b]$. Then its inverse g is defined on the closed interval $[f(a), f(b)]$ and is strictly monotone and continuous.*

Proof.   The assertion that g is defined on the interval $[f(a), f(b)]$ follows from the corollary to Theorem 1 of Section 4. We already have proved above that g is strictly monotone increasing. It therefore remains to prove that $g(y)$ is continuous for each y in $[f(a), f(b)]$.

For any $\varepsilon > 0$, consider the continuous function

$$h(x) = f(x + \varepsilon) - f(x)$$

defined in the interval $[a, b - \varepsilon]$. By Theorem 2 of Section 3, it assumes its minimum δ at a point $\bar{x}$ in the interval. Since $f(x)$ is strictly monotone increasing, $h(\bar{x}) > 0$. Hence $\delta > 0$.

Now take any two points x_1, x_2 in $[a,b]$. If $x_2 \geq x_1 + \varepsilon$, then

$$f(x_2) - f(x_1) \geq f(x_1 + \varepsilon) - f(x_1) \geq \delta.$$

Similarly, if $x_1 \geq x_2 + \varepsilon$, then $f(x_1) - f(x_2) \geq \delta$. Thus

$$|f(x_1) - f(x_2)| \geq \delta \qquad \text{if } |x_1 - x_2| \geq \varepsilon.$$

Setting $f(x_1) = y_1, f(x_2) = y_2$, we conclude that $|y_1 - y_2| \geq \delta$ if $|g(y_1) - g(y_2)| \geq \varepsilon$. Consequently,

$$|g(y_1) - g(y_2)| < \varepsilon \qquad \text{if } |y_1 - y_2| < \delta.$$

This yields the continuity of g.

EXAMPLE 1. Consider $f(x) = x^2$ in $[0, \infty)$. Since f is monotone increasing and continuous, its inverse g is defined on $[0, \infty)$. If $y = x^2$, then $g(y) = x = \sqrt{y}$. Thus g corresponds to any y, $0 \leq y < \infty$, the number $g(y) = \sqrt{y}$.

EXAMPLE 2. *Exponentials.* For any positive number a and for any positive integer n, we define a^n as the product of n numbers, each being a. One can then verify, by induction, that

$$a^m a^n = a^{m+n}, \qquad (a^m)^n = a^{mn}. \tag{8}$$

Consider next the equation

$$x^n = a. \tag{9}$$

Since $f(x) = x^n$ is continuous and since $f(0) = 0 < a, f(1 + a) > a$, we can apply the intermediate value theorem and conclude that Equation (9) has a positive solution $x = x_1$. x_1 is the unique positive solution of (9), for if $x_2 > x_1$ $(x_2 < x_1)$, then $x_2{}^n > x_1{}^n$ $(x_2{}^n < x_1{}^n)$. We write x_1 in the form $a^{1/n}$. We now define

$$a^{m/n} = (a^{1/n})^m, \qquad a^0 = 1, \qquad a^{-m/n} = 1/a^{m/n} \text{ (m,n positive integers).} \tag{10}$$

It can then be shown that

$$a^\alpha a^\beta = a^{\alpha+\beta}, \qquad (a^\alpha)^\beta = a^{\alpha\beta} \tag{11}$$

for all rational numbers α, β.

We shall now assume that $a > 1$. Then $x_1 = a^{1/n}$ is > 1, for if $x_1 \leq 1$, then $a = x_1{}^n \leq 1 < a$. Since $a^{1/n} > 1$, also $a^{m/n} = (a^{1/n})^m > 1$. From the first relation of (11) it now follows that $a^r > a^s$ if $r > s$, r and s being rational numbers. Indeed, $a^r = a^{r-s}a^s > a^s$ since $a^{r-s} > 1$.

Now let x be any irrational number and let $\{r_n\}, \{s_n\}$ be any sequences of rational numbers, monotone decreasing to x. Then

$$\lim_n a^{r_n} = \lim_n a^{s_n}. \tag{12}$$

Indeed, for any n, $s_m < r_n$ if m is sufficiently large. Hence $\lim_m a^{s_m} \le a^{r_n}$. It follows that $\lim_m a^{s_m} \le \lim_n a^{r_n}$. Similarly, one proves the converse inequality.

We have proved that $\lim_n a^{r_n}$ is independent of the particular monotone decreasing sequence of rationals r_n with $\lim r_n = x$. We define

$$a^x = \lim_{n \to \infty} a^{r_n}.$$

This limit a^x is a function of x. a^x is strictly monotone increasing. Indeed, if $x < y$, then there are rationals r,s such that $x < r < s < y$. Let $r_n \searrow x, s_n \searrow y$ as $n \to \infty$, where r_n, s_n are rational numbers. Then $r_n < r$, $s < s_n$ if n is sufficiently large. It follows that

$$a^x = \lim_n a^{r_n} \le a^r < a^s \le \lim_n a^{s_n} = a^y.$$

Using the first relation of (11) with α, β replaced by rationals α_n, β_n, where $\{\alpha_n\}$ decreases to an irrational α and $\{\beta_n\}$ decreases to an irrational β, we find that the first relation of (11) holds also if α and β are irrationals. Similarly, it holds if only α (or if only β) is irrational.

Since $a^{1/n} \to 1$ as $n \to \infty$, for any $\varepsilon > 0$ there is an n_0 such that if $n \ge n_0$

$$0 < a^{1/n} - 1 < \varepsilon, \qquad 0 < 1 - a^{-1/n} < \varepsilon.$$

Let $\delta = 1/n_0$. Since a^x is monotone increasing,

$$|a^x - 1| < \varepsilon \qquad \text{if } -\delta < x < \delta.$$

Thus a^x is continuous at 0. Using this and the first relation of (11), we find that

$$\lim_{x \to \xi} a^x = \lim_{x \to \xi} a^\xi a^{x-\xi} = a^\xi \lim_{x \to \xi} a^{x-\xi} = a^\xi.$$

It follows that a^x is continuous for all x.

We have defined a^x for $a > 1$ and any real x. If $a < 1$, then we define $a^x = (1/a)^{-x}$. If $a = 1$, then we define $a^x = 1$.

The function a^x is called an *exponential function*.

One can easily verify that the first relation in (11) holds also when $0 < a \le 1$. Note that, if $0 < a < 1$, a^x is strictly monotone decreasing.

We shall now prove the second relation in (11) for any real numbers α, β. First we note that the function $f(a) = a^{1/n}$ (n positive integer) is continuous in a, since it is the inverse of the continuous and strictly monotone function $g(x) = x^n$. It follows also that $a^{m/n}$ and $a^{-m/n}$ are continuous in a, for any positive integer m. Now let $\{\alpha_n\}$ and $\{\beta_m\}$ be monotone decreasing sequences of rational numbers, converging to α and β, respectively. Then

$$(a^\alpha)^{\beta_m} = \lim_{n \to \infty} (a^{\alpha_n})^{\beta_m} = \lim_{n \to \infty} a^{\alpha_n \beta_m} = a^{\alpha \beta_m}, \tag{13}$$

by the continuity of $b^{\beta m}$ in b (at $b = a^{\alpha}$) and of a^x in x (at $x = \alpha\beta_m$). Taking $m \to \infty$ in (13), we get

$$(a^{\alpha})^{\beta} = \lim_{m \to \infty} (a^{\alpha})^{\beta m} = \lim_{m \to \infty} a^{\alpha\beta m} = a^{\alpha\beta},$$

by the definition of $(a^{\alpha})^{\beta}$ and by the continuity of a^x at $x = \alpha\beta$.

EXAMPLE 3. *Logarithm.* Let $a > 1$. It can be shown that when x varies on the real line, the numbers a^x vary in the interval $(0,\infty)$ (see Problem 4). From Theorem 3 it follows that the inverse function $g(y)$ is continuous and strictly monotone increasing in the interval $(0,\infty)$. We write $g(y)$ in the form $\log_a y$ and call it the *logarithm of y to the base a.* If a is the number e defined in Section 1.7, then we call it the *natural logarithm* of y and write $\log y$ instead of $\log_e y$. The relations in (7) become

$$a^{\log_a x} = x, \qquad \log_a a^x = x.$$

The function e^x is also denoted by exp x.

The graphs of e^x and $\log x$ are given in Figure 2.1.

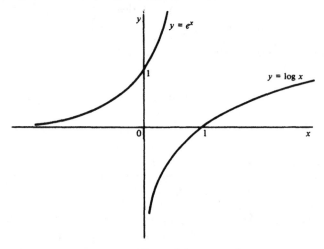

Figure 2.1

PROBLEMS

1. Verify (8) by induction.

2. Let a be a positive real number and let n be a positive integer. Define a cut (A,B) as follows: $x \in A$ if, and only if, either $x < 0$ or $x^n \le a$. Prove that this cut is the positive solution of (9).

3. Prove (11) for all rationals α, β.

4. Show that $\lim\limits_{x \to \infty} a^x = \infty$, $\lim\limits_{x \to -\infty} a^x = 0$, provided $a > 1$.

5. If f and g are monotone increasing and if the composite function $f \circ g$ is defined, then it is also monotone increasing.

6. If f and g are monotone functions on an interval I, is fg also a monotone function on I?

7. Find the inverse of $y = \sqrt{1 + x^2}$, $0 < x < \infty$.

8. The function $y = \sin x$ in $-\pi/2 \le x \le \pi/2$ has an inverse denoted by $\sin^{-1} y$, or arc sin y, with domain $-1 \le y \le 1$. Draw its graph.

9. The function $y = \cos x$ in $0 \le x \le \pi$ has an inverse denoted by $\cos^{-1} y$, or arc cos y, with domain $-1 \le y \le 1$. Draw its graph.

10. The function $y = \tan x$ in $-\pi/2 < x < \pi/2$ has an inverse denoted by $\tan^{-1} y$, or arc tan y, with domain $(-\infty, \infty)$. Draw its graph.

11. Prove: (i) $\log x + \log y = \log(x \cdot y)$; (ii) $\log x^y = y \log x$.

6. SEMICONTINUITY

Let I be an interval and let ξ be a point of I. Let $f(x)$ be a function defined for all $x \in I$, $x \ne \xi$. Suppose there is an $\eta > 0$ such that $f(x)$ is bounded above for $0 < |x - \xi| < \eta$, $x \in I$, that is, $f(x) \le C$ if $0 < |x - \xi| < \eta$. $x \in I$, where C is a constant. We then consider the set G of all the points y obtained in the following manner: There is a sequence $\{x_n\}$ in I, $x_n \ne \xi$ for all n, such that

$$\lim_n x_n = \xi, \qquad \lim_n f(x_n) = y. \tag{1}$$

If the set G is nonempty, then it is bounded above (by the same constant C as above). Denote by A the supremum of the set G. We call A the *limit superior of $f(x)$ as $x \to \xi$*, and write it in any one of the forms

$$\limsup_{x \to \xi} f(x), \qquad \overline{\lim_{x \to \xi}} f(x).$$

If G is empty, then

$$\lim_{x \to \xi} f(x) = -\infty. \tag{2}$$

Indeed, if (2) is not true, then it is not true that for any $N > 0$ there is a $\delta > 0$ such that

$$f(x) < -N \qquad \text{if } 0 < |x - \xi| < \delta, x \in I.$$

Thus there is some $N_0 > 0$ such that for any $\delta = 1/n$ there is a point x_n satisfying

$$f(x_n) \geq -N_0, \qquad 0 < |x_n - \xi| < \frac{1}{n}.$$

If n is sufficiently large, then $|x_n - \xi| < \eta$, so that $f(x_n) \leq C$. It follows that the sequence $\{f(x_n)\}$ is bounded. By the Bolzano–Weierstrass theorem, there is a subsequence $\{f(x_n')\}$ that is convergent to some limit y. Since $\{x_n'\}$ is convergent to ξ, we conclude that $y \in G$. Therefore G is not empty, which contradicts our assumption.

Consider now the case where, for any $\eta > 0$, $f(x)$ is not bounded above on the set given by $0 < |x - \xi| < \eta$, $x \in I$. Then, for any $\eta = 1/n$ there is a point x_n such that $f(x_n) > n$. It follows that

$$\lim_n x_n = \xi, \qquad \lim_n f(x_n) = \infty.$$

In this case we write

$$\limsup_{x \to \xi} f(x) = \infty,$$

or

$$\overline{\lim_{x \to \xi}} f(x) = \infty.$$

Finally, in the previous case where G is empty [and (2) holds], we write

$$\limsup_{x \to \xi} f(x) = -\infty,$$

or

$$\overline{\lim_{x \to \xi}} f(x) = -\infty.$$

THEOREM 1.   *Suppose the limit superior of $f(x)$ as $x \to \xi$ is a real number A. Then*

(a)   *For any $\varepsilon > 0$, there is a $\delta > 0$ such that*

$$f(x) < A + \varepsilon \qquad \text{if } 0 < |x - \xi| < \delta, \qquad x \in I. \tag{3}$$

(b)   *For any $\varepsilon > 0$, $\eta > 0$, there is a point $\bar{x}$ such that*

$$f(\bar{x}) > A - \varepsilon, \qquad 0 < |\bar{x} - \xi| < \eta, \qquad \bar{x} \in I. \tag{4}$$

Proof.   Suppose the assertion (a) is false, Then there is an $\varepsilon_0 > 0$ such that for any $\delta = 1/n$ there is a point x_n satisfying

$$0 < |x_n - \xi| < \frac{1}{n}, \qquad f(x_n) \geq A + \varepsilon_0. \tag{5}$$

Since $f(x)$ is bounded above for $0 < |x - \xi| < \eta$, $x \in I$ (for some $\eta > 0$), it follows that the sequence $\{f(x_n)\}$ is bounded. By the Bolzano–Weierstrass theorem, there is a convergent subsequence $\{f(x_{n_k})\}$. Denote its limit by y. Since $x_{n_k} \to \xi$ as $n_k \to \infty$, the point y belongs to the set G. This implies that $A \geq y$. But from (5) we get $y = \lim f(x_{n_k}) \geq A + \varepsilon_0$, a contradiction.

To prove the assertion (b), we suppose that this assertion is false and derive a contradiction. If (b) is false, then there are positive numbers ε, η such that

$$f(x) \leq A - \varepsilon \qquad \text{if } 0 < |x - \xi| < \eta.$$

But then for any sequence $\{x_n\}$ as in (1), $y \leq A - \varepsilon$. It follows that $A - \varepsilon$ is an upper bound of the set G. But this is impossible since A is the least upper bound of G.

Suppose now that the function $f(x)$ is bounded below in some set given by $0 < |x - \xi| < \eta$, $x \in I$, and define set G as before. If G is nonempty, we denote its infimum by B, and call it the *limit inferior of $f(x)$ as $x \to \xi$.* We denote it by

$$\liminf_{x \to \xi} f(x),$$

or

$$\underline{\lim}_{x \to \xi} f(x).$$

If G is empty, then we can prove that

$$\lim_{x \to \xi} f(x) = \infty.$$

We then also write

$$\liminf_{x \to \xi} f(x) = \infty,$$

or

$$\underline{\lim}_{x \to \xi} f(x) = \infty.$$

Finally, if $f(x)$ is not bounded below in any set given by $0 < |x - \xi| < \eta$, $x \in I$, then there is a sequence $\{x_n\}$ such that

$$x_n \neq \xi, \qquad \lim x_n = \xi, \qquad \lim_n f(x_n) = -\infty.$$

We then write

$$\liminf_{x \to \xi} f(x) = -\infty,$$

or

$$\underline{\lim}_{x \to \xi} f(x) = -\infty.$$

The analog of Theorem 1 is valid. That is, if $B = \liminf_{x \to \xi} f(x)$ is a real number, then, for any $\varepsilon > 0$ there is a $\delta > 0$ such that

$$f(x) > B - \varepsilon \qquad \text{if } 0 < |x - \xi| < \delta, \, x \in I, \tag{6}$$

and for any $\varepsilon > 0, \eta > 0$ there is a point $\bar{x}$ such that

$$f(\bar{x}) < B + \varepsilon, \qquad 0 < |\bar{x} - \xi| < \eta, \qquad \bar{x} \in I. \tag{7}$$

One defines the real number

$$\liminf_{x \to \xi - 0} f(x)$$

analogously to $\liminf_{x \to \xi} f(x)$, replacing set G by the subset G^- obtained by taking in (1) only such sequences $\{x_n\}$ for which $x_n < \xi$. Similarly, one defines

$$\limsup_{x \to \xi + 0} f(x), \qquad \liminf_{x \to \xi - 0} f(x), \qquad \liminf_{x \to \xi + 0} f(x).$$

Next, one defines

$$\limsup_{x \to \infty} f(x)$$

by taking in (1) sequences $\{x_n\}$ that converge to ∞. Similarly, one defines

$$\limsup_{x \to -\infty} f(x), \qquad \liminf_{x \to \infty} f(x), \qquad \liminf_{x \to -\infty} f(x).$$

EXAMPLE. Let $f(x) = \sin 1/x$ for $0 < x < 1$. Then $A = \varlimsup_{x \to 0+} f(x) = 1$. Indeed, since $-1 \le f(x) \le 1, \, -1 \le A \le 1$. Next, if $x_n = 1/(2n\pi + \pi/2)$, then $x_n \to 0$ and $f(x_n) = 1$. This gives $A = 1$. Similarly, one shows that $\varliminf_{x \to 0+} f(x) = -1$.

DEFINITION. A function $f(x)$ defined in an interval I is said to be *lower semicontinuous* at a point ξ if either

$$\liminf_{x \to \xi} f(x) = \infty$$

or

$$\liminf_{x \to \xi} f(x) \ge f(\xi).$$

It is said to be *upper semicontinuous* at a point ξ if either

$$\limsup_{x \to \xi} f(x) = -\infty,$$

or

$$\limsup_{x \to \xi} f(x) \le f(\xi).$$

If f is lower (upper) semicontinuous at each point of I, then we say that it is lower (upper) semicontinuous on I.

EXAMPLE. Let $g(x)$ be a continuous function in $a \le x \le b$ and let $a < \xi < b$. Define

$$f(x) = \begin{cases} g(x) & \text{if } a \le x \le b \text{ and } x \ne \xi, \\ L & \text{if } x = \xi. \end{cases}$$

Then $g(x)$ is lower semicontinuous if $L \le g(\xi)$ and upper semicontinuous if $L \ge g(\xi)$.

We state without proof:

THEOREM 2. *Let* $\lim\sup_{x \to \xi} f(x)$ *and* $\lim\inf_{x \to \xi} f(x)$ *be real numbers. Then,*
(a) $\lim\sup_{x \to \xi} f(x) \ge \lim\inf_{x \to \xi} f(x)$; (b) $\lim\sup_{x \to \xi} f(x) = \lim\inf_{x \to \xi} f(x)$ *if, and only if,* $\lim_{x \to \xi} f(x)$ *exists, and then* $\lim\sup_{x \to \xi} f(x) = \lim_{x \to \xi} f(x)$.

COROLLARY. *A function* $f(x)$ *is both lower semicontinuous and upper semicontinuous at a point* ξ, *if, and only if, it is continuous at the point* ξ.

THEOREM 3. *Let* $f(x)$ *be lower semicontinuous (upper semicontinuous) on a closed, bounded interval. Then* $f(x)$ *is bounded below (above) and has minimum (maximum) in the interval.*

PROBLEMS

1. Prove Theorem 2 and its corollary.

2. Prove Theorem 3 by the method used in the proofs of Theorems 1 and 2 of Section 3.

3. A function f on an interval I is lower semicontinuous at a point $\xi \in I$ if, and only if, $\lim f(x_n) \ge f(\xi)$ for any sequence $\{x_n\}$, $x_n \in I$, $\lim x_n = \xi$.

4. Let f and g be defined and bounded on an interval I and let $\xi \in I$. Prove that

$$\lim\sup_{x \to \xi}(f(x) + g(x)) \le \lim\sup_{x \to \xi} f(x) + \lim\sup_{x \to \xi} g(x); \qquad (8)$$

$$\lim\inf_{x \to \xi}(f(x) + g(x)) \ge \lim\inf_{x \to \xi} f(x) + \lim\inf_{x \to \xi} g(x). \qquad (9)$$

5. Let f and g be defined and bounded on an interval I and let $\xi \in I$. Prove that if $f \geq 0$, $g \geq 0$ on I, then

$$\lim_{x \to \xi} \sup(f(x)g(x)) \leq \left[\lim_{x \to \xi} \sup f(x) \right] \left[\lim_{x \to \xi} \sup g(x) \right].$$

6. Find $\lim\limits_{x \to 0+} \sup f(x)$ and $\lim\limits_{x \to 0-} \inf f(x)$ for the following functions:

(a) $\sin \dfrac{1}{x^2}$;

(b) $\cos \dfrac{1}{x}$;

(c) $\sin \dfrac{1}{x} + \cos \dfrac{1}{x}$;

(d) $a^{1/x}, a > 1$.

7. Let $f(x)$ be defined and bounded above in a neighborhood of a point ξ. Prove

$$\lim_{x \to \xi} \sup f(x) = \lim_{\delta \to 0} \left[\sup_{|x - \xi| \leq \delta} f(x) \right].$$

8. Let $[x]$ be defined as in Problem 1 of Section 2. Find

$$\lim_{x \to \infty} \sup(x - [x]), \qquad \lim_{x \to \infty} \inf(x - [x]).$$

7. UNIFORM CONTINUITY

Let $f(x)$ be a continuous function in an interval I. Thus, for every ξ in I the following property holds. Given any $\varepsilon > 0$ there exists a number $\delta > 0$ such that

$$|f(x) - f(\xi)| < \varepsilon \qquad \text{if } |x - \xi| < \delta, \, x \in I. \qquad (1)$$

The number δ depends, of course, upon ε. But it also may depend on point ξ.

DEFINITION.   If for any $\varepsilon > 0$ there exists a number $\delta > 0$ such that (1) holds for *all* $\xi \in I$, then we say that $f(x)$ is *uniformly continuous* on I.

EXAMPLE.   The function $1/x$ is continuous in the interval $(0, 1]$ but it is not uniformly continuous. Indeed, if it satisfies (1), then for any $0 < \xi < x < \xi + \delta$ we have

$$\varepsilon > \left| \frac{1}{x} - \frac{1}{\xi} \right| = \frac{x - \xi}{x\xi}.$$

Hence $x - \xi < \varepsilon x\xi$, or $x < \xi/(1 - \varepsilon\xi)$. Thus $\xi < x < \xi + \delta$ implies that $\xi < x < \xi/(1 - \varepsilon\xi)$. It follows that $\xi + \delta \leq \xi/(1 - \varepsilon\xi)$, that is, $\delta \leq \varepsilon\xi^2/(1 - \varepsilon\xi)$. Taking $\xi \to 0$, we get $\delta \leq 0$, a contradiction.

THEOREM 1. *Let $f(x)$ be a continuous function in a closed, bounded interval $[a, b]$. Then $f(x)$ is uniformly continuous.*

Proof. We have to prove that for any $\varepsilon > 0$ there is a $\delta > 0$ such that

$$|f(x) - f(y)| < \varepsilon \quad \text{if } |x - y| < \delta \quad \text{and } a \leq x \leq b, a \leq y \leq b. \quad (2)$$

Suppose this is not true. Then there is an $\varepsilon > 0$ for which no such δ exists for which (2) holds. Thus for any $\delta = 1/n$ ($n = 1, 2, \ldots$) there is a pair of points x_n, y_n in $[a, b]$ such that

$$|x_n - y_n| < \frac{1}{n} \quad \text{and } |f(x_n) - f(y_n)| \geq \varepsilon. \quad (3)$$

By the Bolzano–Weierstrass theorem, there is a subsequence $\{x_{n_k}\}$ of $\{x_n\}$ such that $\xi = \lim x_{n_k}$ exists. By the same theorem there is a subsequence of $\{y_{n_k}\}$, say $\{y_{n_{k'}}\}$, such that $\eta = \lim y_{n_{k'}}$ exists. The first inequality in (3) shows that $\xi = \eta$. Since $f(x)$ is continuous,

$$f(\xi) = \lim_k f(x_{n_{k'}}), \qquad f(\xi) = f(\eta) = \lim_k f(y_{n_{k'}}).$$

It follows that

$$\lim_k |f(x_{n_{k'}}) - f(y_{n_{k'}})| = 0.$$

But this contradicts the second inequality in (3).

PROBLEM Suppose $f(x)$ is defined in an open, bounded interval (a,b). Can we extend it into $[a,b]$ such that the extended function is still continuous? Theorem 1 tells us that if this is the case, then the extended function is uniformly continuous. Hence f must be uniformly continuous in (a,b). This is then a necessary condition for such a continuous extension to exist. But this condition is also sufficient! In fact, we have the following theorem:

THEOREM 2. *Let $f(x)$ be a uniformly continuous function in an open bounded interval (a,b). Then $f(a+)$ and $f(b-)$ exist.*

Thus, if we define $f(a) = f(a+)$ and $f(b) = f(b-)$, then the extended f is continuous on $[a,b]$.

Proof. We shall prove that $f(a + 0)$ exists. By our assumptions we know that, for any $\varepsilon > 0$, there is a $\delta > 0$ such that

$$|f(x) - f(y)| < \varepsilon \qquad \text{if } |x - y| < \delta, a < x < b, a < y < b. \qquad (4)$$

Taking for a fixed ε (say $\varepsilon = 1$) a point y_0 satisfying $a < y_0 < a + \delta$, we conclude from (4) that

$$|f(x)| \leq |f(x) - f(y_0)| + |f(y_0)| < \varepsilon + |f(y_0)|$$

if $a < x \leq y_0$. Thus f is bounded in the interval $(a, y_0]$.

Take any sequence $\{x_n\}$ satisfying

$$a < x_n < a + \frac{1}{n}, \qquad a < x_n \leq y_0.$$

By the Bolzano–Weierstrass theorem it follows that there exists a subsequence $\{f(x_{n_k})\}$ of $\{f(x_n)\}$ such that

$$A = \lim_k f(x_{n_k}) \qquad (5)$$

exists.

Now let ε be any positive number. Let δ be a positive number for which (4) holds. If $a < x < a + \delta$ and if $n_k \geq n^*$ where $1/n^* < \delta$, then $|x - x_{n_k}| < \delta$. Hence

$$|f(x) - f(x_{n_k})| < \varepsilon. \qquad (6)$$

On the other hand, by (5),

$$|f(x_{n_k}) - A| < \varepsilon \qquad \text{if } n_k \geq \bar{n} \qquad (7)$$

for some positive integer $\bar{n}$. Combining (6) and (7) with some fixed n_k, we find that

$$|f(x) - A| < 2\varepsilon \qquad \text{if } a < x < a + \delta.$$

Since ε is arbitrary, it follows that

$$f(a + 0) = \lim_{x \to a+} f(x)$$

exists.

The proof that $f(b - 0)$ exists is similar.

PROBLEMS

1. Let $f(x)$ be a continuous function in the interval $[0,\infty)$. Prove that if $\lim_{x \to \infty} f(x)$ exists (as a real number), then $f(x)$ is uniformy continuous in $[0,\infty)$.

2. Give an example of a function $f(x)$ that is uniformly continuous in $(-\infty,\infty)$, but for which $\lim_{x \to \infty} f(x)$ does not exist.

3. The functions given below are continuous in the closed, bounded intervals $[a,b]$ indicated. By Theorem 1 they are uniformly continuous. Find, for any $\varepsilon > 0$, a corresponding δ such that (2) holds:

(a) $\dfrac{1}{\sqrt{x+1}}$ in $0 \le x \le 1$;

(b) $\sqrt{x}$ in $0 \le x \le 2$;

(c) $x \operatorname{sgn} x$ in $-1 \le x \le 1$;

(d) $\sin(x^2)$ in $0 \le x \le 3$.

4. Show that the function $\sin 1/x$ is not uniformly continuous in the interval $(0,1)$.

8. FUNCTIONS OF BOUNDED VARIATION

Let $f(x)$ be a function defined in a closed bounded interval $[a,b]$. A set of points x_i $(i = 0,1,2,...,n)$ satisfying

$$x_0 = a < x_1 < x_2 < \cdots < x_{n-1} < x_n = b$$

is called a *partition* of $[a,b]$. For any such partition, we form the sum

$$K = \sum_{i=1}^{n} |f(x_i) - f(x_{i-1})| . \tag{1}$$

Denote by G the set of all the numbers K obtained when we take all possible partitions of $[a,b]$. If G is a bounded set (or, equivalently, if G is bounded above), then we say that f is of *bounded variation*. We then call the supremum of G the *variation* of f in $[a,b]$, and denote it by

$$\bigvee_a^b f.$$

EXAMPLE 1.   The function $f(x) = \sin x$, $a \le x \le b$, is of bounded variation. Indeed, since

$$|\sin \alpha - \sin \beta| = \left| 2 \cos \frac{\alpha + \beta}{2} \sin \frac{\alpha - \beta}{2} \right| \le 2 \left| \sin \frac{\alpha - \beta}{2} \right| \le |\alpha - \beta|,$$

we have

$$\sum_{i=1}^{n} |f(x_i) - f(x_{i-1})| \le \sum_{i=1}^{n} |x_i - x_{i-1}| = b - a.$$

EXAMPLE 2.   The function

$$f(x) = \begin{cases} \sin \dfrac{1}{x} & \text{if } 0 < x \le 1, \\ 0 & \text{if } x = 0 \end{cases}$$

is not of bounded variation. Indeed, for any $A > 0$ let k be a positive integer satisfying $2k > A$, and let x_i $(0 \le i \le 2k + 1)$ consist of 0,1 and the points

$$\frac{1}{2n\pi \pm \pi/2} \qquad \text{for } n = 1,2,\ldots,k.$$

Then

$$\sum_{i=1}^{2k+1} |f(x_i) - f(x_{i-1})| \ge \sum_{n=1}^{k} \left[\sin\left(2n\pi + \frac{\pi}{2}\right) - \sin\left(2n\pi - \frac{\pi}{2}\right) \right] = 2k > A.$$

THEOREM 1.   *If f is of bounded variation in $[a,b]$ and in $[b,c]$, then f is also of bounded variation in $[a,c]$ and*

$$\bigvee_a^c f = \bigvee_a^b f + \bigvee_b^c f. \tag{2}$$

Proof.   Take any partition

$$x_0 = a < x_1 < x_2 < \cdots < x_n = c$$

of $[a,c]$. Then there is an interval $[x_{k-1},x_k]$ containing the point b. We have

$$\sum_{i=1}^{n} |f(x_i) - f(x_{i-1})| \le \left\{ \sum_{i=1}^{k-1} |f(x_i) - f(x_{i-1})| + |f(b) - f(x_{k-1})| \right\}$$
$$+ \left\{ |f(x_k) - f(b)| + \sum_{i=k-1}^{n} |f(x_i) - f(x_{i-1})| \right\}.$$

The sum in the first braces on the right is less than or equal to $\bigvee_a^b f$, and the sum in the second braces on the right is less than or equal to $\bigvee_b^c f$. Hence each number on the left-hand side is bounded by

$$\bigvee_a^b f + \bigvee_b^c f.$$

Since $\{x_0,x_1,\ldots,x_n\}$ is an arbitrary partition of $[a,c]$, we conclude that f is of bounded variation in $[a,c]$, and

$$\bigvee_a^c f \le \bigvee_a^b f + \bigvee_b^c f. \tag{3}$$

To prove the converse of (3), we use the fact that for any $\varepsilon > 0$ there is a partition

$$y_0 = a < y_1 < \cdots < y_m = b$$

of $[a,b]$ such that

$$\bigvee_a^b f \leq \sum_{i=1}^m |f(y_i) - f(y_{i-1})| + \varepsilon.$$

There is also a partition

$$z_0 = b < z_1 < \cdots < z_h = c$$

of $[b,c]$ such that

$$\bigvee_b^c f \leq \sum_{i=1}^h |f(z_i) - f(z_{i-1})| + \varepsilon.$$

Consider the partition $\{x_0, x_1, \ldots, x_n\}$ of $[a,c]$ given by the points

$$y_0 = a < y_1 < \cdots < y_m < z_1 < \cdots < z_h = c.$$

We have

$$\sum_{i=1}^n |f(x_i) - f(x_{i-1})| = \sum_{i=1}^m |f(y_i) - f(y_{i-1})|$$

$$+ \sum_{i=1}^h |f(z_i) - f(z_{i-1})|$$

$$\geq \bigvee_a^b f + \bigvee_b^c f - 2\varepsilon.$$

Hence

$$\bigvee_a^c f \geq \bigvee_a^b f + \bigvee_b^c f - 2\varepsilon.$$

Since ε is arbitrary, the converse of the inequality (3) follows. This completes the proof of (2).

Observe that a monotone increasing function in $[a,b]$ is of bounded variation, with variation $f(b) - f(a)$. Indeed, for any partition $\{x_0, x_1, \ldots, x_n\}$ of $[a,b]$,

$$\sum_{i=1}^n |f(x_i) - f(x_{i-1})| = \sum_{i=1}^n [f(x_i) - f(x_{i-1})] = f(b) - f(a).$$

We next mention the fact (see Problem 1) that if f and g are of bounded variation, then $f - g$ is also of bounded variation.

It follows that the difference of two monotone increasing functions in $[a,b]$ is of bounded variation. It is remarkable that the converse is also true. This is the content of the next theorem.

THEOREM 2. *Let $f(x)$ be a function of bounded variation in a closed, bounded interval $[a,b]$. Then $f(x)$ can be written as a difference of two monotone increasing functions.*

Proof.   Consider the function

$$\pi(x) = \bigvee_a^x f. \tag{4}$$

If $y > x$ then, by Theorem 1,

$$\pi(y) = \bigvee_a^y f = \bigvee_a^x f + \bigvee_x^y f \geq \bigvee_a^x f = \pi(x).$$

Thus $\pi(x)$ is monotone increasing. Defining

$$\mu(x) = \pi(x) - f(x) \tag{5}$$

it remains to show that $\mu(x)$ is monotone increasing.

Let $y > x$. Then

$$\mu(y) = \pi(y) - f(y) = \pi(x) + \bigvee_x^y f - f(y)$$

$$= [\pi(x) - f(x)] + \left\{ \bigvee_x^y f - [f(y) - f(x)] \right\} \geq \mu(x),$$

since

$$f(y) - f(x) \leq \bigvee_x^y f.$$

THEOREM 3. *If $f(x)$ is continuous and of bounded variation in a closed, bounded interval $[a,b]$, then the function $\bigvee_a^x f$ is a continuous function.*

Proof.   Fix a point x_0, $a \leq x_0 < b$. For any $\varepsilon > 0$ there is a partition

$$x_0 < x_1 < \cdots < x_n = b$$

of $[x_0,b]$ such that

$$\sum_{i=1}^n |f(x_i) - f(x_{i-1})| > \bigvee_{x_0}^b f - \varepsilon. \tag{6}$$

We may assume that x_1 is sufficiently close to x_0 so that

$$|f(x_1) - f(x_0)| < \varepsilon, \tag{7}$$

for otherwise we insert a new point of partition x_0' in (x_0,x_1) and near x_0 such that $|f(x_0') - f(x_0)| < \varepsilon$. The introduction of this new point will not change the inequality in (6); in fact, the sum on the left can only increase with new points of partition.

From (6) and (7) we get

$$\bigvee_{x_0}^{b} f < \varepsilon + \sum_{i=1}^{n} |f(x_i) - f(x_{i-1})| < 2\varepsilon + \sum_{i=2}^{n} |f(x_i) - f(x_{i-1})| \le 2\varepsilon + \bigvee_{x_1}^{b} f. \quad (8)$$

Using (2) and the notation (4), we get the formula: $\bigvee_{x}^{y} f = \pi(y) - \pi(x)$ if $a \le x < y \le b$. Applying this formula to the extreme sides of (8), we find that

$$\pi(x_1) - \pi(x_0) < 2\varepsilon. \quad (9)$$

We have thus shown that for any $\varepsilon > 0$ there is a point $x_1 > x_0$ such that (9) holds. Since $\pi(x)$ is monotone increasing, it follows that

$$0 \le \pi(x) - \pi(x_0) < 2\varepsilon \qquad \text{if } x_0 < x < x_1.$$

Hence $\pi(x_0 +)$ exists and equals $\pi(x_0)$.

Similarly, one can prove that if $a < x_0 \le b$, then $\pi(x_0 -) = \pi(x_0)$. This completes the proof of the theorem.

From Theorems 2 and 3 we get:

COROLLARY. *If $f(x)$ is continuous and of bounded variation in a closed, bounded interval $[a,b]$, then it can be written as a difference of two continuous, monotone increasing functions in $[a,b]$.*

PROBLEMS

1. If f and g are functions of bounded variation in an interval $[a,b]$, then $f \pm g$ is also a function of bounded variation, and

$$\bigvee_{a}^{b} (f \pm g) \le \bigvee_{a}^{b} f + \bigvee_{a}^{b} g.$$

2. If f and g are as in the previous problem, then fg is of bounded variation in $[a,b]$.

3. If f and g are as in Problem 1 and if $|g(x)| \ge \alpha > 0$ for all x in $[a,b]$, then f/g is of bounded variation in $[a,b]$.

4. The continuous function

$$g(x) = \begin{cases} x \sin \dfrac{1}{x} & \text{if } 0 < x \le 1, \\ 0 & \text{if } x = 0, \end{cases}$$

is not of bounded variation in $[0,1]$. [*Hint:* Use the fact that $\sum\limits_{n=1}^{m} 1/n \to \infty$ if $m \to \infty$.]

5. If $|f(x) - f(y)| \le M |x - y|$ for all x, y in a closed, bounded interval $[a,b]$, then f is of bounded variation.

6. The function

$$g(x) = \begin{cases} x^2 \sin \dfrac{1}{x} & \text{if } 0 < x \le 1, \\ 0 & \text{if } x = 0 \end{cases}$$

is of bounded variation in $[0,1]$. [*Hint:* Show that $|g(x_i) - g(x_{i-1})| \le 3|x_i - x_{i-1}|$.]

7. If a function $f(x)$ is of bounded variation in $[a,b]$, then the function $\cos(f(x))$ is also of bounded variation in $[a,b]$.

3

DIFFERENTIABLE FUNCTIONS

1. DEFINITION OF THE DERIVATIVE

Let $f(x)$ be a function defined in a neighborhood (a,b) of a point x_0. Then, for any $h \neq 0$, $|h|$ sufficiently small, the quotient

$$\frac{f(x_0 + h) - f(x_0)}{h}$$

is well defined. Suppose this quotient has a limit as $h \to 0$, that is,

$$\lim_{h \to 0} \frac{f(x_0 + h) - f(x_0)}{h} \tag{1}$$

exists. Then we say that f is *differentiable* at x_0 and its *derivative* at x_0 is equal to the limit in (1). We denote the derivative of $f(x)$ at x_0 by either one of the symbols

$$\frac{df(x_0)}{dx}, \qquad f'(x_0).$$

We also say that $f(x)$ *has a derivative $f'(x_0)$* at x_0.

Note that $f(x)$ is differentiable at x_0 with derivative $f'(x_0)$ if, and only if, the function

$$\phi(h) = \begin{cases} \dfrac{f(x_0 + h) - f(x_0)}{h} & \text{if } h \neq 0, |h| \text{ small,} \\ f'(x_0) & \text{if } h = 0 \end{cases} \tag{2}$$

is continuous at $h = 0$. Therefore we can use Theorems 1 and 2 of Section 2.1 to conclude:

THEOREM 1.   *A function $f(x)$ defined on an interval (a,b) is differentiable at a point x_0 of (a,b) with derivative $f'(x_0)$ if, and only if, for any sequence $\{h_n\}$ of nonzero numbers, with $\lim_{h} h_n = 0$,*

$$\lim_{n \to \infty} \frac{f(x_0 + h_n) - f(x_0)}{h_n} = f'(x_0). \tag{3}$$

Suppose now that $f(x)$ is defined in a half-open interval $[a,b)$. If

$$\lim_{h \to 0+} \frac{f(a + h) - f(a)}{h} \tag{4}$$

exists, then we say that $f(x)$ is *right differentiable* at a, and we call the limit in (4) the *right derivative* of $f(x)$ at a. We denote it by any one of the symbols

$$\frac{df(a + 0)}{dx}, \qquad f'(a + 0), \qquad f'(a+).$$

Similarly, if $f(x)$ is defined in $(a,b]$, and if

$$\lim_{h \to 0-} \frac{f(b + h) - f(b)}{h} \tag{5}$$

exists, then we say that $f(x)$ is *left differentiable* at b, and we call the limit in (5) the *left derivative* of $f(x)$ at b. We denote it by any one of the symbols

$$\frac{df(b - 0)}{dx}, \qquad f'(b - 0), \qquad f'(b-).$$

DEFINITION.   Let $f(x)$ be a function defined on an interval I. If $f(x)$ is differentiable at each point of I, then we say that f is *differentiable on I* (or *in I*). We then denote by f', or df/dx, the function whose value at each point x of I is the number $f'(x)$. We call f' the *derivative of f*.

When dealing with a derivative $f'(x)$ of a function $f(x)$, we shall *not* use the notation $f'(a + 0)$ to denote the limit: $\lim_{x \to a+0} f'(x)$. The reason is that we have already used $f'(a + 0)$ to denote something else, namely, the right derivative of $f(x)$ at a. In general, $\lim_{x \to a+0} f'(x)$ may not exist when $f'(a + 0)$ exists. However, as will be shown in Section 5, if $\lim_{x \to a+0} f'(x)$ exists, then $f'(a + 0)$ also exists and the two numbers are equal.

EXAMPLE 1. $f(x) = x^2$ is differentiable at each point x_0. Indeed, if $x \neq x_0$,

$$\frac{f(x) - f(x_0)}{x - x_0} = \frac{x^2 - x_0^2}{x - x_0} = x + x_0.$$

Hence, with $h = x - x_0$,

$$\lim_{h \to 0} \frac{f(x_0 + h) - f(x_0)}{h} = \lim_{x \to x_0} \frac{f(x) - f(x_0)}{x - x_0} = \lim_{x \to x_0} (x + x_0)$$

$$= x_0 + x_0 = 2x_0.$$

Thus $f'(x_0)$ exists and it is equal to $2x_0$.

EXAMPLE 2. For any positive integer n, x^n is differentiable and

$$\frac{d}{dx} x^n = nx^{n-1}. \tag{6}$$

The proof for $n = 1$ is rather immediate. We therefore shall consider from now on only the case where $n \geq 2$. Write

$$\frac{x^n - x_0^n}{x - x_0} = x^{n-1} + x^{n-2}x_0 + \cdots + xx_0^{n-2} + x_0^{n-1}. \tag{7}$$

We want to show that for any $\varepsilon > 0$ there is a $\delta > 0$ such that

$$\left| \frac{x^n - x_0^n}{x - x_0} - nx_0^{n-1} \right| < \varepsilon \qquad \text{if } 0 < |x - x_0| < \delta. \tag{8}$$

If we take $\delta < 1$, then

$$|x| \leq |x - x_0| + |x_0| < \delta + |x_0| < 1 + |x_0|.$$

Hence, if $0 \le i \le n - 2$,

$$
\begin{aligned}
|x^{n-1-i}x_0{}^i - x_0{}^{n-1}| &= |x^{n-1-i} - x_0{}^{n-1-i}|\,|x_0|^i \\
&= |x - x_0|\,|x^{n-1-i-1} + x^{n-1-i-2}x_0 \\
&\quad + \cdots + x_0{}^{n-1-i-1}|\,|x_0|^i \qquad (9)\\
&\le |x - x_0|\,n(1 + |x_0|)^{n-2-i}|x_0|^i \\
&< |x - x_0|\,n(1 + |x_0|)^{n-2}.
\end{aligned}
$$

Combining (7) with (9), we get

$$
\begin{aligned}
\left| \frac{x^n - x_0{}^n}{x - x_0} - nx_0{}^{n-1} \right| &\le |(x^{n-1} - x_0{}^{n-1}) + (x^{n-2}x_0 - x_0{}^{n-1}) \\
&\quad + \cdots + (xx_0{}^{n-2} - x_0{}^{n-1})| \\
&< n^2(1 + |x_0|)^{n-2}|x - x_0| \le n^2(1 + |x_0|)^{n-2}\delta \le \varepsilon
\end{aligned}
$$

if $\delta = \min(1, \varepsilon/n^2(1 + |x_0|)^{n-2})$.

It is interesting to note that once we have derived (7), we can proceed to establish (6) much more quickly than before if we use the fact that polynomials are continuous functions. Indeed, the polynomial occurring on the right-hand side of (7) conver ̣ ̣s to

$$
x_0{}^{n-1} + x_0{}^{n-1}x_0 + \cdots + x_0 x_0{}^{n-2} + x_0{}^{n-1} = nx_0{}^{n-1}
$$

as $x \to x_0$. Hence the left-hand side of (7) has a limit, equal to $nx_0{}^{n-1}$.

EXAMPLE 3. We shall prove that $\sin x$ is differentiable and $(\sin x)' = \cos x$. First consider Figure 3.1, where $\overline{OC} = 1$ and ADC is an arc of length y on the circle with center O. The area of triangle OAC is $(1/2)\sin y$. It is less

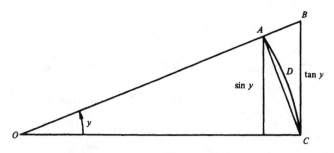

Figure 3.1

than the area of the sector $OADC$, which is equal to $(1/2) y$. The latter is less than the area $(1/2) \tan y$ of triangle OBC. Thus

$$\frac{1}{2} \sin y < \frac{1}{2} y < \frac{1}{2} \tan y = \frac{1}{2} \frac{\sin y}{\cos y},$$

or

$$\cos y < \frac{\sin y}{y} < 1.$$

Since $\cos y \to 1$ as $y \to 0$, we conclude that

$$\lim_{y \to 0} \frac{\sin y}{y} = 1. \tag{10}$$

Using (10) and the continuity of the function $\cos x$, we have

$$\frac{\sin(x + h) - \sin x}{h} = \frac{2}{h} \sin \frac{h}{2} \cos\left(x + \frac{h}{2}\right) \to \cos x \qquad \text{as } h \to 0.$$

EXAMPLE 4.   From the relations

$$\frac{\cos h - 1}{h} = \frac{\cos^2 h - 1}{h(\cos h + 1)} = -\frac{\sin h}{h} \frac{\sin h}{1 + \cos h}$$

we find that

$$\lim_{h \to 0} \frac{\cos h - 1}{h} = 0. \tag{11}$$

Using (10) and (11), we have

$$\frac{\cos(x + h) - \cos x}{h} = \frac{\cos x \cos h - \cos x}{h} - \frac{\sin x \sin h}{h}$$

$$= \cos x \frac{\cos h - 1}{h} - \sin x \frac{\sin h}{h} \to -\sin x.$$

Hence $(\cos x)'$ exists and equals $-\sin x$.

EXAMPLE 5.   Find the derivative of $f(x) = \sqrt{x}, x > 0$.

$$\frac{f(x + h) - f(x)}{h} = \frac{\sqrt{x + h} - \sqrt{x}}{h} = \frac{(x + h) - x}{h(\sqrt{x + h} + \sqrt{x})} = \frac{1}{\sqrt{x + h} + \sqrt{x}}.$$

Since the right-hand side is a continuous function of h, it converges to $1/(2\sqrt{x})$ as $h \to 0$. Thus

$$\frac{d}{dx}\sqrt{x} = \frac{1}{2\sqrt{x}}.$$

THEOREM 2. *If $f(x)$ is differentiable at a point x_0, then it is also continuous at x_0.*

Proof. By assumption,

$$\lim_{x \to x_0} \frac{f(x) - f(x_0)}{x - x_0} = f'(x_0).$$

It follows that for $\varepsilon = 1$ there is a $\delta_0 > 0$ such that

$$\left| \frac{f(x) - f(x_0)}{x - x_0} - f'(x_0) \right| < \varepsilon_0 \qquad \text{if } 0 < |x - x_0| < \delta_0.$$

Consequently

$$\frac{|f(x) - f(x_0)|}{|x - x_0|} < |f'(x_0)| + 1 \qquad \text{if } 0 < |x - x_0| < \delta_0.$$

Hence

$$|f(x) - f(x_0)| < C|x - x_0| \qquad \text{if } |x - x_0| < \delta_0 \quad (C = |f'(x_0)| + 1).$$

Thus for any $\varepsilon > 0$, if

$$\delta = \min\left(\delta_0, \frac{\varepsilon}{C}\right),$$

then $|f(x) - f(x_0)| < \varepsilon$. This completes the proof.

PROBLEMS

1. Find the derivatives of the following functions:

(a) $\dfrac{1}{x^2}$;

(b) $\dfrac{1}{x^n}$;

(c) $\dfrac{1}{\sqrt{x}}$;

(d) $\dfrac{1}{\sqrt{x^2 + x + 1}}$;

(e) $x \sin x$;

(f) $\dfrac{\sin x}{x}$.

2. Show that the function

$$f(x) = \begin{cases} x^2 + x + 1 & \text{if } x \geq 1, \\ 4x - 1 & \text{if } x < 1 \end{cases}$$

is continuous but not differentiable at $x = 1$. Show that it has right and left derivatives at $x = 1$.

3. If $f(x)$ is a monotone increasing function and if its derivative $f'(x)$ exists, then $f'(x) \geq 0$.

4. Let $y = f(x)$ be a function with the graph given in Figure 3.2. At a point

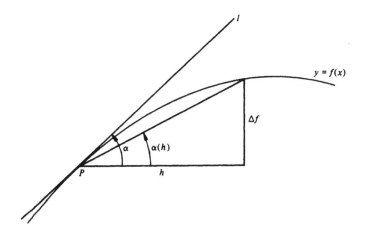

Figure 3.2

$P = (x_0, f(x_0))$ draw a tangent line l to the curve of the graph. The quotient

$$\frac{\Delta f}{h} = \frac{f(x_0 + h) - f(x_0)}{h}$$

is equal to $\tan \alpha(h)$, $\alpha(h)$ as in Figure 3.2. By definition of l, $\alpha = \lim_{h} \alpha(h)$, where α is the angle between l and the x-axis. It follows that $f'(x_0)$ is equal to $\tan \alpha$. The tangent line l is thus given by the equation

$$y - f(x_0) = f'(x_0)(x - x_0).$$

Find the tangent line at x_0 to each of the curves:

(a) $y = x^2$ at $x_0 = 2$;   (b) $y = \sin x$ at $x_0 = \pi/2$;

(c) $y = \cos x$ at $x_0 = \pi/2$;   (d) $y = \dfrac{1}{\sqrt{x+5}}$ at $x_0 = 0$;

(e) $y = x\sqrt{x+1}$ at $x_0 = 1$;   (f) $y = \sin(x^2)$ at $x_0 = 0$.

2. OPERATIONS WITH DIFFERENTIABLE FUNCTIONS

THEOREM 1. *Let f and g be functions defined in a neighborhood of a point x_0. If $f'(x_0)$ and $g'(x_0)$ exist, then $f + g$, $f - g$, and fg are differentiable at x_0, and*

$$\frac{d(f \pm g)(x_0)}{dx} = \frac{df(x_0)}{dx} \pm \frac{dg(x_0)}{dx}, \tag{1}$$

$$\frac{d(fg)(x_0)}{dx} = \frac{df(x_0)}{dx} g(x_0) + f(x_0) \frac{dg(x_0)}{dx}. \tag{2}$$

Proof. By assumption,

$$\frac{f(x_0 + h) - f(x_0)}{h} \to f'(x_0) \qquad \text{as } h \to 0,$$

$$\frac{g(x_0 + h) - g(x_0)}{h} \to g'(x_0) \qquad \text{as } h \to 0. \tag{3}$$

It follows that

$$\frac{[f(x_0 + h) \pm g(x_0 + h)] - [f(x_0) \pm g(x_0)]}{h}$$

$$= \frac{f(x_0 + h) - f(x_0)}{h} \pm \frac{g(x_0 + h) - g(x_0)}{h} \to f'(x_0) + g'(x_0)$$

as $h \to 0$. This proves (1).

To prove the assertion regarding fg, write

$$\frac{f(x_0 + h)g(x_0 + h) - f(x_0)g(x_0)}{h}$$

$$= \frac{f(x_0 + h) - f(x_0)}{h} g(x_0 + h) + f(x_0) \frac{g(x_0 + h) - g(x_0)}{h}.$$

Since $g(x_0 + h) \to g(x_0)$ as $h \to 0$ and since (3) holds, the first term on the right converges to $f'(x_0)g(x_0)$. The second term clearly converges to $f(x_0)g'(x_0)$. This completes the proof.

THEOREM 2. *Let f and g be functions defined in a neighborhood I of a point x_0. Assume that $g(x) \neq 0$ on I and that $f'(x_0)$ and $g'(x_0)$ exist. Then $f(x)/g(x)$ is differentiable at x_0 and*

$$\left(\frac{f(x)}{g(x)}\right)' = \frac{f'(x)g(x) - f(x)g'(x)}{(g(x))^2} \qquad \text{at } x_0. \tag{4}$$

Proof. Write

$$\frac{f(x_0 + h)}{g(x_0 + h)} - \frac{f(x_0)}{g(x_0)} = \frac{f(x_0 + h) - f(x_0)}{g(x_0 + h)} - f(x_0)\frac{g(x_0 + h) - g(x_0)}{g(x_0)g(x_0 + h)}. \tag{5}$$

Since the quotient $\varphi(h)/\psi(h)$ of continuous functions with nonvanishing denominator is continuous, we have

$$-f(x_0)\frac{[g(x_0 + h) - g(x_0)]/h}{g(x_0)g(x_0 + h)} \to -f(x_0)\frac{g'(x_0)}{(g(x_0))^2} \qquad \text{as } h \to 0.$$

Similarly,

$$\frac{[f(x_0 + h) - f(x_0)]/h}{g(x_0 + h)} \to \frac{f'(x_0)}{g(x_0)}.$$

From these relations and from (5) we obtain the assertion (4).

The next theorem deals with the derivative of a composite function. It is called the *chain rule.*

THEOREM 3. *Let $g(x)$ be defined in a neighborhood of a point x_0. Let $f(x)$ be defined in a neighborhood of $y_0 = g(x_0)$. Denote by k the composite function $k(x) = f(g(x))$. If $g'(x_0)$ exists and if $f'(y_0)$ exists, then $k'(x_0)$ exists and*

$$k'(x_0) = f'(y_0)g'(x_0) \qquad (y_0 = g(x_0)). \tag{6}$$

Proof.   There are two cases:

(a)   There is a δ sufficiently small such that $g(x_0) \neq g(x_0 + h)$ for all $0 < |h| < \delta$.

(b)   There is a sequence $\{h_n{}^*\}$, $h_n{}^* \neq 0$, $h_n{}^* \to 0$ as $n \to \infty$, such that $g(x_0 + h_n{}^*) = g(x_0)$.

In case (a) we can write, for any sequence $\{h_n\}$ with $0 < |h_n| < \delta$ and $\lim h_n = 0$,

$$\frac{k(x_0 + h_n) - k(x_0)}{h_n} = \frac{f(g(x_0 + h_n)) - f(g(x_0))}{h_n}$$
$$= \frac{f(g(x_0 + h_n)) - f(g(x_0))}{g(x_0 + h_n) - g(x_0)} \frac{g(x_0 + h_n) - g(x_0)}{h_n}. \tag{7}$$

The sequence $\{g(x_0 + h_n)\}$ converges to $g(x_0)$. Hence, by Theorem 1 of Section 1,

$$\frac{f(g(x_0 + h_n)) - f(g(x_0))}{g(x_0 + h_n) - g(x_0)} \to f'(g(x_0)) \qquad \text{as } n \to \infty.$$

From this and (7) we get

$$\lim_{n \to \infty} \frac{k(x_0 + h_n) - k(x_0)}{h_n} = f'(g(x_0))g'(x_0). \tag{8}$$

Since $\{h_n\}$ is an arbitrary sequence, Theorem 1 of Section 1 implies that $k'(x_0)$ exists and equals $f'(g(x_0))g'(x_0)$. This completes the proof in case (a).

If (b) holds, then

$$g'(x_0) = \lim_{n \to \infty} \frac{g(x_0 + h_n{}^*) - g(x_0)}{h_n{}^*} = 0. \tag{9}$$

We shall prove that for any sequence $\{h_n\}$, $h_n \neq 0$, $\lim h_n = 0$,

$$\frac{k(x_0 + h_n) - k(x_0)}{h_n} \to 0 \qquad \text{as } n \to \infty. \tag{10}$$

This will prove (by means of Theorem 1 of Section 1) that $k'(x_0)$ exists and is equal to $0 = f'(y_0)g'(x_0)$.

We divide $\{h_n\}$ into two subsequences $\{h_n'\}$ and $\{h_n''\}$, such that

$$g(x_0 + h_n') \neq g(x_0),$$
$$g(x_0 + h_n'') = g(x_0).$$

One of the subsequences may be finite (or even empty). It is enough to prove (10) separately for $h_n = h_n'$ and for $h_n = h_n''$.

If $h_n = h_n'$ (and $\{h_n'\}$ is infinite), then we can proceed as in (a) (with h_n replaced by h_n'), and take $n \to \infty$. We then obtain (8). Recalling that (9) holds, that is, $g'(x_0) = 0$, we obtain (10). If $h_n = h_n''$ (and $\{h_n''\}$ is infinite), then

$$k(x_0 + h_n'') - k(x_0) = f(g(x_0 + h_n'')) - f(g(x_0)) = 0.$$

Therefore, (10) is obviously true.

EXAMPLE 1.   We shall prove that for any positive integers n,m:

$$(x^{-n})' = -nx^{-n-1}, \tag{11}$$

$$(x^{1/n})' = \frac{1}{n} x^{(1/n)-1} = \frac{1}{n} x^{(1-n)/n}, \tag{12}$$

$$(x^{m/n})' = \frac{m}{n} x^{(m/n)-1}. \tag{13}$$

Differentiating $x^n x^{-n} = 1$ and using Theorem 1, we get

$$nx^{n-1}x^{-n} + x^n(x^{-n})' = 0.$$

Hence

$$(x^{-n})' = -n\frac{x^{n-1}x^{-n}}{x^n} = -nx^{-n-1}.$$

Next, differentiating $x = (x^{1/n})^n$ as a composite function $[f(y) = y^n$ and $y = x^{1/n}]$, and using the chain rule, we get

$$1 = n(x^{1/n})^{n-1}(x^{1/n})'.$$

Hence

$$(x^{1/n})' = \frac{1}{n}\frac{1}{(x^{1/n})^{(n-1)}} = \frac{1}{n}x^{-(1/n)(n-1)} = \frac{1}{n}x^{(1-n)/n}.$$

Finally, to prove (13), we differentiate $x^{m/n}$ as a composite function $f(y) = y^m$ where $y = x^{1/n}$. We find

$$(x^{m/n})' = my^{m-1}\cdot(x^{1/n})' = mx^{(m-1)/n}\left(\frac{1}{n}\right)x^{(1-n)/n} = \frac{m}{n}x^{(m-n)/n} = \frac{m}{n}x^{(m/n)-1}.$$

PROBLEMS

1. Find the derivatives of the following functions:

    (a) $x^{-m/n}$;

    (b) $x \sin 1/x, \ x > 0$;

    (c) $\cos(\sqrt{1 + x^2})$;

    (d) $\sin^m x$;

    (e) $\tan x$;

    (f) $\dfrac{\sin x + 1}{\sin x - 1}$;

    (g) $\dfrac{x + \sqrt{x}}{x - \sqrt{x}}$;

    (h) $\dfrac{x^2 + 2x - 7}{x^3 + x + 1}$.

2. Find which of the following functions has right derivative at $x = 0$:

$$f(x) = \begin{cases} x \sin \dfrac{1}{x} & \text{if } x > 0, \\ 0 & \text{if } x = 0; \end{cases} \qquad f(x) = \begin{cases} x^2 \sin \dfrac{1}{x} & \text{if } x > 0, \\ 0 & \text{if } x = 0. \end{cases}$$

3. Find the derivatives of the following functions:

    (a) $\dfrac{1}{\sqrt{1 - x^2}}$;

    (b) $\dfrac{1}{\sqrt{1 + x^2}}$;

    (c) $\dfrac{\sqrt{1 - x^2}}{\sqrt{1 + x^2}}$;

    (d) $\tan^5 x$.

3. INVERSE FUNCTIONS

Let $f(x)$ be a strictly monotone function in an open interval I and assume that its derivative $f'(x)$ exists. Let $x_0 \in I$ and suppose that $f'(x_0) \neq 0$. Denote by g the inverse function to f. If

$$y_0 = f(x_0), \qquad y_0 + k = f(x_0 + h),$$

then

$$g(y_0) = x_0, \qquad g(y_0 + k) = x_0 + h.$$

By Theorem 2 of Section 1, f is continuous in I. By Theorem 3 of Section 2.5, g is then continuous and strictly monotone. Furthermore, if h varies in a set $-h_1 < h < h_2$, $h \neq 0$, where h_1, h_2 are sufficiently small, then k varies in some set $-k_1 < k < k_2$, $k \neq 0$, the correspondence $h \leftrightarrow k$ is one to one, and $h \to 0$ if, and only if, $k \to 0$. Writing

$$\frac{g(y_0 + k) - g(y_0)}{k} = \frac{(x_0 + h) - x_0}{f(x_0 + h) - f(x_0)} = \frac{1}{\dfrac{f(x_0 + h) - f(x_0)}{h}}$$

we see that if $k \to 0$, then the function on the right converges to $1/f'(x_0)$. We conclude that $g'(y_0)$ exists and equals $1/f'(x_0)$. We sum up:

THEOREM 1.   *Let $f(x)$ be a strictly monotone, differentiable function in an open interval I. Then, for each point x_0 where $f'(x_0) \neq 0$, the inverse function $g(y)$ is differentiable at $y_0 = f(x_0)$, and*

$$g'(y_0) = \frac{1}{f'(x_0)}. \tag{1}$$

EXAMPLE 1.   We shall prove that the natural logarithm $\log x$ is differentiable and

$$\frac{d}{dx}(\log x) = \frac{1}{x}. \tag{2}$$

First we need to show that

$$\lim_{y \to 0+} (1 + y)^{1/y} = e. \tag{3}$$

If $\dfrac{1}{n+1} < y \leq \dfrac{1}{n}$ where n is a positive integer, then

$$a_n \equiv \left(1 + \frac{1}{n+1}\right)^n < (1 + y)^{1/y} < \left(1 + \frac{1}{n}\right)^{n+1} \equiv b_n.$$

Since $a_n \to e$, $b_n \to e$ as $n \to \infty$, it follows that for any $\varepsilon > 0$ there is an n_0 such that $a_n > e - \varepsilon$, $b_n < e + \varepsilon$ if $n \geq n_0$. Hence, if $0 < y < 1/n_0$,

$$|(1 + y)^{1/y} - e| < \varepsilon.$$

This proves (3).

Similarly, one proves that

$$\lim_{y \to 0-} (1 + y)^{1/y} = e. \tag{4}$$

Here one uses the fact that

$$\left(1 - \frac{1}{n+1}\right)^{-n} \to e, \quad \left(1 - \frac{1}{n}\right)^{-(n+1)} \to e \quad \text{as } n \to \infty. \tag{5}$$

We now can prove (2) easily. Since $x > 0$,

$$\lim_{h \to 0} \frac{\log(x + h) - \log x}{h} = \lim_{h \to 0} \frac{\log(x + h)/x}{h} = \frac{1}{x} \lim_{h/x \to 0} \frac{\log(1 + h/x)}{h/x}$$

$$= \frac{1}{x} \lim_{h/x \to 0} \log\left(1 + \frac{h}{x}\right)^{x/h} = \frac{1}{x} \log e = \frac{1}{x}$$

by (3) and (4) with $y = h/x$, and the continuity of the function log.

EXAMPLE 2.   We shall show that the exponential function e^x is differentiable, and

$$\frac{d}{dx} e^x = e^x. \tag{6}$$

This function is the inverse of $x = \log y$. Thus it is differentiable and

$$\frac{d}{dx} e^x = \frac{1}{(d/dy)\log y} = \frac{1}{1/y} = y = e^x.$$

EXAMPLE 3.   Consider the function $y = a^x$. Taking the natural logarithm of both sides we get

$$\log y = \log a^x = x \log a. \tag{7}$$

This shows that the (composite) function $\log y(x)$ (where $y(x) = a^x$) is differentiable. Hence also the function

$$e^{\log y(x)} = y(x) = a^x$$

is differentiable. From (7) we get

$$\frac{d}{dx}(\log y) = \log a.$$

Differentiating on the left by the chain rule, we find that

$$\log a = \frac{1}{y}\frac{dy}{dx} = \frac{1}{y}(a^x)'.$$

Hence

$$(a^x)' = a^x \log a. \tag{8}$$

EXAMPLE 4.   For any real α, the function $y = x^\alpha$ is defined and differentiable if $x > 0$, and

$$(x^\alpha)' = \alpha x^{\alpha-1}. \tag{9}$$

The proof is similar to the proof of (8). One first takes the logarithm and obtains

$$\log y = \alpha \log x.$$

It follows that $\log y$ is differentiable, as a (composite) function of x. Hence $e^{\log y} = y = x^\alpha$ is also differentiable. Next,

$$\frac{\alpha}{x} = \alpha(\log x)' = \frac{d}{dx}\log y = \frac{1}{y}\frac{dy}{dx}.$$

Therefore,

$$(x^\alpha)' = \frac{\alpha}{x} y = \alpha x^{\alpha-1}.$$

EXAMPLE 5.   Consider the function $y = \sin^{-1} x$. By Theorem 1, this function is differentiable and (since $x = \sin y$)

$$(\sin^{-1} x)' = \frac{1}{dx/dy} = \frac{1}{\cos y}.$$

Since y varies in the interval $-\frac{\pi}{2} \le y < \frac{\pi}{2}$, $\cos y$ is nonnegative. Hence

$$\cos y = \sqrt{1 - \sin^2 y} = \sqrt{1 - x^2},$$

and we get

$$\frac{d}{dx}\sin^{-1} x = \frac{1}{\sqrt{1 - x^2}}. \tag{10}$$

PROBLEMS

Prove that

$$\frac{d}{dx}\cos^{-1}x = -\frac{1}{\sqrt{1-x^2}}; \qquad \frac{d}{dx}\tan^{-1}x = \frac{1}{1+x^2}.$$

2. The *hyperbolic sine* is the function

$$y = \sinh x = \frac{e^x - e^{-x}}{2}$$

and the *hyperbolic cosine* is the function

$$y = \cosh x = \frac{e^x + e^{-x}}{2}.$$

Prove that (i) $(\sinh x)' = \cosh x$, $(\cosh x)' = \sinh x$; (ii) $\cosh^2 y - \sinh^2 y = 1$; (iii) $x = \sinh y$ has inverse $y = \sinh^{-1}x$, and $y = \sinh^{-1}x = \log(x + \sqrt{x^2 + 1})$.

3. Find the derivatives of the following functions:

(a) x^x;

(b) $x^{s\ln x}$;

(c) $e^{\sqrt{1+x^2}}$;

(d) $\sin^{-1}\sqrt{1-x^2}$;

(e) $a^{x^2 + \log(3 + \sin x)}$;

(f) $\tan^{-1}\dfrac{\sin x}{1 + \cos x}$;

(g) $\log(x + \sqrt{x^2 + 1})$;

(h) $\tan^{-1}\dfrac{1 + x}{1 - x}$.

4. Find whether $f'(0+)$ exists for each of the following functions:

(a) $f(x) = \begin{cases} \dfrac{1}{1 + \sqrt{x}} & \text{if } x = 0, \\ 1 & \text{if } x > 0; \end{cases}$

(b) $f(x) = \begin{cases} \sin\sqrt{x} & \text{if } x > 0, \\ 0 & \text{if } x = 0; \end{cases}$

(c) $f(x) = \begin{cases} x^\alpha & \text{if } x > 0 \\ 0 & \text{if } x = 0 \end{cases}$ where $\alpha > 0$.

5. Find the derivatives of the following functions:

(a) $x^{(\sin x)/x}$;

(b) $\exp\left(\dfrac{x^2 + 1}{x^2 - 1}\right)$;

(c) $\log(\log x)$;

(d) $x^{(x^x)}$;

(e) $(x^x)^x$;

(f) $\log(\log(\log\sqrt{1 + x^2}))$.

6. Let f and g be differentiable functions, and $f(x) > 0$. Prove the formula:

$$\frac{d}{dx}\{[f(x)]^{g(x)}\} = [f(x)]^{g(x)}g'(x)\log f(x) + g(x)[f(x)]^{g(x)-1}f'(x).$$

4. HIGHER DERIVATIVES

DEFINITION.   Given a function $f(x)$ in some interval I, we denote its derivative by $f'(x)$ or df/dx. The derivative of $f'(x)$ is called the *second derivative* of $f(x)$ and is denoted by

$$f''(x), \quad \text{or} \quad \frac{d^2f}{dx^2}.$$

The derivative of $f''(x)$ is called the *third derivative* of $f(x)$ and is denoted by

$$f'''(x), \quad \text{or} \quad \frac{d^3f}{dx^3}.$$

If we differentiate fn times, we get the nth *derivative* of $f(x)$. It is denoted by

$$f^{(n)}(x), \quad \text{or} \quad \frac{d^nf}{dx^n}.$$

PROBLEMS

1. Compute the nth derivative of each of the following functions:

(a) x^n;   (b) $\sin 3x$;   (c) $\frac{1}{x}$;

(d) $\sqrt{1+x}$;   (e) e^x;   (f) x^{2n};

(g) $\log x$;   (h) $\cos(x+1)$.

2. Prove *Leibnitz' rule*: If $f(x)$ and $g(x)$ have derivatives of all orders $\leq n$, then $f(x)g(x)$ has derivatives of all orders $\leq n$ and

$$\frac{d^n}{dx^n}[f(x)g(x)] = f^{(n)}(x)g(x) + \binom{n}{1}f^{(n-1)}(x)g'(x) + \cdots$$

$$+ \binom{n}{k}f^{(n-k)}(x)g^{(k)}(x) + \cdots + \binom{n}{1}f'(x)g^{(n-1)}(x) + f(x)g^{(n)}(x).$$

[*Hint*: Use induction on n.]

3. Prove that a function $y = e^{\lambda x}$ (λ constant) satisfies
$$a_0 y^{(n)} + a_1 y^{(n-1)} + a_2 y^{(n-2)} + \cdots + a_{n-1} y' + a_n y = 0$$
($a_0, a_1, \ldots, a_n$ constants) if, and only if, λ satisfies:
$$a_0 \lambda^n + a_1 \lambda^{n-1} + a_2 \lambda^{n-2} + \cdots + a_{n-1} \lambda + a_n = 0.$$

4. Use Leibnitz' rule in order to find the nth derivative of each of the following functions:

(a) xe^x;   (b) $x \sin x$;

(c) $x\sqrt{1 + x}$;   (d) $e^x \sin x$.

5. THE MEAN VALUE THEOREM

THEOREM 1 (Rolle's Theorem). *Let $f(x)$ be a continuous function in a closed, bounded interval $[a,b]$. Assume also that $f'(x)$ exists for all x in the open interval (a,b). If $f(a) = f(b)$, then there exists a point ξ in (a,b) such that $f'(\xi) = 0$.*

Proof. Since $f(x)$ is continuous in $[a,b]$, it assumes both its maximum M and its minimum m. If $M = f(a) = m$, then $f(x)$ is constant and the assertion of the theorem is obvious (with any ξ in (a,b)). Thus we may suppose that either $M > f(a)$ or $m < f(a)$. Consider the first case and let ξ be a point where f attains the value M. Thus $a < \xi < b$ and
$$f(\xi) \geq f(x) \qquad \text{for all } a \leq x \leq b.$$
It follows that
$$\frac{f(\xi + h) - f(\xi)}{h} \leq 0 \qquad \text{if } h > 0, \geq 0 \qquad \text{if } h < 0.$$
Hence
$$f'(\xi) = f'(\xi + 0) \leq 0, \qquad f'(\xi) = f'(\xi - 0) \geq 0.$$
Thus $f'(\xi) = 0$.

If $M = f(a)$ but $m < f(a)$, then we prove in the same way that at the point ξ where $f(\xi) = m$, $f'(\xi) = 0$.

One of the most important theorems in calculus is the *mean value theorem*.

THEOREM 2 (Mean Value Theorem). *Let $f(x)$ be a continuous function in a closed, bounded interval $[a,b]$. Assume that $f'(x)$ exists for all x in (a,b). Then there is a point ξ in (a,b) such that*
$$\frac{f(b) - f(a)}{b - a} = f'(\xi). \tag{1}$$

Geometrically, (1) can be stated as follows: There is a tangent line l to $y = f(x)$ at some point $(\xi, f(\xi))$ such that its slope (that is, the tangent of the angle between the x-axis and l) is equal to the slope of the line connecting $(a, f(a))$ to $(b, f(b))$; see Figure 3.3.

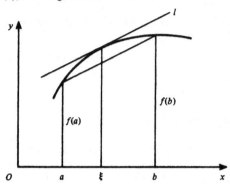

Figure 3.3

Proof. If $f(a) = f(b)$, then the assertion follows from Rolle's theorem. If $f(a) \neq f(b)$, then we introduce the function

$$g(x) = f(x) - Cx, \qquad \text{where } C = \frac{f(b) - f(a)}{b - a}.$$

It satisfies: $g(a) = g(b)$. Hence, by Rolle's theorem, there is a point ξ in (a,b) such that $g'(\xi) = 0$, that is,

$$f'(\xi) = C = \frac{f(b) - f(a)}{b - a}.$$

The assertion (1) also can be written in the form:

$$f(x + h) - f(x) = hf'(x + \theta h) \qquad (0 < \theta < 1) \tag{2}$$

where $x = a$, $x + h = b$.

We shall now give some applications based on the mean value theorem.

THEOREM 3.   *Let $f(x)$ be a differentiable function in an interval I. If $f'(x) \equiv 0$ in I, then $f(x) \equiv$ const.*

Proof. For any two points x, y in I, if $x > y$, then

$$f(x) - f(y) = (x - y)f'(y + \theta(x - y)) = 0 \qquad (0 < \theta < 1).$$

Thus $f(x) = f(y)$. This proves the assertion.

Note that the converse is obvious, namely, the derivative of a constant function is zero.

THEOREM 4.   *Let $f(x)$ be a differentiable function in an interval I. If $f'(x) \geq 0$ for all x in I, then f is monotone increasing. If, further, $f'(x) > 0$ for all x in I, then f is strictly monotone increasing.*

Proof.   If $x > y$, then

$$f(x) - f(y) = (x - y)f'(z) \geq 0$$

for some $y < z < x$. This proves the first assertion. If $f'(x) > 0$ for all x, then, whenever $x > y$,

$$f(x) - f(y) = (x - y)f'(z) > 0$$

for some $y < z < x$.

Similarly, one can show that if $f'(x) \leq 0 \, (f'(x) < 0)$ for all x in I, then $f(x)$ is monotone decreasing (strictly monotone decreasing).

THEOREM 5.   *Let $f(x)$ be a continuous function in a bounded interval $[a,b)$ and assume that $f'(x)$ exists for all $a < x < b$, and*

$$A = \lim_{x \to a+} f'(x) \qquad (3)$$

exists. Then $f'(a + 0)$ exists and equals A.

Proof.   By the mean value theorem, if $0 < h < b - a$,

$$f(a + h) - f(a) = hf'(a + \theta h) \qquad (0 < \theta < 1).$$

Hence

$$\frac{f(a + h) - f(a)}{h} - A = f'(a + \theta h) - A.$$

From (3) it follows that for any $\varepsilon > 0$ there is a $\delta > 0$ such that

$$|f'(x) - A| < \varepsilon \qquad \text{if } a < x < a + \delta.$$

Therefore, if $0 < h < \delta$,

$$\left| \frac{f(a + h) - f(a)}{h} - A \right| < \varepsilon.$$

Thus $f'(a + 0)$ exists and is equal to A.

PROBLEMS

1.   If $f(x)$ is defined in (a,b) and if $f'(x)$ exists and is bounded in (a,b), then $f(a + 0)$ exists. [*Hint:* Show that $f(x)$ is uniformly continuous in (a,b) and use Theorem 2 of Section 2.7.]

2. Compute the derivative of the function $f(x) = (1 - 1/x)^x$, $x > 1$, and conclude that $f(x)$ is monotone increasing.

3. Do the same to the function $x^{\sqrt{1+x^2}}$.

4. The equation

$$1 - x + \frac{x^2}{2} - \frac{x^3}{3} + \cdots + (-1)^n \frac{x^n}{n} = 0$$

has one real root if n is odd, and no real root if n is even.

5. Any equation

$$a_0 x^n + a_1 x^{n-1} + a_2 x^{n-2} + \cdots + a_{n-1}x + a_n = 0 \ (a_0 \neq 0),$$

where $a_0, a_1, \ldots, a_n$ are real constants, has at least one real root if n is odd.

6. If the equation in the previous problem (with real $a_0, \ldots, a_n$) has m real and distinct roots, then the equation

$$na_0 x^{n-1} + (n-1)a_1 x^{n-2} + (n-2)a_2 x^{n-3} + \cdots + a_{n-1} = 0$$

has at least $m - 1$ real and distinct roots.

7. Let $f(x)$ be a continuous function for $x \geq 0$, $f(0) = 0$. If $f'(x) > 0$ for all $x > 0$, then $f(x) > 0$ for all $x > 0$.

8. Let $f(x)$ be defined in a closed, bounded interval I. Assume that $f'(x)$ and $f''(x)$ exist for all x in I. Assume also that there is a sequence $\{x_n\}$ of mutually distinct points of I such that $\lim x_n = y$ exists and such that $f(x_n) = 0$ for all n. Prove that $f(y) = f'(y) = f''(y) = 0$.

9. If in the previous problem $f(x)$ has derivatives of all orders in I, then $f^{(n)}(y) = 0$ for all $n \geq 0$.

10. Prove the inequality $\sqrt{1 + x} < 1 + (1/2)x$ for $x > 0$. [*Hint*: $f(x) = 1 + (1/2)x - \sqrt{1 + x}$ satisfies: $f(0) = 0$, $f'(x) > 0$.]

11. Prove the inequalities

$$\frac{m(x-1)}{x^{1-m}} < x^m - 1 < m(x-1), \qquad 0 < m < 1, \qquad x > 1.$$

12. Prove that $x > \log(1 + x) > x/(1 + x)$ if $x > 0$.

13. Prove the following theorem of *Darboux*: *Suppose $f'(x)$ exists for all $a \leq x \leq b$, and $f'(a) \neq f'(b)$. Then for any γ between $f'(a)$ and $f'(b)$, there is a point c in (a,b) such that $f'(c) = \gamma$.* Note that this is similar to the intermediate value theorem; however, $f'(x)$ need not be continuous. [*Hint*: Let $f'(a) < \gamma < f'(b)$. $g(x) = f(x) - \gamma(x - a)$ attains its maximum

at some point $c \in [a,b]$. $c \neq a$ since $g'(a) = f'(a) - \gamma < 0$. Similarly, $c \neq b$. Hence $c \in (a,b)$, and then $g'(c) = 0$.]

14. Compute the derivative of $f(x) = (1 + 1/x)^x$, $x > 0$, and conclude that $f(x)$ is monotone increasing.

6. THE SECOND MEAN VALUE THEOREM

The following theorem will be of fundamental importance in the subsequent sections.

THEOREM 1 (Second Mean Value Theorem). *Let* $f(x)$ *and* $g(x)$ *be continuous functions in a closed, bounded interval* $[a,b]$. *Assume that* $f'(x)$ *and* $g'(x)$ *exist for all* $a < x < b$, *and that they do not both vanish at any point* x. *Assume finally that* $g(a) \neq g(b)$. *Then, there is a point* ξ *in* (a,b) *such that*

$$\frac{f(b) - f(a)}{g(b) - g(a)} = \frac{f'(\xi)}{g'(\xi)}. \tag{1}$$

Note that in the special case where $g(x) = x$, Theorem 1 reduces to the mean value theorem.

Proof. Consider the function

$$k(x) = f(x) - Cg(x), \qquad \text{where } C = \frac{f(b) - f(a)}{g(b) - g(a)}.$$

It satisfies: $k(a) = k(b)$. Hence, by Rolle's theorem, there is a point ξ in (a,b) such that

$$k'(\xi) = f'(\xi) - Cg'(\xi) = 0. \tag{2}$$

$g'(\xi) \neq 0$ since, otherwise, both $g'(\xi)$ and $f'(\xi)$ vanish, and this contradicts one of our assumptions. Since $g'(\xi) \neq 0$, (2) yields (1).

In many applications of the second mean value theorem the function $g(x)$ satisfies $g'(x) \neq 0$ for all $x \in (a,b)$. Thus, the condition that $f'(x)$ and $g'(x)$ do not simultaneously vanish at any point x is satisfied.

7. L'HOSPITAL'S RULE

This rule roughly states the following: If $f(a) = g(a) = 0$ or if $f(a) = g(a) = \infty$, and if, further,

$$A = \lim_{x \to a} \frac{f'(x)}{g'(x)}$$

exists, then also

$$\lim_{x \to a} \frac{f(x)}{g(x)}$$

exists and equals A. Here A may be either a real number or $\pm \infty$. We shall include all these cases in one theorem.

In what follows, I is an open interval and a is either one of its endpoints We shall be concerned with limits $\lim_{x \to a}$. Thus if a is the left (right) endpoint of I, then such limits are, in effect, $\lim_{x \to a-}$ ($\lim_{x \to a+}$). We shall also allow a to be ∞ or $-\infty$, that is, when $a = \infty$ ($a = -\infty$) the interval I is unbounded from the right (left) and $\lim_{x \to a}$ is taken to mean $\lim_{x \to \infty}$ ($\lim_{x \to -\infty}$).

We shall impose the following conditions:

    (i) $f(x)$ and $g(x)$ are continuous in I and either

$$\lim_{x \to a} f(x) = \lim_{x \to a} g(x) = 0, \tag{1}$$

or

$$\lim_{x \to a} f(x) = \lim_{x \to a} g(x) = \infty. \tag{2}$$

    (ii) $f'(x)$ and $g'(x)$ exist for all x in I.
    (iii) $g(x)$ and $g'(x)$ do not vanish for all x in I.
    (iv) The limit

$$A = \lim_{x \to a} \frac{f'(x)}{g'(x)} \tag{3}$$

    exists, where A is either a real number, or $A = \infty$, or $A = -\infty$.

We can now state *l'Hospital's rule*:

THEOREM 1. *Let $f(x)$ and $g(x)$ satisfy the conditions* (i)–(iv). *Then*

$$\lim_{x \to a} \frac{f(x)}{g(x)} \tag{4}$$

exists, and equals A.

Proof. Consider first the case where (1) holds. We begin with the subcase where a is a real number. To be specific, we will assume that a is the left endpoint of I. From the second mean value theorem we get, for any $x \in I$,

$$\frac{f(x)}{g(x)} = \frac{f(x) - f(a+)}{g(x) - g(a+)} = \frac{f'(\xi)}{g'(\xi)}$$

where ξ lies between a and x. If $x \to a$, then $\xi \to a$, and then

$$\frac{f'(\xi)}{g'(\xi)} \to A.$$

It follows that

$$\frac{f(x)}{g(x)} \to A \qquad \text{if } x \to a.$$

Consider next the subcase where $a = \infty$. We introduce the functions

$$\phi(t) \quad f\left(\frac{1}{t}\right), \qquad \psi(t) = g\left(\frac{1}{t}\right)$$

for $0 < t < b$, for some $b > 0$. We have

$$\phi'(t) = -\frac{1}{t^2} f'\left(\frac{1}{t}\right), \qquad \psi'(t) = -\frac{1}{t^2} g'\left(\frac{1}{t}\right).$$

It is easily seen that φ and ψ satisfy all the conditions of the previous case with $a = 0$, and

$$\lim_{t \to 0} \frac{\phi'(t)}{\psi'(t)} = \lim_{t \to 0} \frac{f'(1/t)}{g'(1/t)} = \lim_{x \to \infty} \frac{f'(x)}{g'(x)} = A.$$

It follows that

$$\lim_{x \to \infty} \frac{f(x)}{g(x)} = \lim_{t \to 0} \frac{f(1/t)}{g(1/t)} = \lim_{t \to 0} \frac{\phi(t)}{\psi(t)}$$

exists and equals A. The case where $a = -\infty$ can be treated in the same way.

Consider now the case where (2) holds. As before, we first treat the subcase where a is a real number. We take I to be of the form $a < x < a + h$. (If $I = (a - h, a)$, the proof will be similar.)

By the second mean value theorem, if $a < x < b < a + h$,

$$\frac{f(x) - f(b)}{g(x) - g(b)} = \frac{f(x)}{g(x)} \frac{1 - f(b)/f(x)}{1 - g(b)/g(x)} = \frac{f'(\xi)}{g'(\xi)} \qquad \text{where } a < x < \xi < b.$$

Hence

$$\frac{f(x)}{g(x)} = \frac{f'(\xi)}{g'(\xi)} \frac{1 - g(b)/g(x)}{1 - f(b)/f(x)}, \tag{5}$$

provided x is sufficiently close to a, so that $1 - f(b)/f(x) \neq 0$. We now consider three cases:

    (i)   *A is a real number.* For any $\varepsilon > 0$ we can take b sufficiently close to a so that

$$\left| \frac{f'(\xi)}{g'(\xi)} - A \right| < \varepsilon. \tag{6}$$

With b fixed, choose δ so small that if $a < x < a + \delta$, then

$$\left| \frac{1 - g(b)/g(x)}{1 - f(b)/f(x)} - 1 \right| \leq \frac{|g(b)/g(x)| + |f(b)/f(x)|}{1 - |f(b)/f(x)|} < \varepsilon. \qquad (7)$$

Here we use the assumptions that $|f(x)| \to \infty$, $|g(x)| \to \infty$ as $x \to a$. Using (6) and (7) in (5), we find that

$$\left| \frac{f(x)}{g(x)} - A \right| < C\varepsilon \qquad \text{if } a < x < a + \delta,$$

where C is a constant independent of ε. Since ε is arbitrary, the assertion (4) follows.

(ii) $A = \infty$. For any $N > 0$ we can choose b sufficiently close to a so that

$$\frac{f'(\xi)}{g'(\xi)} > N.$$

With b fixed, now choose δ sufficiently small so that

$$\frac{1 - g(b)/g(x)}{1 - f(b)/f(x)} > 1 - \frac{1}{N} \qquad \text{if } a < x < a + \delta.$$

Then (5) yields

$$\frac{f(x)}{g(x)} > N\left(1 - \frac{1}{N}\right) > N - 1$$

if $a < x < a + \delta$. Since N is arbitrary, it follows that (4) holds (with $A = \infty$).

(iii) $A = -\infty$. This case is reduced to case (ii), by considering $-f/g$ instead of f/g.

It remains to consider the subcase where (4) holds and $a = \infty$ (or $a = -\infty$). This can be reduced to the case where $a = 0$ by introducing the functions $\phi(t)$, $\psi(t)$ as before.

Remark 1.   Suppose f and g satisfy all the conditions of Theorem 1 with the possible exception of (iv), and suppose that

$$\lim_{x \to a} f'(x) = \lim_{x \to a} g'(x) = \beta, \qquad \beta = 0 \text{ or } \beta = \infty.$$

If $f''(x)$ and $g''(x)$ exist, and if

$$B = \lim_{x \to a} \frac{f''(x)}{g''(x)}$$

exists, then we conclude (by Theorem 1) that the limit in (3) also exists, and $B = A$. We can now apply Theorem 1 once more and deduce that

$$\lim_{x \to a} \frac{f(x)}{g(x)} = \lim_{x \to a} \frac{f'(x)}{g'(x)} = \lim_{x \to a} \frac{f''(x)}{g''(x)}. \tag{8}$$

Remark 2. Theorem 1 remains true if a is an interior point of the open interval I. The functions $f(x), g(x)$ need not be defined at point a. The proof follows by applying Theorem 1 separately to each of the two subintervals of I lying to the right and to the left of a.

EXAMPLE 1.

$$\lim_{x \to 0} \frac{\log(1 + x)}{x} = \lim_{x \to 0} \frac{1}{1 + x} = 1.$$

EXAMPLE 2.

$$\lim_{x \to \infty} \frac{2x^2 + x - 1}{3x^2 - 2x + 5} = \lim_{x \to \infty} \frac{4x + 1}{6x - 2} = \lim_{x \to \infty} \frac{4}{6} = \frac{2}{3}.$$

EXAMPLE 3.

$$\lim_{x \to \infty} \frac{x^n}{e^x} = \lim_{x \to \infty} \frac{nx^{n-1}}{e^x} = \lim_{x \to \infty} \frac{n(n-1)x^{n-2}}{e^x} = \lim_{x \to \infty} \frac{n!}{e^x} = 0.$$

EXAMPLE 4. For any $\alpha > 0$,

$$\lim_{x \to \infty} \frac{\log x}{x^\alpha} = \lim_{x \to \infty} \frac{1/x}{\alpha x^{\alpha - 1}} = \lim_{x \to \infty} \frac{1}{\alpha x^\alpha} = 0.$$

EXAMPLE 5.

$$\lim_{x \to 0} \frac{\tan x - x}{x - \sin x} = \lim_{x \to 0} \frac{\sec^2 x - 1}{1 - \cos x} = \lim_{x \to 0} \frac{2 \sec^2 x \tan x}{\sin x}$$

$$= \lim_{x \to 0} 2 \sec^3 x = 2.$$

L'Hospital's rule is concerned with $\lim(f/g)$ when either $\lim f = \lim g = 0$ or $\lim f = \lim g = \infty$. We shall refer to these cases as $0/0$ and ∞/∞, respectively. There are other indeterminate cases. For instance, if $\lim f = 0$ and $\lim g = \infty$, then we cannot know a priori anything about $\lim(fg)$. We denote this situation briefly by $0 \cdot \infty$. We can reduce it to the situation occurring in l'Hospital's rule by writing

$$fg = \frac{f}{1/g} \quad \text{or} \quad fg = \frac{g}{1/f}.$$

EXAMPLE 6.

$$\lim_{x \to 0+} x \log x = \lim_{x \to 0+} \frac{\log x}{1/x} = \lim_{x \to 0+} \frac{1/x}{(-1/x^2)} = \lim_{x \to 0+} (-x) = 0.$$

If $\lim f = \infty$, $\lim g = \infty$, then we cannot know a priori anything about $\lim(f - g)$. We denote this indeterminate case by $\infty - \infty$. Here we may try to write

$$f - g = (fg)\left(\frac{1}{g} - \frac{1}{f}\right)$$

and reduce it to the case $0 \cdot \infty$.

If $\lim f = 1$, $\lim g = \infty$, then $\lim f^g$ is an indeterminate case 1^∞. If $\lim f = 0$, $\lim g = 0$, then $\lim f^g$ is an indeterminate case 0^0. In both cases, we first compute the limit of $\log(f^g) = g \log f$ and then use the relation

$$\lim f^g = \exp\{\lim(g \log f)\}.$$

EXAMPLE 7.

$$\lim_{x \to 0+} \log x^x = \lim_{x \to 0+} x \log x = 0.$$

Hence

$$\lim_{x \to 0+} x^x = e^0 = 1.$$

PROBLEMS

1. Compute the limit of each of the following functions, as $x \to 0$:

(a) $\dfrac{1}{x} - \dfrac{1}{\sin x}$;

(b) $(\cos x)^{1/x^2}$;

(c) $\dfrac{e^{-1/x^2}}{x}$;

(d) $\dfrac{\tan x - \sin x}{x^3}$;

(e) $\dfrac{\sin(\pi \cos x)}{x \sin x}$;

(f) $\dfrac{\cos\left((\pi/2)\cos x\right)}{\sin^2 x}$;

(g) $\dfrac{e - (1 + x)^{1/x}}{x}$;

(h) $\dfrac{a^x - b^x}{x}\ (a > 0, b > 0)$;

(i) $\dfrac{\sin x - x \cos x}{x - \sin x}$;

(j) $\dfrac{1}{x} \log \dfrac{1 + x}{1 - x}$.

2. Evaluate the following limits:

(a) $\displaystyle\lim_{x \to \infty} \dfrac{\log(1 + e^{2x})}{x}$;

(b) $\displaystyle\lim_{x \to 0} \left(\dfrac{\sin x}{x}\right)^{1/x^2}$;

(c) $\displaystyle\lim_{x \to \infty} \dfrac{\log(1 + xe^{2x})}{x^2}$;

(d) $\displaystyle\lim_{x \to \infty} (x\sqrt{x^2 + a^2} - x), a \neq 0$;

(e) $\displaystyle\lim_{x \to \infty} x^{1/x}$;

(f) $\displaystyle\lim_{x \to 1} x^{1/(1 - x)}$;

(g) $\displaystyle\lim_{x \to \infty} \dfrac{\log(\log x)}{\log(x - \log x)}$;

(h) $\displaystyle\lim_{x \to \pi/2} (\sin x)^{\tan x}$.

3. Evaluate the following limits:

(a) $\displaystyle\lim_{x \to 0} \dfrac{1 - \cos x}{a^{x^2} - b^{x^2}} (a > 0, b > 0)$;

(b) $\displaystyle\lim_{x \to 0} \dfrac{1 - a^x}{1 - b^x} (a > 0, b > 0)$;

(c) $\displaystyle\lim_{x \to 0} \dfrac{\sin x - x}{\arcsin x - x}$;

(d) $\displaystyle\lim_{x \to 0} \dfrac{(e^x - 1)\sin x}{\cos x - \cos^2 x}$;

(e) $\displaystyle\lim_{x \to \infty} x \log \dfrac{x + 1}{x - 1}$;

(f) $\displaystyle\lim_{x \to (\pi/2)-} (\tan x)^{\tan x}$.

4. Let $f(x)$ be defined in an open interval I and assume that $f'(x)$ and $f''(x)$ exist in I and $f'(x) \neq 0$. Let $a \in I$. Prove

(a) $\displaystyle\lim_{h \to 0} \dfrac{f(a + 2h) - 2f(a + h) + f(a)}{h^2} = f''(a)$;

(b) $\displaystyle\lim_{h \to 0} \left[\dfrac{1}{f(a + h) - f(a)} - \dfrac{1}{hf'(a)}\right] = -\dfrac{f''(a)}{2(f'(a))^2}$.

8. TAYLOR'S THEOREM

TAYLOR'S THEOREM. *Let $f(x)$ be a function defined on a closed, bounded interval $a \leq x \leq b$. Assume that its derivatives $f'(x), f''(x),...,f^{(n+1)}(x)$*

exist for all $a \le x \le b$. Then, for any x, $a < x \le b$, there is a point ξ, $a < \xi < x$, such that

$$f(x) = f(a) + \frac{f'(a)}{1!}(x - a) + \frac{f''(a)}{2!}(x - a)^2 + \cdots + \frac{f^{(n)}(a)}{n!}(x - a)^n$$

$$+ \frac{f^{(n+1)}(\xi)}{(n+1)!}(x - a)^{n+1}. \tag{1}$$

Proof. We shall prove (1) for $x = c$ where $a < c \le b$. Consider the function

$$F(x) = f(c) - f(x) - f'(x)(c - x) - \frac{f''(x)}{2!}(c - x)^2 - \cdots - \frac{f^{(n)}(x)}{n!}(c - x)^n. \tag{2}$$

In computing $F'(x)$ we find that all the inner terms cancel out, so that

$$F'(x) = -\frac{f^{(n+1)}(x)}{n!}(c - x)^n. \tag{3}$$

Next we introduce the function

$$G(x) = \frac{(c - x)^{n+1}}{(n+1)!}. \tag{4}$$

It is clear that

$$G'(x) = -\frac{(c - x)^n}{n!}. \tag{5}$$

Applying the second mean value theorem to $F(x)$, $G(x)$ in the interval $a \le x \le c$, and noting that $F(c) = G(c) = 0$, we get

$$\frac{F(a)}{G(a)} = \frac{F'(\xi)}{G'(\xi)}, \qquad a < \xi < c.$$

Hence

$$F(a) = \frac{F'(\xi)}{G'(\xi)} G(a).$$

Substituting F, F', G, G' from (2)–(5) into the last relation, Equation (1) for $x = c$ follows.

Formula (1) is called *Taylor's formula*. It can also be written in the form:

$$f(a + h) = f(a) + hf'(a) + \frac{h^2}{2!}f''(a) + \cdots + \frac{h^n}{n!}f^{(n)}(a)$$

$$+ \frac{h^{n+1}}{(n+1)!}f^{(n+1)}(a + \theta h), \tag{6}$$

for some $0 < \theta < 1$.

The term

$$R_{n+1} = \frac{f^{(n+1)}(\xi)}{(n+1)!} (x-a)^{n+1} \qquad (7)$$

occurring in (1) is called the *Lagrange form* of the *remainder*.

If we follow the proof of Theorem 1 but take a different function $G(x)$, namely, $G(x) = c - x$, then we arrive at the relation:

$$F(a) = \frac{f^{(n+1)}(\xi)}{n!} (c - \xi)^n (c-a).$$

This yields Taylor's formula with a different form for the remainder, namely,

$$R_{n+1} = \frac{f^{(n+1)}(\xi)}{n!} (x - \xi)^n (x-a). \qquad (8)$$

It is called the *Cauchy form* of the remainder.

The importance of Taylor's theorem is that it enables us to approximate a given function by polynomials that have a simple form. How precise this approximation is will depend on the size of the remainder.

EXAMPLE. If $f(x) = \sin x$ and $a = 0$, then $f^{(n)}(x)$ is either $\pm \sin x$ or $\pm \cos x$, so that (by (7))

$$|R_{n+1}| \le \frac{|x|^{n+1}}{(n+1)!}.$$

The right-hand side converges to 0 if $n \to \infty$.

PROBLEMS

1. Show that, for any positive x,

$$e^x = 1 + x + \frac{x^2}{2} + \cdots + \frac{x^n}{n!} + R_{n+1},$$

where $|R_{n+1}| < e^x \dfrac{|x|^{n+1}}{(n+1)!}.$

2. Show that

$$\frac{1}{\sqrt{1-x}} = 1 + \frac{1}{2}x + \frac{1 \cdot 3}{2 \cdot 4}x^2 + \cdots + \frac{1 \cdot 3 \cdots (2n-1)}{2 \cdot 4 \cdots (2n)}x^n + R_{n+1},$$

where

$$|R_{n+1}| < \frac{1 \cdot 3 \cdots (2n+1)}{2 \cdot 4 \cdots (2n+2)} |x|^{n+1} \qquad \text{if } -1 < x < 0,$$

$$|R_{n+1}| < \frac{1 \cdot 3 \cdots (2n+1)}{2^{n+1} n!} \frac{x^{n+1}}{(1-x)^{3/2}} \qquad \text{if } 0 < x < 1.$$

3. Find the remainder R_{n+1}, when $a = 0$, for each of the following functions, and prove that $R_{n+1} \to 0$ if $n \to \infty$:

    (a) $\log(1 + x)$;   (b) $\sqrt{1 + x}$;   (c) $\sinh x$;   (d) $x \cos x$.

4. Let $P(x)$ be a polynomial. If $P(\alpha) = P'(\alpha) = \cdots = P^{(k)}(\alpha) = 0$, then there is a polynomial $Q(x)$ such that $P(x) = (x - \alpha)^{k+1} Q(x)$.

9. COMPUTATION OF EXTREMUM

Let $f(x)$ be a function defined on an interval I. Suppose c is a point of I such that

$$f(c) \geq f(x) \tag{1}$$

for all x satisfying

$$x \in I \quad \text{and} \quad c - \delta < x < c + \delta, \tag{2}$$

where δ is some positive number. Then we say that $f(x)$ has a *local maximum* (or a *relative maximum*) $f(c)$ at c. If instead of (1) we have

$$f(c) \leq f(x) \tag{3}$$

for all x as in (2), then we say that $f(x)$ has a *local minimum* (or a *relative minimum*) $f(c)$ at c. We say that $f(x)$ has a *local* (or *relative*) *extremum* $f(c)$ at c (and that c is a *locally extreme* point) if it either has a local maximum $f(c)$ or a local minimum $f(c)$ at c. For brevity, we shall refer to locally extreme points as extreme points.

In this section we shall use Taylor's theorem in order to compute the extreme points c. Unless otherwise stated, we shall always assume, from now on, that I is an open interval.

THEOREM 1. *If $f'(x)$ exists for all $x \in I$, and if c is an extreme point of f, then $f'(c) = 0$.*

Proof. Suppose $f(c)$ is a local maximum. Then

$$\frac{f(c+h) - f(c)}{h} \leq 0 \qquad \text{if } h > 0, \text{ and } \geq 0 \text{ if } h < 0.$$

It follows that

$$0 \le f'(c) \le 0,$$

that is, $f'(c) = 0$. If $f(c)$ is a local minimum, then

$$\frac{f(c + h) - f(c)}{h} \ge 0 \qquad \text{if } h > 0, \text{ and } \le 0 \text{ if } h < 0.$$

This again yields $f'(c) = 0$.

Theorem 1 shows that in order to find the extreme points x we have to solve the equation

$$f'(x) = 0. \tag{4}$$

A point x satisfying (4) is called a *critical point* of f.

Suppose we find a solution $x = c$ of (4). Is it necessarily an extreme point? The answer is negative. Indeed, for the function $f(x) = x^3$, $f'(x) = 3x^2 = 0$ if $x = 0$, but 0 is not an extreme point.

In the following theorem we give sufficient conditions for a solution of (4) to be an extreme point.

THEOREM 2. *Assume that $f'(x)$, $f''(x)$,..., $f^{(n+1)}(x)$ exist in the open interval I and that $f^{(n+1)}(x)$ is continuous. Let c be a point of I for which*

$$f'(c) = f''(c) = \cdots = f^{(n)}(c) = 0, \tag{5}$$

$$f^{(n+1)}(c) \ne 0. \tag{6}$$

Then (i) *if n is even then c is not an extreme point, and* (ii) *if n is odd then c is an extreme point: a local maximum if $f^{(n+1)}(c) < 0$ and a local minimum if $f^{(n+1)}(c) > 0$.*

Proof. From Taylor's formula (6) of Section 8, with $a = c$, we get

$$f(c + h) = f(c) + \frac{h^{n+1}}{(n+1)!} f^{(n+1)}(c + \theta h). \tag{7}$$

Since $f^{(n+1)}(c) \ne 0$ and since $f^{(n+1)}(x)$ is continuous, $f^{(n+1)}(c + \theta h) > 0$ (< 0) for all h with $|h|$ sufficiently small, say $|h| < h_0$, if $f^{(n+1)}(c) > 0$ ($f^{(n+1)}(c) < 0$). Suppose now that n is even. Then

$$\frac{h^{n+1}}{(n+1)!} f^{(n+1)}(c + \theta h)$$

is positive when h varies in one of the two intervals $(0, h_0)$, $(-h_0, 0)$ and is negative when h varies in the other interval. It follows (from (7)) that the same is true of $f(c + h) - f(c)$. Hence $f(c)$ is neither a local maximum nor a local minimum.

Suppose next that n is odd. If $f^{(n+1)}(c) > 0$ and $0 < |h| < h_0$, then we see from (7) that

$$f(c + h) - f(c) > 0.$$

Thus $f(c)$ is a relative minimum. Similarly, if $f^{(n+1)}(c) < 0$, then

$$f(c + h) - f(c) < 0 \quad \text{if} \quad 0 < |h| < h_0.$$

Thus $f(c)$ is a relative maximum. This completes the proof.

Notice that if case (i) holds, the function $g(x) = f'(x)$ satisfies

$$g'(c) = \cdots = g^{(n-1)}(c) = 0, \qquad g^{(n)}(c) \neq 0.$$

Hence c is an extreme point for $g(x)$, and $g(c) = 0$. When $f'(c) = 0$ and $f'(c)$ is an extreme value of $f'(x)$, then we say that c is a *point of inflection*. For example, 0 is a point of inflection of x^3.

From Theorem 2 we deduce: If (5) and (6) hold, then either c is a point of extreme or it is a point of inflection.

If in (1) $f(c) > f(x)$ for all x as in (2) with $x \neq c$, then we say that $f(c)$ is a *strict local maximum*. Similarly, we define a *strict local minimum*, a *strict extreme*, and a *point of strict inflection*. Theorem 2 gives sufficient conditions for strict extreme and strict inflection.

Theorems 1 and 2 cannot be applied in case c is an endpoint of a closed interval I. However, if $f'(x) \geq 0$ for all x near c, or if $f'(x) \leq 0$ for all x near c, then f is monotone near c. Hence c will be either a local minimum or a local maximum.

Theorem 2 cannot always be applied (even if $f(x)$ has derivatives of all orders). It is possible to have a function $f(x)$ that has an extreme at a point c, and at the same time

$$f^{(n)}(c) = 0 \quad \text{for} \quad n = 1, 2, 3, \ldots . \tag{8}$$

For example, $f(x) = e^{-1/x^2}$ if $x \neq 0$ and $f(0) = 0$ satisfies (8) at $c = 0$ (for proof see Section 5.7), and 0 is a minimum point of $f(x)$.

PROBLEMS

Find the critical points of the following functions (in $(-\infty, \infty)$), and determine whether they are local maximum points, local minimum points, or points of inflection:

(a) $x^2(x - 1)^2$;

(b) $(4x + 3)/(x^2 + 1)$;

(c) $2x + \log(1 + x^2)$;

(d) $\sin(x^4)$;

(e) $x/(a^2 + x^2)$, $a \neq 0$;

(f) $\cosh x$;

(g) $x^n e^{-x^2}$ (n positive integer);

(h) $e^{x^2} + e^{-x^2}$.

2. Prove the inequality

$$x^m(a - x)^n \le m^m n^n a^{m+n}/(m + n)^{m+n} \qquad \text{for } 0 \le x \le a.$$

Here m and n are positive integers and $a > 0$.

3. Find the extreme values of $f(x) = a \sin x + b \cos x$ $(a^2 + b^2 > 0)$.

4. Find the extreme points of $f(x) = (x - a)^m(x - c)^n$, where m,n are positive integers and $c > a$.

5. Prove that $e^x < 1/(1 - x)$ if $x < 1$, $x \ne 0$.

6. Show that $\log(1 + x) < x$ if $x > -1$, $x \ne 0$.

4

INTEGRATION

1. THE DARBOUX INTEGRAL

Let $f(x)$ be a bounded function defined in a closed, bounded interval $a \le x \le b$. A set of points $\{x_0, x_1, \ldots x_n\}$ is called a *partition* of $[a,b]$ if

$$x_0 = a < x_1 < x_2 < \cdots < x_{n-1} < x_n = b. \tag{1}$$

The *mesh* of the partition is the largest among the numbers

$$x_1 - x_0, x_2 - x_1, \ldots, x_n - x_{n-1}.$$

We denote it by $\max(x_i - x_{i-1})$.

Given any partition (1), we define

$$M_i = \underset{x_{i-1} \le x \le x_i}{\text{l.u.b.}} f(x), \qquad m_i = \underset{x_{i-1} \le x \le x_i}{\text{g.l.b.}} f(x),$$

and form the sums

$$S = \sum_{i=1}^{n} M_i(x_i - x_{i-1}), \tag{2}$$

$$s = \sum_{i=1}^{n} m_i(x_i - x_{i-1}). \tag{3}$$

S is called an *upper Darboux sum* and s is called a *lower Darboux sum*.

Since f is a bounded function, the set G^0 of all the numbers S is bounded. We denote by J the infimum of this set. Similarly, we denote by G_0 the set of all the numbers s. The supremum of G_0 is denoted by I.

A partition

$$y_0 = a < y_1 < y_2 < \cdots < y_{m-1} < y_m = b \tag{4}$$

is said to be *more refined* (or *finer*) than the partition (1) if every point x_j occurring in (1) occurs also in (4).

We shall compare the upper sum (2) with the upper sum

$$S' = \sum_{i=1}^{m} M_i'(y_i - y_{i-1}) \tag{5}$$

corresponding to the partition (4).

THEOREM 1.   *If partition (4) is more refined than partition (1), then*

$$S' \leq S. \tag{6}$$

Proof.   Consider a term $M_i(x_i - x_{i-1})$ occurring in the sum in (2). Then $x_{i-1} = y_{j-1}$ for some j and $x_i = y_k$ for some k. If $k = j$, then

$$M_i(x_i - x_{i-1}) = M_j'(y_k - y_{j-1}). \tag{7}$$

If, however, $k > j$, then

$$x_{i-1} = y_{j-1} < y_j < \cdots < y_{k-1} < y_k = x_i.$$

We then have

$$M_i \geq M_j', \, M_i \geq M_{j+1}', \ldots, M_i \geq M_k'.$$

It follows that

$$M_i(x_i - x_{i-1}) \geq M_j'(y_j - y_{j-1})$$
$$+ M_{j+1}'(y_{j+1} - y_j) + \cdots + M_k'(y_k - y_{k-1}). \tag{8}$$

Thus, for any i, $1 \leq i \leq n$, either (7) or (8) holds. Summing over i, we get (6).

THEOREM 2.   *For any $\varepsilon > 0$ there is a $\delta > 0$ such that for any partition* (1),

$$0 \leq S - J < \varepsilon \qquad \text{if } \max(x_i - x_{i-1}) < \delta. \tag{9}$$

We express this assertion also in the form:

$$\sum_{i=1}^{n} M_i(x_i - x_{i-1}) \to J \qquad \text{if } \max(x_i - x_{i-1}) \to 0. \tag{10}$$

Proof. Since J is the infimum of the set G, for any $\varepsilon > 0$ there is a partition (4) such that the corresponding upper Darboux sum (5) satisfies

$$J \le S' < J + \frac{\varepsilon}{2}. \tag{11}$$

Consider now any partition (1) with mesh $< \delta$. Let

$$z_0 = a < z_1 < z_2 < \cdots < z_{r-1} < z_r = b \tag{12}$$

be the partition whose points consist of the points of both (1) and (4). Denote by S the upper Darboux sum corresponding to partition (1) (S is given by (2)) and denote by S'' the upper Darboux sum corresponding to partition (12), that is,

$$S'' = \sum_{i=1}^{r} M_i''(z_i - z_{i-1}). \tag{13}$$

By Theorem 1,

$$S'' \le S'. \tag{14}$$

We shall compare S'' with S. Take any term $M_i(x_i - x_{i-1})$ in S. Then there are points z_{j-1} and z_k in (12) such that $z_{j-1} = x_{i-1}$ and $z_k = x_i$. If $k = j$, then

$$M_i(x_i - x_{i-1}) = M_s''(z_j - z_{j-1}). \tag{15}$$

If $k > j$, then

$$x_{i-1} = z_{j-1} < z_j < z_{j+1} < \cdots < z_{k-1} < z_k = x_i.$$

The points $z_j, \ldots, z_{k-1}$ belong to partition (12) but do not belong to partition (1). Denote by M an upper bound on $|f|$. Then

$$M_i(x_i - x_{i-1}) < \delta M. \tag{16}$$

The case $k > j$ can occur at most m times, since partition (12) has at most m additional points over partition (1) (m as in (4)). Consequently, by (15) and (16),

$$S = \sum_{i=1}^{n} M_i(x_i - x_{i-1}) \le \sum_{i=1}^{r} M_i''(z_i - z_{i-1}) + m\delta M = S'' + m\delta M.$$

Combining this with (14), and using (11), we get

$$J \le S < J + \frac{\varepsilon}{2} + m\delta M < J + \varepsilon \qquad \text{if } \delta < \frac{\varepsilon}{2mM}.$$

This completes the proof of (9).

THEOREM 3.   *If partition* (4) *is more refined than partition* (1), *then*

$$s' \geq s,$$

where s' and s are the lower Darboux sums corresponding to partitions (4) *and* (1), *respectively.*

THEOREM 4.   *For any $\varepsilon > 0$, there is a $\delta > 0$ such that for any partition* (1)

$$-\varepsilon < s - I \leq 0 \qquad if \quad \max(x_i - x_{i-1}) < \delta.$$

We express this assertion also in the form:

$$\sum_{i=1}^{n} m_i(x_i - x_{i-1}) \to I \qquad \text{if } \max(x_i - x_{i-1}) \to 0.$$

The proofs of Theorems 3 and 4 are similar to the proofs of Theorems 1 and 2.

THEOREM 5.   *The following inequality holds:*

$$J \geq I. \tag{17}$$

Proof.   Let s be any lower Darboux sum as in (2). Let S' be any upper Darboux sum as in (5). Denote by S'' the upper Darboux sum given by (13), where the partition (12) is obtained by adding the points of partition (4) to partition (1). Denote by s'' the corresponding lower Darboux sum. By Theorem 1,

$$S'' \leq S'.$$

By Theorem 3,

$$s'' \geq s.$$

Since $S'' \geq s''$, it follows that $S' \geq s$. Hence

$$J = \inf_{S' \in G^0} S' \geq s.$$

Since this is true for any s,

$$J \geq \sup_{G_0} s = I.$$

DEFINITION. J is called the *upper Darboux integral*, and I is called the *lower Darboux integral*. If $I = J$, then f is said to be *Darboux integrable* over (on, or in) the interval $[a,b]$, and the number I is then called the *Darboux integral* of f over (on, or in) $[a,b]$. We shall denote it by

$$(D) \int_a^b f(x)\,dx.$$

THEOREM 6.   *Let $f(x)$ be a bounded function in a closed, bounded interval $a \le x \le b$. Then f is Darboux integrable over $[a,b]$ if, and only if, for any $\varepsilon > 0$ there is a partition of $[a,b]$ for which the upper and lower Darboux sums satisfy*

$$S - s < \varepsilon. \tag{18}$$

Proof.   If (18) holds, then

$$J - I < S - s < \varepsilon.$$

Since ε is arbitrary, $J \le I$. Since, by Theorem 5, $J \ge I$, we find that $J = I$. Hence f is Darboux integrable.

Suppose conversely that f is Darboux integrable. Then for any $\varepsilon > 0$, there is a $\delta > 0$ such that for any partition (1) with mesh $< \delta$, the corresponding upper and lower Darboux sums, S and s, satisfy

$$s > I - \frac{\varepsilon}{2}, \quad S < I + \frac{\varepsilon}{2}.$$

This implies (18).

PROBLEMS

1.   Consider the function

$$f(x) = \begin{cases} 0 & \text{if } x \text{ is irrational,} \\ 1 & \text{if } x \text{ is rational.} \end{cases}$$

Prove that f is not Darboux integrable over $[0,1]$.

2.   Consider the function

$$f(x) = \begin{cases} \alpha & \text{if } -\infty < x \le 0, \\ \beta & \text{if } 0 < x < \infty, \end{cases}$$

where α, β are real numbers. Show that $f(x)$ is Darboux integrable over any bounded intreval $[a,b]$ and compute $(D) \int_a^b f(x)\,dx$.

3.   Show that the function $[x]$ is Darboux integrable over any bounded interval $[a,b]$, and compute $(D) \int_a^b f(x)\,dx$.

2. THE RIEMANN INTEGRAL

Let $f(x)$ be a bounded function in a closed, bounded interval $[a,b]$. Take any partition

$$x_0 = a < x_1 < x_2 < \cdots < x_{n-1} < x_n = b \tag{1}$$

and any points $\xi_1, \xi_2, \ldots, \xi_n$ such that $x_{i-1} \le \xi_i \le x_i$. Form the sum

$$T = \sum_{i=1}^{n} f(\xi_i)(x_i - x_{i-1}). \tag{2}$$

Suppose that there is a number K such that the following is true: For any $\varepsilon > 0$ there is a $\delta > 0$ such that

$$|T - K| < \varepsilon \qquad \text{if } \max(x_i - x_{i-1}) < \delta, \tag{3}$$

no matter what the points x_i of the partition and the points ξ_i in $[x_{i-1}, x_i]$ are. Then we write

$$\sum_{i=1}^{n} f(\xi_i)(x_i - x_{i-1}) \to K \qquad \text{if } \max(x_i - x_{i-1}) \to 0. \tag{4}$$

DEFINITION.   If there is a number K such that (4) holds, then f is said to be *Riemann integrable* over (on, or in) the interval $[a,b]$. The number K is then called the *Riemann integral* of f over (on, or in) $[a,b]$, and it is denoted by

$$(R) \int_a^b f(x) \, dx.$$

The function f is called the *integrand* of the Riemann integral. The points a and b are called the *lower and upper endpoints* of the integral.

If $f(x) \ge 0$ for all $a \le x \le b$, then both the Darboux and the Riemann integrals give, intuitively, the area that is bounded by the curves $y = f(x)$, $y = 0$, $x = a$, and $x = b$. It is therefore natural to expect that the two integrals are the same. This is indeed the case:

THEOREM 1.   *Let $f(x)$ be a bounded function in a closed, bounded interval $a \le x \le b$. Then f is Riemann integrable if, and only if, f is Darboux integrable, and, in this case,*

$$(R) \int_a^b f(x) \, dx = (D) \int_a^b f(x) \, dx. \tag{5}$$

Proof. If f is Darboux integrable then, for any partition (1), and for
for any ξ_i with $x_{i-1} \le \xi_i \le x_i$,

$$s = \sum_{i=1}^{n} m_i(x_i - x_{i-1}) \le \sum_{i=1}^{n} f(\xi_i)(x_i - x_{i-1}) \le \sum_{i=1}^{n} M_i(x_i - x_{i-1}) = S.$$

For any $\varepsilon > 0$ there is a $\delta > 0$ such that if $\max(x_i - x_{i-1}) < \delta$, then

$$S < I + \varepsilon, \qquad s > I - \varepsilon,$$

where I is the Darboux integral of f over $[a,b]$. It follows that

$$\left| \sum_{i=1}^{n} f(\xi_i)(x_i - x_{i-1}) - I \right| < \varepsilon.$$

This shows that f is Riemann integrable in $[a,b]$, and also that (5) holds.
   Suppose now that f is Riemann integrable in $[a,b]$, and denote by K its
Riemann integral over $[a,b]$. Then, for any $\varepsilon > 0$ there is a $\delta > 0$ such that
for any partition (1) with mesh $< \delta$,

$$K - \varepsilon < \sum_{i=1}^{n} f(\xi_i)(x_i - x_{i-1}) < K + \varepsilon \tag{6}$$

where ξ_i are any points satisfying $x_{i-1} \le \xi_i \le x_i$.
   Let

$$M_i = \operatorname*{l.u.b.}_{x_{i-1} \le x \le x_i} f(x).$$

Then there is a point ξ_i' in $[x_{i-1},x_i]$ such that

$$f(\xi_i') > M_i - \frac{\varepsilon}{b - a}.$$

It follows that

$$\sum_{i=1}^{n} f(\xi_i')(x_i - x_{i-1}) > \sum_{i=1}^{n} M_i(x_i - x_{i-1}) - \varepsilon.$$

Using (6) with $\xi_i = \xi_i'$, we get

$$S = \sum_{i=1}^{n} M_i(x_i - x_{i-1}) < K + 2\varepsilon.$$

Similarly, we get

$$s = \sum_{i=1}^{n} m_i(x_i - x_{i-1}) > K - 2\varepsilon.$$

It follows that

$$J \le S < K + 2\varepsilon < s + 4\varepsilon \le I + 4\varepsilon.$$

Since ε is arbitrary, $J \le I$. Since also $J \ge I$, we conclude that $J = I$, so that f is Darboux integrable in $[a,b]$.

From now on we shall speak only of Riemann integrals. We shall denote the Riemann integral of $f(x)$ over $[a,b]$ by

$$\int_a^b f(x) \, dx.$$

In proving various properties for Riemann integrals we often make use of Theorem 1. This theorem tells us that it is sufficient to prove the corresponding properties in which the Riemann integral is replaced everywhere by the Darboux integral.

As an example, we quote the following result, which is a consequence of Theorem 6 of Section 1.

THEOREM 2. *A bounded function f on a bounded interval $[a,b]$ is Riemann integrable if, and only if, for any $\varepsilon > 0$ there is a partition of $[a,b]$ for which the corresponding upper and lower Darboux sums, S and s, satisfy*

$$S - s < \varepsilon.$$

Some authors call the Darboux integral, as defined in Section 1, the Riemann integral, or the Riemann–Darboux integral. Theorem 1 then takes the form: f is Riemann integrable if, and only if,

$$\sum_{j=1}^n f(\xi_j)(x_j - x_{j-1}) \to \int_a^b f(x) \, dx \qquad \text{if } \max(x_i - x_{i-1}) \to 0,$$

where $\int_a^b f(x) \, dx$ is the Riemann integral. The sum in (2) is called the *Riemann sum*.

In what follows, when we speak of a Riemann integrable function in $[a,b]$, we always understand that $[a,b]$ is a bounded interval.

PROBLEMS

1. A function $f(x)$ in $[a,b]$ is called a *step function* if there is a partition (1) of $[a,b]$ and constants c_i such that $f(x) = c_i$ if $x_{i-1} < x < x_i$. Prove that such a function is Riemann integrable and

$$\int_a^b f(x) \, dx = \sum_{i=1}^n c_i(x_i - x_{i-1}).$$

2. Let $f(x)$ and $g(x)$ be integrable functions over $[a,b]$ and let α be a point of $[a,b]$. If $f(x) = g(x)$ for all $x \ne \alpha$, then

$$\int_a^b f(x) \, dx = \int_a^b g(x) \, dx.$$

3. Generalize the result of the preceding problem to the case where $f(x) = g(x)$ for all $x \neq \alpha_1,...,x \neq \alpha_m$ where $\alpha_1,...,\alpha_m$ are some points in the interval $[a,b]$.

3. PROPERTIES OF THE INTEGRAL

We shall often refer to Riemann integrable functions, briefly, as integrable functions.

THEOREM 1.   *If $f(x)$ is integrable over $[a,b]$, then it is integrable also over any subinterval $[\alpha,\beta]$.*

Proof.   By Theorems 2 and 4 of Section 1, for any $\varepsilon > 0$ there is a $\delta > 0$ such that for any partition of $[a,b]$ with mesh $< \delta$, the corresponding Darboux sums S and s satisfy

$$0 \leq I - s < \frac{\varepsilon}{2}, \qquad 0 \leq S - J < \frac{\varepsilon}{2},$$

Since $I = J$, we get $S - s < \varepsilon$. Now take a particular partition of $[a,b]$

$$x_0 = a < x_1 < x_2 < \cdots < x_{j-1} < x_j < \cdots < x_{k-1} < x_k < \cdots < x_{n-1} < x_n = b$$

with mesh $< \delta$, such that $x_{j-1} = \alpha$, $x_k = \beta$. Then

$$\alpha = x_{j-1} < x_j \cdots < x_{k-1} < x_k = \beta$$

is a partition of $[\alpha,\beta]$. Denote by S^* and s^* the upper and lower Darboux sums corresponding to this partition of $[\alpha,\beta]$. Then

$$S - s = \sum_{i=1}^{n} (M_i - m_i)(x_i - x_{i-1})$$

$$= S^* - s^* + \sum_{i=1}^{j-1} (M_i - m_i)(x_i - x_{i-1})$$

$$+ \sum_{i=k+1}^{n} (M_i - m_i)(x_i - x_{i-1}) \geq S^* - s^*.$$

It follows that $S^* - s^* < \varepsilon$. Since ε is arbitrary, Theorem 2 of Section 2 implies that f is integrable over $[\alpha,\beta]$.

THEOREM 2.   *If $f(x)$ is integrable over $[a,b]$ and over $[b,c]$, then it is integrable over $[a,c]$ and*

$$\int_a^b f(x)\, dx + \int_b^c f(x)\, dx = \int_a^c f(x)\, dx. \tag{1}$$

Proof.   For any $\varepsilon > 0$ there is a partition

$$a = x_0 < x_1 < \cdots < x_{k-1} < x_k = b \tag{2}$$

of $[b,c]$ such that the corresponding Darboux sums

$$S_1 = \sum_{i=1}^{k} M_i(x_i - x_{i-1}), \qquad s_1 = \sum_{i=1}^{k} m_i(x_i - x_{i-1})$$

satisfy

$$S_1 - s_1 < \frac{\varepsilon}{2}.$$

Similarly, there is a partition

$$b < x_{k+1} < x_{k+2} < \cdots < x_{n-1} < x_n = c \tag{3}$$

of $[b,c]$ such that the corresponding Darboux sums

$$S_2 = \sum_{i=k+1}^{n} M_i(x_i - x_{i-1}), \qquad s_2 = \sum_{i=k+1}^{n} m_i(x_i - x_{i-1})$$

satisfy

$$S_2 - s_2 < \frac{\varepsilon}{2}.$$

Consider the partition of $[a,c]$ given by the points of (2) and (3). Its upper and lower Darboux sums, S and s, are given by

$$S = S_1 + S_2, \qquad s = s_1 + s_2.$$

Therefore $S - s < \varepsilon$. Using Theorem 2 of Section 2, we conclude that f is integrable. If we now take sequences of partitions (2) and (3) with mesh tending to 0, then from the relation $S = S_1 + S_2$ we obtain, in the limit, relation (1).

THEOREM 3.   *Let $f(x)$ and $g(x)$ be integrable functions over $[a,b]$. Then $f + g$ and $f - g$ are also integrable over $[a,b]$, and*

$$\int_a^b [f(x) \pm g(x)]\, dx = \int_a^b f(x)\, dx \pm \int_a^b g(x)\, dx. \tag{4}$$

Proof.   For any $\varepsilon > 0$ there are positive numbers δ_1 and δ_2 such that for any partition (2) of $[a,b]$, and for any points ξ_i, $x_{i-1} \le \xi_i \le x_i$,

$$\left| \sum_{i=1}^{k} f(\xi_i)(x_i - x_{i-1}) - \int_a^b f(x)\, dx \right| < \frac{\varepsilon}{2} \qquad \text{if } \max(x_i - x_{i-1}) < \delta_1,$$

and

$$\left| \sum_{i=1}^{k} g(\xi_i)(x_i - x_{i-1}) - \int_a^b g(x)\,dx \right| < \frac{\varepsilon}{2} \quad \text{if } \max(x_i - x_{i-1}) < \delta_2.$$

Hence, if $\max(x_i - x_{i-1}) < \delta$ where $\delta = \min(\delta_1, \delta_2)$, then

$$\left| \sum_{i=1}^{k} [f(\xi_i) \pm g(\xi_i)](x_i - x_{i-1}) - \left\{ \int_a^b f(x)\,dx \pm \int_a^b g(x)\,dx \right\} \right| < \varepsilon.$$

From the definition of the Riemann integral it follows that $f \pm g$ is integrable and (4) holds.

THEOREM 4.   *Let $f(x)$ and $g(x)$ be integrable functions over $[a,b]$. If $f(x) \le g(x)$ for all x in $[a,b]$, then*

$$\int_a^b f(x)\,dx \le \int_a^b g(x)\,dx. \tag{5}$$

Proof.   For any partition (2) of $[a,b]$, and for any $\xi_i \in [x_{i-1}, x_i]$,

$$\sum_{i=1}^{k} f(\xi_i)(x_i - x_{i-1}) \le \sum_{i=1}^{k} g(\xi_i)(x_i - x_{i-1}).$$

Now take a sequence of partitions with mesh converging to 0.

THEOREM 5.   *If f is integrable over $[a,b]$, then for any constant λ, λf is also integrable over $[a,b]$, and*

$$\int_a^b \lambda f(x)\,dx = \lambda \int_a^b f(x)\,dx. \tag{6}$$

The proof is left to the student.

Let $f(x)$ be a bounded function on $[a,b]$. We introduce two functions:

$$f^+(x) = \begin{cases} f(x) & \text{if } f(x) \ge 0, \\ 0 & \text{if } f(x) < 0, \end{cases}$$

$$f^-(x) = \begin{cases} -f(x) & \text{if } f(x) \le 0, \\ 0 & \text{if } f(x) > 0. \end{cases}$$

f^+ is called the *positive part* of f and f^- is called the *negative part* of f. We can write f in the form:

$$f(x) = f^+(x) - f^-(x). \tag{7}$$

The function $|f|$, defined by $|f|(x) = |f(x)|$, clearly satisfies

$$|f(x)| = f^+(x) + f^-(x). \tag{8}$$

THEOREM 6.   *If f is integrable over* [a,b] *then also* f^+, f^- *and* $|f|$ *are integrable over* [a,b], *and*

$$\left| \int_a^b f(x)\, dx \right| \leq \int_a^b |f(x)|\, dx. \tag{9}$$

Proof.   For every subinterval $[\alpha, \beta]$ of $[a,b]$, if $f(x) > 0$ for at least one x in $[\alpha, \beta]$, then $\sup_{\alpha \leq x \leq \beta} f^+(x) = \sup_{\alpha \leq x \leq \beta} f(x)$ and $\inf_{\alpha \leq x \leq \beta} f^+(x) \geq \inf_{\alpha \leq x \leq \beta} f(x)$ so that

$$\sup_{\alpha \leq x \leq \beta} f^+(x) - \inf_{\alpha \leq x \leq \beta} f^+(x) \leq \sup_{\alpha \leq x \leq \beta} f(x) - \inf_{\alpha \leq x \leq \beta} f(x).$$

This inequality holds also in case $f(x) \leq 0$ for all x in $[\alpha, \beta]$, for then $f^+(x) = 0$ for all x in $[\alpha, \beta]$. Hence, for any partition of $[a,b]$, the Darboux sums S, s of f and the Darboux sums S^+, s^+ of f^+ satisfy the inequality

$$S^+ - s^+ \leq S - s.$$

Using this and Theorem 2 of Section 2, we find that f^+ is integrable over $[a,b]$. From (7) and Theorem 3 it follows that also f^- is integrable. From (8) and Theorem 3 it then follows that $|f|$ is integrable. The inequality (9) is obtained by applying Theorem 4 to f and $g = |f|$ and to $-f$ and $g = |f|$.

The next theorem is called the *mean value theorem for integrals.*

THEOREM 7.   *Let $f(x)$ be integrable over* [a,b] *and let*

$$M = \text{l.u.b. } f(x), \qquad m = \text{g.l.b. } f(x).$$
$$\quad\quad a \leq x \leq b \qquad\qquad\quad a \leq x \leq b$$

Then

$$\int_a^b f(x)\, dx = \mu(b - a) \qquad \text{for some } m \leq \mu \leq M. \tag{10}$$

If, furthermore, $f(x)$ is continuous, then

$$\int_a^b f(x)\, dx = f(\xi)(b - a) \qquad \text{for some } a < \xi < b. \tag{11}$$

Proof.   Since $m \leq f(x) \leq M$ for all x in $[a,b]$, Theorem 4, coupled with the fact that $\int_a^b dx = b - a$, yields

$$m(b - a) \leq \int_a^b f(x)\, dx \leq M(b - a).$$

Thus, if we define μ by

$$\mu = \frac{1}{b - a} \int_a^b f(x)\, dx,$$

then $m \leq \mu \leq M$. This proves (10). If $f(x)$ is continuous, then $M = f(\alpha)$, $m = f(\beta)$ for some points α and β in $[a,b]$. From the intermediate value theorem it then follows that if $\alpha \neq \beta$, there is a point ξ lying between α and β such that $f(\xi) = \mu$. This proves (11) in case $\alpha \neq \beta$. If $\alpha = \beta$, that is, if $M = m$, then (11) is certainly true with any ξ in (a,b).

PROBLEMS

1. Prove Theorem 5.

2. Show that the function

$$f(x) = \begin{cases} 1 & \text{if } x \text{ is irrational,} \\ -1 & \text{if } x \text{ is rational} \end{cases}$$

   is not integrable, but that $|f|$ is integrable.

3. If f is integrable on $[a,b]$ and $\int_a^b f(x)\, dx > 1$, then there exists a point ξ in (a,b) such that $f(\xi) > 1/(b - a)$.

4. If f is integrable and continuous over $[a,b]$, and if $\int_\alpha^\beta f(x)\, dx \geq 0$ for any subinterval (α,β) of (a,b), then $f \geq 0$ in $[a,b]$. [*Hint:* If $f(\xi) < 0$, $a < \xi < b$, then $f(x) < f(\xi)/2$ in some neighborhood of ξ.]

5. If f is integrable and continuous in $[a,b]$, and if $\int_\alpha^\beta f(x)\, dx = 0$ for any subinterval (α,β) of (a,b), then $f(x) = 0$ for all $a \leq x \leq b$.

6. If $f(x)$ and $g(x)$ are integrable and continuous functions over $[a,b]$, and if $\int_\alpha^\beta f(x)\, dx \geq \int_\alpha^\beta g(x)\, dx$ for all subintervals (α,β) of (a,b), then $f(x) \geq g(x)$ for all $a \leq x \leq b$.

4. EXAMPLES OF INTEGRABLE FUNCTIONS

THEOREM 1. *Any continuous function $f(x)$ on a closed, bounded interval $[a,b]$ is integrable over $[a,b]$.*

Proof. By Theorem 1 of Section 2.7, a continuous function $f(x)$ on $[a,b]$ is uniformly continuous. Thus, for any $\varepsilon > 0$ there is a $\delta > 0$ such that

$$|f(x') - f(x'')| < \varepsilon \qquad \text{if } |x' - x''| < \delta.$$

Take any partition

$$x_0 = a < x_1 < x_2 < \cdots < x_{n-1} < x_n = b \tag{1}$$

with $\max(x_i - x_{i-1}) < \delta$. We have

$$M_i - m_i = \max_{x_{i-1} \leq x \leq x_i} f(x) - \min_{x_{i-1} \leq x \leq x_i} f(x) = f(\xi_i^1) - f(\xi_i^2)$$

for some ξ_i^1, ξ_i^2 in $[x_{i-1},x_i]$. Since $|\xi_i^1 - \xi_i^2| < \delta$, it follows that $M_i - m_i < \varepsilon$.

Consequently,

$$\sum_{i=1}^{n} M_i(x_i - x_{i-1}) - \sum_{i=1}^{n} m_i(x_i - x_{i-1}) < (b - a)\varepsilon.$$

Using Theorem 2 of Section 2 we conclude that $f(x)$ is integrable over $[a,b]$.

THEOREM 2.   *Any monotone function $f(x)$ on a closed, bounded interval $[a,b]$ is integrable over $[a,b]$.*

Proof.   Suppose $f(x)$ is monotone increasing. Then, for any partition (1) of $[a,b]$,

$$M_i = f(x_i), \qquad m_i = f(x_{i-1}).$$

Hence, if $\max(x_i - x_{i-1}) < \delta$,

$$\sum_{i=1}^{n} M_i(x_i - x_{i-1}) - \sum_{i=1}^{n} m_i(x_i - x_{i-1})$$

$$= \sum_{i=1}^{n} f(x_i)(x_i - x_{i-1}) - \sum_{i=1}^{n} f(x_{i-1})(x_i - x_{i-1})$$

$$= \sum_{i=1}^{n} [f(x_i) - f(x_{i-1})](x_i - x_{i-1}) \le \delta \sum_{i=1}^{n} [f(x_i) - f(x_{i-1})]$$

$$= \delta[f(b) - f(a)] < \varepsilon,$$

if $\delta < \varepsilon/[f(b) - f(a)]$, where ε is a given positive number. Since ε is arbitrary, Theorem 2 of Section 2 implies that $f(x)$ is integrable on $[a,b]$.
If $f(x)$ is monotone decreasing, then the proof is similar.

COROLLARY.   *Any function $f(x)$ of bounded variation on a closed, bounded interval $[a,b]$ is integrable over $[a,b]$.*
Indeed, by Theorem 2 of Section 2.8, $f(x) = g_1(x) - g_2(x)$ where g_1 and g_2 are monotone increasing. By Theorem 2, g_1 and g_2 are integrable over $[a,b]$. Now use Theorem 3 of Section 3 to conclude that $f(x)$ is integrable over $[a,b]$.

THEOREM 3.   *Let $f(x)$ be a bounded function on a closed, bounded interval $[a,b]$. Assume that f is continuous at all the points of $[a,b]$ with the exception of points of a subset S having the following property: For any $\varepsilon > 0$, S is contained in a finite number of open intervals $I_1,...,I_N$ whose lengths $l_1,...,l_N$ satisfy: $l_1 + \cdots + l_N < \varepsilon$. Then f is integrable over $[a,b]$.*

Proof.   For any $\varepsilon > 0$, let $I_1,\ldots,I_N$ be open intervals as in the assumption of the theorem. If two of these intervals intersect, then their union is an open interval. If we replace these two intervals by their union, then the new set of intervals has all the properties as in the assumption of the theorem. Continuing in this way step by step, we arrive at a finite number of open intervals $I'_1,\ldots,I'_N$, no two of which intersect, such that every point of S is contained in at least one of these intervals and the sum of their lengths is $< \varepsilon$. For simplicity of notation we assume that $I'_1 = I_1,\ldots,I'_N = I_N$. We may also suppose that the intervals I_j are so indexed that $I_j = \{x ; a_j < x < b_j\}$ and

$$a_1 < b_1 < a_2 < b_2 < \cdots < a_N < b_N, a < b_1, a_N < b.$$

If $a_1 < a$, we redefine I_1, taking it to be the interval (a,b_1). If $b_N > b$, we redefine I_N, taking it to be the interval (a_N,b).
     Let

$$J_1 = [b_1,a_2], \qquad J_2 = [b_2,a_3], \ldots, J_{N-1} = [b_{N-1},a_N].$$

If $a < a_1$, then we define $J_0 = [a,a_1]$. If $b_N < b$, then we define $J_N = [b_N,b]$. In each closed interval J_k, f is continuous. Hence, by Theorem 1, there is a partition of J_k for which the upper and lower Darboux sums, S_k and s_k, satisfy

$$S_k - s_k < \frac{\varepsilon}{N + 1}.$$

     For any partition of the closure of I_j, the upper and lower Darboux sums. $S_j{}^0$ and $s_j{}^0$, satisfy

$$S_j{}^0 - s_j{}^0 < 2M(b_j - a_j)$$

where M is an upper bound on $|f|$. Setting

$$S = \sum_k S_k + \sum_j S_j{}^0, \qquad s = \sum_k s_k + \sum_j s_j{}^0,$$

we have

$$S - s < (N + 1)\frac{\varepsilon}{N + 1} + 2M \sum (b_j - a_j) < \varepsilon + 2M\varepsilon = (1 + 2M)\varepsilon.$$

     Since S and s are the upper and lower Darboux sums for a particular partition of $[a,b]$, it follows from Theorem 2 of Section 2 that f is integrable.

     COROLLARY.   *A bounded function having a finite number of points of discontinuity is integrable.*
     Indeed, if the points of discontinuity are $\alpha_1,\ldots,\alpha_m$, then for any ε we take $I_1,\ldots,I_m$ with $I_j = \{x ; \alpha_j - \varepsilon/2m < x < \alpha_j + \varepsilon/2m\}$. We then see that the set of discontinuities of f is contained in the intervals I_j, and the sum of their lengths is ε.

DEFINITION.   Let S be a set having the following property: For any $\varepsilon > 0$, there is a countable number of open intervals I_j such that (i) each point of S is contained in at least one of the intervals I_j, and (ii) the lengths l_j of the intervals I_j satisfy $l_1 + l_2 + \cdots + l_n < \varepsilon$ for any n. We then say that S has *measure zero*.

The following remarkable theorem characterizes those bounded functions which are Riemann integrable.

THEOREM 4.   *A bounded function on a closed, bounded interval is Riemann integrable if, and only if, the set of its points of discontinuity has measure zero.*

Note that Theorem 3 is contained in Theorem 4.

We shall not give the proof of Theorem 4 here.

THEOREM 5.   *If f and g are integrable functions over* $[a,b]$, *then fg is also integrable over* $[a,b]$.

Proof.   Suppose first that $f \geq 0$, $g \geq 0$. Set $h = fg$. For any partition (1) of $[a,b]$, let

$$M_i = \sup_{x_{i-1} \leq x \leq x_i} f(x), \qquad m_i = \inf_{x_{i-1} \leq x \leq x_i} f(x),$$

$$\overline{M}_i = \sup_{x_{i-1} \leq x \leq x_i} g(x), \qquad \overline{m}_i = \inf_{x_{i-1} \leq x \leq x_i} g(x),$$

$$\overline{\overline{M}}_i = \sup_{x_{i-1} \leq x \leq x_i} h(x), \qquad \overline{\overline{m}}_i = \inf_{x_{i-1} \leq x \leq x_i} h(x).$$

Denote by S, $\overline{S}$, and $\overline{\overline{S}}$ the upper Darboux sums corresponding to f, g, and h, respectively. Denote by s, $\overline{s}$, and $\overline{\overline{s}}$ the lower Darboux sums corresponding to f, g, and h, respectively. Finally, let M be any upper bound on $|f|$ and let $\overline{M}$ be any upper bound on $|g|$. Since $f \geq 0$, $g \geq 0$,

$$M_i \overline{M}_i \geq \overline{\overline{M}}_i, \qquad m_i \overline{m}_i \leq \overline{\overline{m}}_i.$$

Hence

$$\overline{\overline{S}} = \sum_{i=1}^{n} \overline{\overline{M}}_i (x_i - x_{i-1}) \leq \sum_{i=1}^{n} M_i \overline{M}_i (x_i - x_{i-1}),$$

$$\overline{\overline{s}} = \sum_{i=1}^{n} \overline{\overline{m}}_i (x_i - x_{i-1}) \geq \sum_{i=1}^{n} m_i \overline{m}_i (x_i - x_{i-1}).$$

It follows that

$$\bar{S} - \underset{\bar{s}}{} \leq \sum_{i=1}^{n} (M_i \bar{M}_i - m_i \bar{m}_i)(x_i - x_{i-1})$$

$$= \sum_{i=1}^{n} (M_i - m_i)\bar{M}_i(x_i - x_{i-1}) + \sum_{i=1}^{n} m_i(\bar{M}_i - \bar{m}_i)(x_i - x_{i-1})$$

$$\leq \bar{M}(S - s) + M(\bar{S} - \bar{s}).$$

Since f is integrable, for any $\varepsilon > 0$ there is a $\delta_1 > 0$ such that, for any partition (1) with mesh $< \delta_1$,

$$S - s < \frac{\varepsilon}{2\bar{M}}.$$

Since g is integrable, for any $\varepsilon > 0$ there is a $\delta_2 > 0$ such that, for any partition (1) with mesh $< \delta_2$,

$$\bar{S} - \bar{s} < \frac{\varepsilon}{2M}.$$

If we then take a partition (1) with mesh $< \delta$, where $\delta = \min(\delta_1, \delta_2)$, then we get

$$\bar{S} - \bar{s} < \bar{M}\frac{\varepsilon}{2\bar{M}} + M\frac{\varepsilon}{2M} = \varepsilon.$$

Theorem 2 of Section 2 now implies that fg is integrable.

So far we have considered only the case $f \geq 0$, $g \geq 0$. For general f, g write

$$f = f^+ - f^-, \qquad g = g^+ - g^-$$

so that

$$fg = f^+g^+ + f^-g^- - f^+g^- - f^-g^+.$$

By Theorem 6 of Section 3, f^+, f^- and g^+, g^- are integrable. Applying the special case of Theorem 5 proved above, it follows that f^+g^+, f^-g^-, f^+g^-, and f^-g^+ are integrable. Theorem 3 of Section 3 now shows that fg is integrable.

THEOREM 6.   *If f and g are integrable over $[a,b]$ and if $|g(x)| \geq m > 0$ for all $a \leq x \leq b$, then f/g is integrable over $[a,b]$.*

Proof.   In view of Theorem 5, it suffices to prove that $1/g$ is integrable. Take any partition (1) and set

$$M_i = \sup_{x_{i-1} \leq x \leq x_i} g(x), \qquad m_i = \inf_{x_{i-1} \leq x \leq x_i} g(x).$$

If $m_i > 0$ or if $M_i < 0$, then

$$\sup_{x_{i-1} \leq x \leq x_i} \frac{1}{g(x)} = \frac{1}{m_i}, \qquad \inf_{x_{i-1} \leq x \leq x_i} \frac{1}{g(x)} = \frac{1}{M_i}.$$

It follows that

$$\sup_{x_{i-1} \leq x \leq x_i} \frac{1}{g(x)} - \inf_{x_{i-1} \leq x \leq x_i} \frac{1}{g(x)} = \frac{M_i - m_i}{M_i m_i} \leq \frac{M_i - m_i}{m^2}.$$

On the other hand, if $m_i < 0 < M_i$, then, for $\varepsilon = (M_i - m_i)/m^2$, there are some η_i, ξ_i in $x_{i-1} \leq x \leq x_i$ such that

$$\sup_{x_{i-1} \leq x \leq x_i} \frac{1}{g(x)} - \inf_{x_{i-1} \leq x \leq x_i} \frac{1}{g(x)} \leq \left| \frac{1}{g(\eta_i)} - \frac{1}{g(\xi_i)} \right| + \varepsilon$$

$$= \frac{|g(\xi_i) - g(\eta_i)|}{|g(\eta_i)(g\xi_i)|} + \varepsilon \leq \frac{1}{m^2}(M_i - m_i) + \varepsilon = \frac{2}{m^2}(M_i - m_i).$$

Denote by S and s the upper and lower Darboux sums for the function $1/g$. Then

$$S - s \leq \frac{2}{m^2} \sum_{i=1}^{n}(M_i - m_i)(x_i - x_{i-1}).$$

The right-hand side is, up to a factor, the difference of Darboux upper and lower sums; since g is integrable, it can be made less than ε if $\max(x_i - x_{i-1})$ is sufficiently small. Hence, $1/g$ is integrable.

PROBLEMS

1. A countable set S has measure zero. [*Hint:* Let $S = \{x_1, x_2, \ldots\}$. For any $\varepsilon > 0$, S is contained in the intervals $I_n = (x_n - \varepsilon/2^n, x_n + \varepsilon/2^n)$.]

2. Let f be a bounded function in $[0,1]$, continuous at all the points x except for $x = 1/n$, $n = 1,2,\ldots$. Use Theorem 3 to show that f is integrable.

3. Evaluate $\int_0^1 f(x)\, dx$, where

$$f(x) = \begin{cases} 0 & \text{if } x = \dfrac{1}{n},\ n = 1,2,\ldots, \\ 1 & \text{if } x \neq \dfrac{1}{n}. \end{cases}$$

4. If f is integrable over $[a,b]$ and $f(x) \geq m > 0$, then $\sqrt{f}$ is integrable over $[a,b]$.

5. Extend the result of the preceding problem to the case where $m = 0$.

6. Compute the following integrals:

  (a) $\displaystyle\int_a^b x\,dx;$ (b) $\displaystyle\int_0^2 [x]x\,dx;$

  (c) $\displaystyle\int_0^1 x^2\,dx;$ (d) $\displaystyle\int_0^1 e^x\,dx.$

7. A function $f(x)$ is said to be *even* if $f(-x) = f(x)$. It is said to be *odd* if $f(-x) = -f(x)$. Show that if f is integrable and even on $[-1,1]$, then $\int_{-1}^1 f(x)dx = 2\int_0^1 f(x)dx$. Show that if f is integrable and odd on $[-1,1]$, then $\int_{-1}^1 f(x)\,dx = 0$.

8. Prove that a monotone function on an interval $[a,b]$ has at most a countable number of discontinuities. [*Hint:* Suppose $f(x)$ is monotone increasing. The set S_n of points x where $f(x+0) > f(x-0) + 1/n$ is finite. The set of points of discontinuities of f is contained in the union of the sets S_n.]

5. THE FUNDAMENTAL THEOREM OF CALCULUS

Let $f(x)$ be an integrable function over $[a,b]$. For any subinterval (α,β) of (a,b), we define

$$\int_\beta^\alpha f(x)\,dx = -\int_\alpha^\beta f(x)\,dx. \qquad (1)$$

We also define

$$\int_\alpha^\alpha f(x)\,dx = 0. \qquad (2)$$

One can then verify the formula

$$\int_\alpha^\beta f(x)\,dx + \int_\beta^\gamma f(x)\,dx = \int_\alpha^\gamma f(x)\,dx, \qquad (3)$$

for any three numbers α,β,γ in $[a,b]$.

Fix a point c in $[a,b]$, and consider the integral

$$g(x) = \int_c^x f(t)\,dt \qquad (4)$$

for each x in $[a,b]$. This is a function of x. It is called an *indefinite integral* of f. For different values of c, we get different indefinite integrals. Since, however, by (3),

$$\int_c^x f(t)\, dt - \int_{c'}^x f(t)\, dt = \int_c^{c'} f(t)\, dt,$$

any two indefinite integrals differ by a constant.

THEOREM 1. *An indefinite integral of an integrable function f is a continuous function.*

Proof. Let M be an upper bound on $|f|$. Then the indefinite integral $g(x)$ given by (4) satisfies

$$|g(x) - g(y)| = \left| \int_c^x f(t)\, dt - \int_c^y f(t)\, dt \right| = \left| \int_x^y f(t)\, dt \right|$$

$$\leq \left| \int_x^y |f(t)|\, dt \right| \leq M\,|x - y|.$$

Hence, for any $\varepsilon > 0$,

$$|g(x) - g(y)| < \varepsilon \qquad \text{if } |x - y| < \delta,$$

provided $\delta < \varepsilon/M$.

The *fundamental theorem of calculus* is the following:

THEOREM 2. *At each point $x_0 \in (a,b)$ where $f(x)$ is continuous, the derivative $g'(x_0)$ of the indefinite integral $g(x)$ exists, and it is equal to $f(x_0)$. Thus*

$$\frac{d}{dx} \int_c^x f(t)\, dt = f(x) \tag{5}$$

at each point x where f is continuous.

Proof. Using (3) we have

$$\frac{g(x_0 + h) - g(x_0)}{h} = \frac{1}{h} \left[\int_c^{x_0 + h} f(t)\, dt - \int_c^{x_0} f(t)\, dt \right] = \frac{1}{h} \int_{x_0}^{x_0 + h} f(t)\, dt.$$

Hence

$$\frac{g(x_0 + h) - g(x_0)}{h} - f(x_0) = \frac{1}{h} \int_{x_0}^{x_0 + h} [f(t) - f(x_0)]\, dt. \tag{6}$$

Since $f(x)$ is continuous at x_0, for any $\varepsilon > 0$, there is a $\delta > 0$ such that

$$|f(t) - f(x_0)| < \varepsilon \qquad \text{if } |t - x_0| < \delta.$$

Therefore, if $|h| < \delta$, then

$$\left| \frac{1}{h} \int_{x_0}^{x_0+h} [f(t) - f(x_0)] \, dt \right| \leq \frac{1}{|h|} \left| \int_{x_0}^{x_0+h} |f(t) - f(x_0)| \, dt \right|$$

$$< \frac{\varepsilon}{|h|} \left| \int_{x_0}^{x_0+h} dt \right| = \varepsilon.$$

From (6) it then follows that

$$\left| \frac{g(x_0 + h) - g(x_0)}{h} - f(x_0) \right| < \varepsilon \qquad \text{if } |h| < \delta.$$

This completes the proof.

If in the foregoing proof we restrict h to take only positive values (or only negative values), we obtain the following result:

COROLLARY 1.   *If $x_0 \in [a,b]$ and if $f(x)$ is right continuous (or left continuous) at x_0, then $g(x)$ has right derivative (or left derivative) at x_0, and $g'(x_0 + 0) = f(x_0 + 0)$ [or $g'(x_0 - 0) = f(x_0 - 0)$].*

DEFINITION.   A function $F(x)$ is said to be a *primitive* of a function $f(x)$, if $F'(x) = f(x)$.

Notice that if $F(x)$ and $G(x)$ are primitives of $f(x)$, then $F'(x) - G'(x) = 0$. From Theorem 3 of Section 3.5 it follows that $G(x) = F(x) + C$, where C is a constant.

From Theorem 2 and Corollary 1 we immediately deduce:

COROLLARY 2.   *If $f(x)$ is continuous in $[a,b]$, then any indefinite integral $g(x)$ of f is a primitive function. In fact, $g'(x) = f(x)$ for all $a < x < b$, and $g'(a + 0) = f(a)$, $g'(b - 0) = f(b)$.*

The fundamental theorem of calculus, or, rather, Corollary 2, gives a method for computing integrals. Suppose f is continuous in $[a,b]$ and suppose we know a primitive function $F(x)$ of f. Since the indefinite integral

$$G(x) = \int_a^x f(t) \, dt$$

is also a primitive function of f,

$$G(x) - F(x) = C, \; C \text{ constant.}$$

Setting $x = a$ and $x = b$, we get the equations

$$G(a) - F(a) = C$$

$$G(b) - F(b) = C.$$

Thus

$$G(b) - G(a) = F(b) - F(a).$$

Since $G(a) = 0$ and $G(b) = \int_a^b f(t)\, dt$, we conclude:

THEOREM 3.   *Let $f(x)$ be a continuous function in $[a,b]$, and let $F(x)$ be a primitive of f. Then*

$$\int_a^b f(x)\, dx = \int_a^b F'(x)\, dx = F(b) - F(a). \tag{7}$$

One often writes the right-hand side of (7) in the form

$$\left[F(x) \right]_a^b, \qquad \text{or } F(x) \Big|_a^b.$$

Theorem 3 provides a powerful method for computing integrals. Instead of trying to compute them from the definition of the integral, all one has to do is find a primitive function.

EXAMPLE 1.   The function $x^{n+1}/(n+1)$ is a primitive of x^n, n being a positive number. Hence

$$\int_a^b x^n\, dx = \frac{b^{n+1}}{n+1} - \frac{a^{n+1}}{n+1}.$$

EXAMPLE 2.   The function $\sin x$ is a primitive of $\cos x$. Hence

$$\int_0^{\pi/2} \cos x\, dx = \sin \frac{\pi}{2} - \sin 0 = 1.$$

EXAMPLE 3.   If $-1 < \alpha < \beta < 1$, then

$$\int_\alpha^\beta \frac{dx}{\sqrt{1-x^2}} = \left[\sin^{-1} x \right]_\alpha^\beta = \sin^{-1} \beta - \sin^{-1} \alpha.$$

PROBLEMS

1.   Compute the following integrals:

(a) $\displaystyle\int_1^e \frac{1}{x}\, dx$;

(b) $\displaystyle\int_1^2 \frac{dx}{x^n}$, n positive integer;

(c) $\int_{-1}^{\sqrt{3}} \dfrac{dx}{1 + x^2}$;

(d) $\int_{0}^{\sqrt{\pi}} x \sin(x^2)\, dx$;

(e) $\int_{0}^{1} x e^{x^2}\, dx$;

(f) $\int_{0}^{a} (x - 2)^5\, dx$.

2. Let $f(x)$ be continuous in $[a,b]$. Prove that the function $H(x) = \int_{x}^{b} f(t)\, dt$ satisfies $H'(x) = -f(x)$.

3. Let $f(x)$ be a continuous function in $[a,b]$. Let $\phi(x)$ be a function having a continuous derivative in $[a,b]$. Assume also that $a \le \phi(x) \le b$. Prove that the function $k(x) = \int_{a}^{\phi(x)} f(t)\, dt$ is differentiable in (a,b), and $k'(x) = \phi'(x) f(\phi(x))$.

6. METHODS OF INTEGRATION

Theorem 3 of Section 5 will be used to derive two important formulas for evaluating integrals. The first formula, called the *change of variables formula*, is given in the following theorem.

THEOREM 1.   *Let $f(x)$ be a continuous function in a closed interval $[a,b]$. Let $\phi(t)$ be a continuous function defined in a closed interval $[c,d]$ and assume that $\phi'(t)$ exists and is continuous in $[c,d]$. Finally, assume that the values of $\phi(t)$ lie in the interval $[a,b]$, and that $\phi(c) = a$, $\phi(d) = b$. Then*

$$\int_{c}^{d} f(\phi(t))\phi'(t)\, dt = \int_{a}^{b} f(x)\, dx. \tag{1}$$

Proof.   Let

$$g(x) = \int_{a}^{x} f(y)\, dy,$$

and consider the function

$$F(t) = g(\phi(t)).$$

It satisfies

$$F'(t) = g'(\phi(t))\phi'(t) = f(\phi(t))\phi'(t).$$

Hence

$$\int_{c}^{d} f(\phi(t))\phi'(t)\, dt = \int_{c}^{d} F'(t)\, dt = F(d) - F(c)$$

$$= g(\phi(d)) - g(\phi(c)) = g(b) - g(a) = \int_{a}^{b} f(y)\, dy.$$

The change of variables formula (1) often is written also in the abbreviated form

$$\int_c^d f(x)\frac{dx}{dt}\,dt = \int_a^b f(x)\,dx. \tag{2}$$

EXAMPLE 1. In order to compute

$$\int_e^{e^2} \frac{\log t}{t}\,dt,$$

substitute $x = \phi(t) = \log t$. Then $dx/dt = \phi'(t) = 1/t$ and

$$\int_e^{e^2} \frac{\log t}{t}\,dt = \int_1^2 x\,dx = \left[\frac{x^2}{2}\right]_1^2 = \frac{3}{2}.$$

EXAMPLE 2. In order to compute

$$\int_0^1 t\sqrt{1 - t^2}\,dt,$$

substitute $x = \phi(t) = 1 - t^2$. Then $dx/dt = \phi'(t) = -2t$. Hence

$$\int_0^1 t\sqrt{1 - t^2}\,dt = -\frac{1}{2}\int_0^1 \sqrt{x}\,\frac{dx}{dt}\,dt = -\frac{1}{2}\int_1^0 \sqrt{x}\,dx$$

$$= \frac{x^{3/2}}{3}\Big|_0^1 = \frac{1}{3}.$$

The next theorem gives another formula for computing integrals. It is called the *method of integration by parts.*

THEOREM 2. *Let $f(x)$ and $g(x)$ be functions having continuous derivatives on a closed interval $[a,b]$. Then*

$$\int_a^b f(x)g'(x)\,dx = f(x)g(x)\Big|_a^b - \int_a^b f'(x)g(x)\,dx. \tag{3}$$

Proof. From the relation

$$\frac{d}{dx}(f(x)g(x)) = f'(x)g(x) + f(x)g'(x)$$

and Theorem 3 of Section 5, it follows that

$$\int_a^b [f'(x)g(x) + f(x)g'(x)]\,dx = f(x)g(x)\Big|_a^b.$$

This yields (3).

EXAMPLE 3.   To compute the integral

$$\int_0^1 xe^x \, dx,$$

we use (3) with $f(x) = x$, $g'(x) = e^x$. Since $g(x) = e^x$, we get

$$\int_0^1 xe^x \, dx = xe^x \Big|_0^1 - \int_0^1 e^x \, dx = \left[xe^x - e^x \right]_0^1 = 1.$$

EXAMPLE 4.   Compute the integral

$$I = \int_0^1 \sqrt{1 + x^2} \, dx.$$

Here we take $g'(x) = 1$, $f(x) = \sqrt{1 + x^2}$. Since $g(x) = x$, $f'(x) = x/\sqrt{1 + x^2}$, we get

$$I = \int_0^1 \sqrt{1 + x^2} \, dx = x\sqrt{1 + x^2} \Big|_0^1 - \int_0^1 \frac{x^2}{\sqrt{1 + x^2}} \, dx.$$

Writing

$$-\int_0^1 \frac{x^2}{\sqrt{1 + x^2}} \, dx = -\int_0^1 \frac{1 + x^2}{\sqrt{1 + x^2}} \, dx + \int_0^1 \frac{dx}{\sqrt{1 + x^2}}$$

$$= -\int_0^1 \sqrt{1 + x^2} \, dx + \tan^{-1} x \Big|_0^1,$$

we find that

$$2I = \left[x\sqrt{1 + x^2} + \tan^{-1} x \right]_0^1.$$

Formulas (1) and (3) can be used also to find primitives. We just allow point b to be a variable.

A primitive or an indefinite integral of a function $f(x)$ is usually denoted by

$$\int f(x) \, dx.$$

PROBLEMS

1. Compute the following indefinite integrals:

    (a) $\int \tan^{-1}x \, dx$;

    (b) $\int x^2 e^x \, dx$;

    (c) $\int x^n \log x \, dx, n \neq -1$;

    (d) $\int x \tan^{-1}x \, dx$;

    (e) $\int x^2 \sin ax \, dx$;

    (f) $\int \sin^{-1}x \, dx$;

    (g) $\int \log(a^2 + x^2) \, dx$;

    (h) $\int \sin(\log x) \, dx$;

    (i) $\int e^{-x} \log(1 + e^x) \, dx$;

    (j) $\int \dfrac{dx}{\sqrt{x+1} + \sqrt{x}}$.

2. Compute the following indefinite integrals:

    (a) $\int \dfrac{dx}{x^2 + 2x + 5}$;

    (b) $\int \dfrac{x \, dx}{x^2 + 4x - 3}$;

    (c) $\int \dfrac{x + 2}{x^2 + 4x - 5} \, dx$;

    (d) $\int \dfrac{dx}{(x - 1)(x^2 + 1)}$.

3. Prove

$$\int \cos^n x \, dx = \frac{\cos^{n-1} x \sin x}{n} - \frac{n - 1}{n} \int \cos^{n-2} x \, dx + C;$$

$$\int e^{ax} \sin bx \, dx = e^{ax} \frac{a \sin bx - b \cos bx}{a^2 + b^2} + C.$$

4. Prove

$$\int_0^{\pi/2} \sin^n x \, dx = \begin{cases} \dfrac{(n - 1)(n - 3) \cdots 4 \cdot 2}{n(n - 2) \cdots 5 \cdot 3} & \text{if } n \text{ is odd,} \\[3mm] \dfrac{(n - 1)(n - 3) \cdots 3 \cdot 1}{n(n - 2) \cdots 4 \cdot 2} \dfrac{\pi}{2} & \text{if } n \text{ is even.} \end{cases}$$

5. When the integrand is a function of $\sin x$ and $\cos x$, one often substitutes $z = \tan(x/2)$. Show that this gives

$$\cos x = \frac{1 - z^2}{1 + z^2}, \qquad \sin x = \frac{2z}{1 + z^2}, \qquad \frac{dx}{dz} = \frac{2}{1 + z^2}.$$

Use this substitution to show that

$$\int \frac{dx}{\sin x} = \log \left| \tan \frac{x}{2} \right|.$$

Prove also that

$$\int \frac{dx}{\cos x} = \log \left| \tan\left(\frac{x}{2} + \frac{\pi}{4}\right) \right|.$$

6.   When the integrand is a function of $\sin^2 x$ and $\cos^2 x$, one often substitutes $z = \tan x$. Show that this gives

$$\sin^2 x = \frac{z^2}{1 + z^2}, \qquad \cos^2 x = \frac{1}{1 + z^2}, \qquad \frac{dx}{dz} = \frac{1}{1 + z^2}.$$

Use this substitution in order to prove that

$$\int_0^{\pi/2} \frac{dx}{a^2 - b^2 \sin^2 x} = \frac{\pi/2}{a\sqrt{a^2 - b^2}}, \qquad \text{provided } a > b.$$

7. MEAN VALUE THEOREMS

THEOREM 1. *Let $f(x)$ and $g(x)$ be integrable functions over $[a,b]$. Assume that $g(x) \geq 0$ if $a \leq x \leq b$, and denote by M and m the l.u.b. and g.l.b. of f, respectively. Then*

$$\int_a^b f(x)g(x)\, dx = \mu \int_a^b g(x)\, dx \qquad \text{where } m \leq \mu \leq M. \tag{1}$$

If, furthermore, f is continuous, then there is a point ξ, $a < \xi < b$, such that

$$\int_a^b f(x)g(x)\, dx = f(\xi) \int_a^b g(x)\, dx. \tag{2}$$

This theorem is called the *first mean value theorem* (for integrals). In the special case where $g(x) = 1$, it reduces to the mean value theorem for integrals.

Proof.   Since

$$mg(x) \leq f(x)g(x) \leq Mg(x),$$

$$m \int_a^b g(x)\, dx \leq \int_a^b f(x)g(x)\, dx \leq M \int_a^b g(x)\, dx. \tag{3}$$

Consequently, if $\int_a^b g(x)\, dx \neq 0$ and if we define

$$\mu = \frac{\int_a^b f(x)g(x)\, dx}{\int_a^b g(x)\, dx},$$

then (1) follows. If $\int_a^b g(x)\, dx = 0$, then from (3) we conclude that $\int_a^b f(x)g(x)\, dx = 0$, so that (1) holds for any μ. If $f(x)$ is continuous on $[a,b]$, then $f(\alpha) = m$, $f(\beta) = M$ for some α, β in $[a,b]$. By the intermediate value theorem, there is a point ξ between α and β such that $f(\xi) = \mu$. This, together with (1), gives (2).

The next theorem is called the *second mean value theorem* (for integrals).

THEOREM 2.   *Let $f(x)$ and $g(x)$ be integrable functions over $[a,b]$ and assume that $g(x) \geq 0$, and that $f(x)$ is monotone. Then there is a point ξ, $a < \xi < b$, such that*

$$\int_a^b f(x)g(x)\,dx = f(a)\int_a^\xi g(x)\,dx + f(b)\int_\xi^b g(x)\,dx. \tag{4}$$

Proof.   The function

$$\psi(t) = f(a)\int_a^t g(x)\,dx + f(b)\int_t^b g(x)\,dx$$

is continuous in $[a,b]$, and

$$\psi(a) = f(b)\int_a^b g(x)\,dx, \qquad \psi(b) = f(a)\int_a^b g(x)\,dx.$$

Since $f(x)$ is monotone, the number μ occurring in the proof of Theorem 1 lies between $f(a)$ and $f(b)$. Therefore, the number

$$\mu\int_a^b g(x)\,dx = \int_a^b f(x)g(x)\,dx \tag{5}$$

lies between $\psi(a)$ and $\psi(b)$. By the intermediate value theorem, there is a point ξ in (a,b) such that $\psi(\xi)$ is equal to either side of (5). This completes the proof of (4).

We shall give an application to Taylor's formula. We assume that $f(x)$ and all its derivatives up to order $n + 1$ exist and are continuous in $[a,b]$. By Theorem 3 of Section 5,

$$f(x) = f(a) + \int_a^x f'(t)\,dt. \tag{6}$$

By integration by parts,

$$\int_a^x f'(t)\,dt = -f'(t)(x - t)\Big|_a^x + \int_a^x f''(t)(x - t)\,dt$$

$$= f'(a)(x - a) + \int_a^x f''(t)(x - t)\,dt.$$

Substituting this into (6), we get

$$f(x) = f(a) + f'(a)(x - a) + \int_a^x f''(t)(x - t)\,dt. \tag{7}$$

Again, by integration by parts,

$$\int_a^x f''(t)(x-t)\,dt = -f''(t)\frac{(x-t)^2}{2}\Big|_a^x + \int_a^x f^{(3)}(t)\frac{(x-t)^2}{2!}\,dt.$$

Substituting this into (7), we get

$$f(x) = f(a) + f'(a)(x-a) + f''(a)\frac{(x-a)^2}{2!} + \int_a^x f^{(3)}(t)\frac{(x-t)^2}{2!}\,dt.$$

Continuing in this way step by step, we arrive at the relation

$$f(x) = f(a) + f'(a)(x-a) + \frac{f''(a)}{2!}(x-a)^2$$

$$+ \dots + \frac{f^{(n)}(a)}{n!}(x-a)^n + \frac{1}{n!}\int_a^x (x-t)^n f^{(n+1)}(t)\,dt. \tag{8}$$

We call this formula *Taylor's formula with integral remainder*. The function

$$R_{n+1} = \frac{1}{n!}\int_a^x (x-t)^n f^{(n+1)}(t)\,dt \tag{9}$$

is called the *integral remainder*.

Using the first mean value theorem (Theorem 1) with f being $f^{(n+1)}$ and with $g(t) = (x-t)^n$, we find that

$$R_{n+1} = \frac{f^{(n+1)}(\xi)}{n!}\int_a^x (x-t)^n\,dt = \frac{f^{(n+1)}(\xi)}{(n+1)!}(x-a)^{n+1}, \text{ where } a < \xi < x.$$

Thus the integral remainder gives rise to the Lagrange form for the remainder.

PROBLEMS

1. If f and g are integrable over $[a,b]$, $g(x) \geq 0$, and $f(x)$ positive and decreasing, then

$$\int_a^b f(x)g(x)\,dx = f(a)\int_a^\xi g(x)\,dx \qquad \text{for some } a < \xi < b.$$

2. Let f be a continuous function in $[a,b]$. If $\int_a^b f(x)\phi(x)\,dx = 0$ for all continuous functions ϕ, then $f(x) = 0$ for all $a \leq x \leq b$.

3. Generalize the result of the preceding problem, taking the functions ϕ to be functions having continuous first derivatives.

8. IMPROPER INTEGRALS

Let $f(x)$ be a function defined in a half-open, bounded interval $[a,b)$. Suppose that for any $\varepsilon > 0$, $f(x)$ is integrable over $[a,b - \varepsilon]$. If

$$\lim_{\varepsilon \to 0} \int_a^{b-\varepsilon} f(x)\, dx \tag{1}$$

exists, then we say that f has *improper integral* over (on, or in) $[a,b)$. The limit (1) is then called the improper integral of $f(x)$ over $[a,b)$, and we write

$$\int_a^b f(x)\, dx = \lim_{\varepsilon \to 0} \int_a^{b-\varepsilon} f(x)\, dx. \tag{2}$$

Similarly, we define the improper integral of f over $(a,b]$ as

$$\int_a^b f(x)\, dx = \lim_{\varepsilon' \to 0} \int_{a+\varepsilon'}^b f(x)\, dx. \tag{3}$$

Here $f(x)$ is integrable over every interval $[a + \varepsilon',b]$, $\varepsilon' > 0$, and the limit in (3) exists.

Next let $f(x)$ be a function defined in an open, bounded interval (a,b). If for some $a < c < b$, $f(x)$ has improper integrals over $(a,c]$ and over $[c,b)$, then we say that f has *improper integral* over (a,b) and write

$$\int_a^b f(x)\, dx = \int_a^c f(x)\, dx + \int_c^b f(x)\, dx. \tag{4}$$

It follows that

$$\int_a^b f(x)\, dx = \lim_{\substack{\varepsilon \to 0 \\ \varepsilon' \to 0}} \int_{a+\varepsilon'}^{b-\varepsilon} f(x)\, dx, \tag{5}$$

where by this relation we mean the following: For any $\eta > 0$, there is a positive number δ, such that if $0 < \varepsilon' < \delta$ and $0 < \varepsilon < \delta$, then

$$\left| \int_{a+\varepsilon'}^{b-\varepsilon} f(x)\, dx - \int_a^b f(x)\, dx \right| < \eta. \tag{6}$$

Conversely, one can show that if the limit in (5) exists, then, for any $a < c < b$, f has improper integrals over $(a,c]$ and over $[c,b)$, and the right-hand sides in (4) and (5) are equal.

EXAMPLE 1.   Consider the function $1/x^\alpha$, $\alpha > 0$ and $\alpha \neq 1$. Its integral over $(\varepsilon,1)$ is

$$\int_\varepsilon^1 \frac{dx}{x^\alpha} = \frac{x^{1-\alpha}}{1-\alpha} \bigg|_\varepsilon^1 = \frac{1}{1-\alpha} - \frac{\varepsilon^{1-\alpha}}{1-\alpha}.$$

If $0 < \alpha < 1$, then

$$\int_0^1 \frac{dx}{x^\alpha} = \lim_{\varepsilon \to 0} \int_\varepsilon^1 \frac{dx}{x^\alpha} = \frac{1}{1 - \alpha}.$$

If, however, $\alpha > 1$, then the improper integral over $(0,1]$ does not exist.

So far we have considered improper integrals over bounded intervals. Suppose now that $f(x)$ is defined in $[a,\infty)$ and is integrable over any interval $[a,N]$. If

$$\lim_{N \to \infty} \int_a^N f(x)\, dx$$

exists, then we say that $f(x)$ has *improper integral* over $[a,\infty)$. It is given by

$$\int_a^\infty f(x)\, dx = \lim_{N \to \infty} \int_a^N f(x)\, dx.$$

Similarly, we define the improper integral

$$\int_{-\infty}^b f(x)\, dx = \lim_{M \to \infty} \int_{-M}^b f(x)\, dx.$$

If $f(x)$ is defined in (a,∞) and if, for some $c > 0$,

$$\int_a^c f(x)\, dx \qquad \text{and} \qquad \int_c^\infty f(x)\, dx$$

exist as improper integrals over $(a,c]$ and $[c,\infty)$, respectively, then we define the improper integral

$$\int_a^\infty f(x)\, dx = \int_a^c f(x)\, dx + \int_c^\infty f(x)\, dx.$$

It is equal to

$$\lim_{\substack{\varepsilon \to 0 \\ N \to \infty}} \int_{a+\varepsilon}^N f(x)\, dx.$$

Similarly, one defines $\int_{-\infty}^\infty f(x)\, dx$ as $\int_{-\infty}^c f(x)\, dx + \int_c^\infty f(x)\, dx$ where c is any real number. We have

$$\int_{-\infty}^\infty f(x)\, dx = \lim_{\substack{N \to \infty \\ M \to \infty}} \int_{-M}^N f(x)\, dx. \tag{7}$$

EXAMPLE 2.   Consider the function $f(x) = 1/x^\alpha$ for $\alpha > 1$. Since

$$\int_1^N \frac{dx}{x^\alpha} = \frac{x^{1-\alpha}}{1-\alpha}\bigg|_1^N = \frac{N^{1-\alpha}}{1-\alpha} - \frac{1}{1-\alpha} \to -\frac{1}{1-\alpha} \qquad \text{if } N \to \infty,$$

we conclude that

$$\int_1^\infty \frac{dx}{x^\alpha}$$

exists and equals $-1/(1 - \alpha)$.

There is yet another concept of an integral that may exist even when the improper integral does not exist. It is called the *Cauchy principal value*. Suppose $f(x)$ is defined in $(-\infty,\infty)$ and is integrable over any interval $[-N,N]$. If

$$\lim_{N \to \infty} \int_{-N}^N f(x) \, dx \tag{8}$$

exists, then this limit is called the Cauchy principal value of $f(x)$ over $(-\infty,\infty)$. We denote it by

$$\int_{-\infty}^\infty f(x) \, dx.$$

If the improper integral $\int_{-\infty}^\infty f(x) \, dx$ exists, that is, if (7) is valid, then clearly also the limit in (8) exists and is equal to the limit in (7). But it may well happen that f has a Cauchy principal value but does not have improper integral.

If $f(x)$ is defined in an interval $[a,b]$, except for one point c, $a < c < b$, and if

$$\lim_{\varepsilon \to 0} \left[\int_a^{c-\varepsilon} f(x) \, dx + \int_{c+\varepsilon}^b f(x) \, dx \right]$$

exists, then we denote this limit by

$$\int_a^b f(x) \, dx$$

and call it the *Cauchy principal value*. Again we note that if the improper integrals

$$\int_a^c f(x) \, dx$$

and

$$\int_c^b f(x) \, dx$$

exist, then the improper integral $\int_a^b f(x) \, dx$, defined by

$$\int_a^b f(x) \, dx = \int_a^c f(x) \, dx + \int_c^b f(x) \, dx,$$

exists and is equal to the Cauchy principal value.

EXAMPLE 3. Since

$$\int_\varepsilon^1 \frac{dx}{x} = \log x \Big|_\varepsilon^1 = -\log \varepsilon \to \infty \qquad \text{as } \varepsilon \to 0,$$

the improper integral $\int_0^1 dx/x$ does not exist. Similarly, the improper integral $\int_{-2}^0 dx/x$ does not exist. However, since

$$\int_{-2}^{-\varepsilon} \frac{dx}{x} + \int_\varepsilon^1 \frac{dx}{x} = \log|x| \Big|_{-2}^{-\varepsilon} + \log x \Big|_\varepsilon^1$$

$$= (\log \varepsilon - \log 2) + (\log 1 - \log \varepsilon) = -\log 2,$$

the Cauchy principal value

$$\int_{-2}^1 \frac{dx}{x}$$

exists and is equal to $-\log 2$.

PROBLEMS

1. Find which of the following improper integrals exist, and evaluate those integrals that do exist:

(a) $\displaystyle\int_1^3 \frac{dx}{x^2 - 1}$;

(b) $\displaystyle\int_{-1}^1 \frac{dx}{\sqrt{1 - x^2}}$;

(c) $\displaystyle\int_0^2 \log x \, dx$;

(d) $\displaystyle\int_0^\infty x e^{-x^2} \, dx$;

(e) $\displaystyle\int_1^\infty \sin x \, dx$;

(f) $\displaystyle\int_1^\infty x^2 e^{-x} \, dx$;

(g) $\displaystyle\int_0^1 \frac{dx}{x \log x}$;

(h) $\displaystyle\int_0^\infty e^{-x} \sin x \, dx$;

(i) $\displaystyle\int_0^\infty \frac{dx}{x\sqrt{1 + x^2}}$;

(j) $\displaystyle\int_{-\infty}^\infty \frac{e^{x/3}}{1 + e^x} \, dx$.

2. If the improper integrals $\int_a^b f(x) \, dx$, $\int_a^b g(x) \, dx$ exist, then the improper integral $\int_a^b (f(x) + g(x)) \, dx$ also exists, and

$$\int_a^b (f(x) + g(x)) \, dx = \int_a^b f(x) \, dx + \int_a^b g(x) \, dx.$$

3. Let

$$f(x) = \begin{cases} \dfrac{1}{1+x} & \text{if } |x| > 2, \\ 0 & \text{if } |x| \leq 2. \end{cases}$$

Show that the integral $\int_{-\infty}^{\infty} f(x)\, dx$ exists as a Cauchy principal value, but not as an improper integral.

5

SEQUENCES AND SERIES OF FUNCTIONS

1. SERIES OF NUMBERS: CRITERIA FOR CONVERGENCE

Let $\{a_n\}$ be a sequence of numbers. We say that $\{a_n\}$ is a *Cauchy sequence* if for any ε there is a positive integer n_0 such that

$$|a_n - a_m| < \varepsilon \qquad \text{whenever } n \geq n_0, m \geq n_0. \tag{1}$$

We write this condition briefly in the form

$$\lim_{\substack{n \to \infty \\ m \to \infty}} |a_n - a_m| = 0. \tag{2}$$

If $\{a_n\}$ is a convergent sequence, then it is also a Cauchy sequence. Indeed, if $\lim a_n = L$, then for any $\varepsilon > 0$ there is a positive integer n_0 such that

$$|a_n - L| < \frac{\varepsilon}{2}, \qquad |a_m - L| < \frac{\varepsilon}{2} \qquad \text{if } n \geq n_0, m \geq n_0.$$

Hence

$$|a_n - a_m| \leq |a_n - L| + |L - a_m| < \frac{\varepsilon}{2} + \frac{\varepsilon}{2} = \varepsilon.$$

The converse is also true. This is stated in the following theorem.

THEOREM 1. *If $\{a_n\}$ is a Cauchy sequence, then $\{a_n\}$ is convergent.*

Proof. Using condition (1) with $\varepsilon = 1$, $m = n_0$, we get

$$|a_n| \le |a_n - a_{n_0}| + |a_{n_0}| < 1 + |a_{n_0}| \qquad \text{if } n \ge n_0.$$

It follows that the sequence $\{a_n\}$ is bounded. By the Bolzano–Weierstrass theorem, there is a subsequence $\{a_{m_i}\}$ that converges to some number L. Thus, for any $\varepsilon > 0$ there is a positive integer n_1 such that

$$|a_{m_i} - L| < \varepsilon \qquad \text{if } m_i \ge n_1. \tag{3}$$

From (1) we also have

$$|a_n - a_{m_i}| < \varepsilon \qquad \text{if } m_i \ge n_0, n \ge n_0.$$

Hence, if $n^* = \max(n_0, n_1)$ and $n \ge n^*$, then

$$|a_n - L| \le |a_n - a_{m_i}| + |a_{m_i} - L| < \varepsilon + \varepsilon = 2\varepsilon.$$

This shows that $\lim a_n$ exists and equals L.

Given a sequence $\{a_n\}$ we form another sequence $\{S_n\}$ where

$$S_n = a_1 + a_2 + \cdots + a_n.$$

The sequence $\{S_n\}$ is called an *infinite series*. The S_n are called *partial sums* of the sequence $\{a_m\}$. If $S = \lim S_n$ exists, then we call it the *sum* of the infinite series. We then say that the infinite series *is convergent* (or *converges*) to S, and write

$$S = a_1 + a_2 + \cdots + a_n + \cdots = \sum_{n=1}^{\infty} a_n.$$

If $\lim S_n$ does not exist, then we say that the series *is divergent* (or *diverges*). If, in particular, $\lim S_n = \infty$, then we say that the series diverges to ∞. We then write

$$\sum_{n=1}^{\infty} a_n = \infty.$$

One often writes the infinite series $\{S_n\}$ in the (more explicit) form $\sum_{n=1}^{\infty} a_n$ or $\sum a_n$.

THEOREM 2. *If an infinite series $\sum a_n$ is convergent, then $\lim a_n = 0$.*

Indeed, if $\lim S_n = S$, then $a_n = S_n - S_{n-1} \to S - S = 0$ as $n \to \infty$.

Applying Theorem 1 and its converse to the sequence $\{S_n\}$, we get the following *Cauchy's criterion for convergence*:

THEOREM 3.   *An infinite series* $\sum a_n$ *is convergent if, and only if, for any* $\varepsilon > 0$ *there is a positive integer* n_0 *such that*

$$|S_n - S_m| = |a_{m+1} + \cdots + a_n| < \varepsilon \qquad \text{for any } n > m \geq n_0. \qquad (4)$$

EXAMPLE.   The series $\sum aq^n$ $(a \neq 0, q \neq 0)$ is called a *geometric series*. If $q \geq 1$, the series diverges since the nth term aq^n does not converge to 0 as $n \to \infty$. If $0 < q < 1$, the series is convergent, by Theorem 3, since

$$|S_n - S_m| = |aq^{m+1} + \cdots + aq^n| = |a| \, q^{m+1}(1 + q + \cdots + q^{n-m-1})$$

$$= |a| \, q^{m+1} \frac{1 - q^{n-m}}{1 - q} < \frac{|a|}{1 - q} \, q^{m+1} \to 0 \qquad \text{if } n \geq m, \, m \to \infty.$$

The partial sum S_n is equal to

$$aq + aq^2 + \cdots + aq^n = aq \, \frac{1 - q^n}{1 - q} = \frac{aq}{1 - q} - \frac{aq^{n+1}}{1 - q}.$$

Since $aq^{n+1}/(1 - q) \to 0$ as $n \to \infty$, we find that $S = \lim S_n = aq/(1 - q)$.

Suppose that the terms a_n of an infinite series $\sum a_n$ are nonnegative for all n sufficiently large, say, $n \geq \bar{n}$. Then the sequence of partial sums S_n $(n \geq \bar{n})$ is monotone increasing. By Theorem 1 of Section 1.7, if $\{S_n\}$ is bounded above, then $\lim S_n$ exists. If $\{S_n\}$ is not bounded above, then (by Theorem 4 of Section 1.7) $\lim S_n = \infty$. We therefore can state:

THEOREM 4.   *If* $a_n \geq 0$ *for all n sufficiently large, then the series* $\sum a_n$ *is convergent if, and only if, the sequence of partial sums* S_n *is bounded above. If* $\sum a_n$ *is not convergent, then it diverges to* ∞.

The next theorem gives a useful criterion for convergence. It is called the *comparison test*.

THEOREM 5.   *Let* $\sum a_n, \sum b_n$ *be two series whose terms satisfy the inequalities:* $0 \leq a_n \leq b_n$ *for all n sufficiently large.*

(i)   *If* $\sum b_n$ *is convergent, then* $\sum a_n$ *is convergent.*
(ii) *If* $\sum a_n$ *is divergent, then* $\sum b_n$ *is divergent.*

Proof.   Let n_0 be such that $0 \leq a_n \leq b_n$ for all $n \geq n_0$. If $\sum b_n$ is convergent, then, for any $m > n_0$,

$$\sum_{n=n_0}^{m} a_n \leq \sum_{n=n_0}^{m} b_n \leq K$$

where K is a constant independent of m. The partial sums S_n of $\sum a_n$ then form a sequence bounded above. By Theorem 4, $\lim S_n$ exists.

Suppose now that $\sum a_n$ is divergent. Then $\lim S_n = \infty$. Thus, for any $N > 0$, there is a positive integer $\bar{n} > n_0$ such that $S_m > N$ if $m > \bar{n}$. But then

$$\sum_{n=n_0}^{m} a_n = S_m - \sum_{n=1}^{n_0-1} a_n > N - \sum_{n=1}^{n_0-1} a_n.$$

It follows that if $m > \bar{n}$, then

$$\sum_{n=1}^{m} b_n = \sum_{n=n_0}^{m} b_n + \sum_{n=1}^{n_0-1} b_n \geq N - \sum_{n=1}^{n_0-1} a_n + \sum_{n=1}^{n_0-1} b_n.$$

Since N is arbitrary,

$$\lim_{m \to \infty} \sum_{n=1}^{m} b_n = \infty.$$

The next criterion is called *Cauchy's root test*:

THEOREM 6.   *Suppose $a_n \geq 0$ for all n. If*

$$\varlimsup_{n \to \infty} \sqrt[n]{a_n} < 1, \tag{5}$$

then the series $\sum a_n$ converges. If

$$\varlimsup_{n \to \infty} \sqrt[n]{a_n} > 1, \tag{6}$$

then the series $\sum a_n$ diverges.

Proof.   Suppose (5) holds. Let q be a number such that

$$\varlimsup_{n \to \infty} \sqrt[n]{a_n} < q < 1.$$

Then, there is a positive integer n_0 such that

$$\sqrt[n]{a_n} < q \qquad \text{for all } n \geq n_0.$$

Hence $a_n < q^n$ if $n \geq n_0$. Since $\sum q^n$ is convergent, the comparison test implies that $\sum a_n$ is convergent.

If (6) holds, then choose p such that

$$\varlimsup_{n \to \infty} \sqrt[n]{a_n} > p > 1.$$

It follows that there is a subsequence $\{a_{n_i}\}$ of $\{a_n\}$ such that

$$\sqrt[n_i]{a_{n_i}} > p \qquad \text{for all } n_i.$$

Hence

$$a_{n_i} > p^{n_i} \to \infty \qquad \text{as } n_i \to \infty.$$

Theorem 2 then shows that $\sum a_n$ is not convergent.

The next test for convergence is called the *ratio test*.

THEOREM 7.   *Suppose $a_n > 0$ for all n. If*

$$\varlimsup_{n \to \infty} \frac{a_{n+1}}{a_n} < 1, \tag{7}$$

then the series $\sum a_n$ is convergent. If

$$\varlimsup_{n \to \infty} \frac{a_{n+1}}{a_n} > 1, \tag{8}$$

then the series is divergent.

Proof.   If (7) holds, then there is a number q such that

$$\varlimsup_{n \to \infty} \frac{a_{n+1}}{a_n} < q < 1.$$

Hence there is a positive integer n_0 such that

$$\frac{a_{n+1}}{a_n} < q \qquad \text{if } n \geq n_0.$$

It follows that for any nonnegative integer m,

$$a_{n_0+m} < qa_{n_0+m-1} < q^2 a_{n_0+m-2} < \cdots < q^m a_{n_0}.$$

Since $\sum a_{n_0} q^m$ is convergent, the comparison test implies that $\sum a_n$ is convergent.

If (8) holds, then

$$\lim_{n \to \infty} \frac{a_{n+1}}{a_n} > p > 1$$

for some number p. Hence, there is a positive integer n_1 such that

$$\frac{a_{n+1}}{a_n} > p \qquad \text{if } n \geq n_1.$$

It follows that for any nonnegative integer, m,

$$a_{n_1+m} > pa_{n_1+m-1} > p^2 a_{n_1+m-2} > \cdots > p^m a_{n_1}.$$

Since $\sum p^m a_{n_1}$ is divergent, $\sum a_n$ is also divergent.

EXAMPLE. The series $\sum\limits_{n=1}^{\infty} n^{\alpha}t^{n-1}$ is convergent if $0 < t < 1, \alpha > 0$.
This follows by both the root test and the ratio test.
The next criterion for convergence is called the *integral test*.

THEOREM 8. *Let $\{a_n\}$ be a monotone decreasing sequence of positive numbers. Let $f(x)$ be a monotone decreasing function defined for $N \leq x < \infty$, N a positive integer, such that $f(n) = a_n$ for all $n \geq N$. Then the series $\sum a_n$ is convergent if, and only if, the improper integral $\int_N^{\infty} f(x)\,dx$ exists.*

Proof. If $n \geq N$, then

$$a_n = f(n) \geq f(x) \geq f(n+1) = a_{n+1} \qquad \text{for all } n < x < n+1.$$

Integrating over the interval $[n, n+1]$, we get

$$a_n \geq \int_n^{n+1} f(x)\,dx \geq a_{n+1}.$$

Hence, for any $m > N$,

$$\sum_{n=N}^{m} a_n \geq \int_N^{m+1} f(x)\,dx \geq \sum_{n=N+1}^{m+1} a_n. \qquad (9)$$

If $\int_N^{\infty} f(x)\,dx$ exists, then the sequence of numbers

$$\int_N^{m+1} f(x)\,dx \qquad (m = 1,2,\dots)$$

is convergent; therefore, it is also bounded. From the second inequality in (9), it follows that the partial sums of $\sum a_n$ then form a bounded sequence. Hence, by Theorem 4, the series $\sum a_n$ converges.

Conversely, suppose that the series $\sum a_n$ is convergent. Then the partial sums form a bounded sequence. The first inequality in (9) then shows that for all $m = 1,2,\cdots,$

$$\int_N^{m+1} f(x)\,dx \leq K, \qquad K \text{ constant.}$$

We deduce that the monotone sequence of numbers

$$\int_N^{m+1} f(x)\,dx$$

is convergent. Denote the limit by L. Then, for any $\varepsilon > 0$ there is a positive integer m_0 such that

$$L - \varepsilon < \int_N^{m} f(x)\,dx < L \qquad \text{if } m \geq m_0.$$

It follows that

$$L - \varepsilon < \int_N^y f(y)\, dy < L$$

for any real y, $y > m_0$. This shows that

$$\int_N^\infty f(x)\, dx = \lim_{y \to \infty} \int_N^y f(x)\, dx$$

exists.

EXAMPLE.   Let $\alpha > 0$, $\alpha \neq 1$, and consider the series $\sum 1/n^\alpha$. We can apply the integral test with $f(x) = 1/x^\alpha$. Since

$$\int_1^M \frac{dx}{x^\alpha} = \frac{x^{1-\alpha}}{1-\alpha}\bigg|_1^M = \frac{M^{1-\alpha}}{1-\alpha} - \frac{1}{1-\alpha},$$

the improper integral $\int_1^\infty x^{-\alpha}\, dx$ exists if, and only if, $\alpha > 1$. Thus $\sum 1/n^\alpha$ is convergent if $\alpha > 1$ and is divergent if $\alpha < 1$.

Consider now the series $\sum 1/n$, called the *harmonic series*. We again use the integral test with $f(x) = 1/x$. Since

$$\int_1^M \frac{dx}{x} = \log x \bigg|_1^M = \log M \to \infty \qquad \text{if } M \to \infty,$$

the harmonic series is divergent.

Remark.   Given a series $\sum_{n=1}^{\infty} a_n$, any series $\sum_{n=k}^{\infty} a_n$ is called a *tail* of the series $\sum_{n=1}^{\infty} a_n$. When we check a series for convergence, or divergence, it is sufficient to check a tail of the series.

PROBLEMS

1. Prove that if $\sum a_n$ and $\sum b_n$ are convergent, then $\sum(a_n + b_n)$ is convergent and $\sum(a_n + b_n) = \sum a_n + \sum b_n$.

2. Prove that if $\sum a_n$ is convergent, then, for any number λ, $\sum(\lambda a_n)$ is convergent to $\lambda \sum a_n$.

3. Prove the following *theorem of Abel*: If $\{a_n\}$ is a monotone decreasing sequence of positive numbers, and if $\sum a_n$ converges, then $na_n \to 0$ if $n \to \infty$. [*Hint:* $na_{2n} \leq a_{n+1} + a_{n+2} + \cdots + a_{2n} \to 0$ if $n \to \infty$. Similarly, $na_{2n+1} \to 0$ if $n \to \infty$.]

4. If $a_n > 0$ for all n and if instead of (8) we assume that $a_{n+1}/a_n \geq 1$ for all n sufficiently large, then $\sum a_n$ is divergent.

5. Determine, for each of the following series, whether it converges or diverges:

(a) $\sum \dfrac{n}{n^2 + 2}$;

(b) $\sum \sin \dfrac{1}{n + 1}$;

(c) $\sum \dfrac{1}{n\sqrt[n]{n}}$;

(d) $\sum \dfrac{1}{(\log n)^\alpha}$, $\alpha > 0$;

(e) $\sum \dfrac{n + 1}{n^2 + 2n - 1}$;

(f) $\sum \dfrac{1}{(\log n)^n}$;

(g) $\sum \dfrac{n!}{n^n}$;

(h) $\sum \left(\dfrac{n}{n + 1}\right)^{n^2}$.

6. Use the integral test to determine whether each of the following series is convergent or divergent:

(a) $\sum \dfrac{1}{n(\log n)^\alpha}$, $\alpha > 0$;

(b) $\sum \dfrac{1}{n \log n (\log \log n)^\alpha}$, $\alpha > 0$.

7. Show that if $\sum a_n$ is a convergent series of positive numbers and if $\{b_n\}$ is a bounded sequence, then the series $\sum a_n b_n$ is convergent.

8. Determine, for each of the following series, whether it converges or diverges:

(a) $\sum \dfrac{1}{n^{1 + 1/n}}$;

(b) $\sum \dfrac{n + \sqrt{n}}{n^2 - n + 1}$;

(c) $\sum (\sqrt{n + 1} - \sqrt{n})$;

(d) $\sum \dfrac{\sqrt{n + 1} - \sqrt{n}}{n}$;

(e) $\sum \left(1 - \dfrac{\log n}{n}\right)^n$;

(f) $\sum (\sqrt[n]{n} - 1)$.

2. SERIES OF NUMBERS: CONDITIONAL AND ABSOLUTE CONVERGENCE

A series $\sum a_n$ is said to be *absolutely convergent* if the series $\sum |a_n|$ is convergent. If $\sum a_n$ is convergent but not absolutely convergent, then it is said to be *conditionally convergent*.

THEOREM 1. *If a series is absolutely convergent, then it is convergent.*

Proof. We have to show that if $\sum |a_n|$ is convergent, then $\sum a_n$ is convergent. Denote the partial sums of $\sum |a_n|$ and $\sum a_n$ by $\tilde{S}_n$ and S_n, respectively. Then, if $n > m$,

$$|S_n - S_m| = |a_{m+1} + \cdots + a_n| \leq |a_{m+1}| + \cdots + |a_m| = |\tilde{S}_n - \tilde{S}_m|.$$

By Theorem 3 of Section 1, for any $\varepsilon > 0$ there is a positive integer N such that

$$|\tilde{S}_n - \tilde{S}_m| < \varepsilon \qquad \text{if } m \geq N.$$

It follows that

$$|S_n - S_m| < \varepsilon \qquad \text{if } n \geq N.$$

Hence, by Theorem 3 of Section 1, the series $\sum a_n$ converges.

In all the convergence criteria of the previous section (except for Theorem 2), the terms a_n are nonnegative for all n sufficiently large. Thus these criteria may be used to establish absolute convergence. We now shall give a criterion for convergence of a series that has infinitely many positive terms and infinitely many negative terms. It is called the *alternating series test.*

THEOREM 2. *Let $\{b_n\}$ be a monotone decreasing sequence of positive numbers, and let $\lim b_n = 0$. Then the series $\sum (-1)^{n-1} b_n$ is convergent.*

Proof. Consider the partial sums

$$S_{2n} = b_1 - b_2 + b_3 - b_4 + \cdots + b_{2n-1} - b_{2n}.$$

Since

$$S_{2(n+1)} - S_{2n} = b_{2n+2} - b_{2n+1} \geq 0,$$

the sequence $\{S_{2n}\}$ is monotone increasing. This sequence is also bounded above since

$$S_{2n} = b_1 - (b_2 - b_3) - (b_4 - b_5) - \cdots - (b_{2n-2} - b_{2n-1}) - b_{2n} \leq b_1.$$

It follows that

$$\lim_{n \to \infty} S_{2n} \tag{1}$$

exists.

Consider next the partial sums

$$S_{2n+1} = b_1 - b_2 + b_3 - b_4 + \cdots + b_{2n-1} - b_{2n} + b_{2n+1}.$$

Since

$$S_{2(n+1)+1} - S_{2n+1} = -b_{2n+2} + b_{2n+3} \leq 0,$$

the sequence $\{S_{2n+1}\}$ is monotone decreasing. It is also bounded below. Indeed,

$$S_{2n+1} = (b_1 - b_2) + (b_3 - b_4) + \cdots + (b_{2n-1} - b_{2n}) + b_{2n+1} \geq b_{2n+1} \geq 0.$$

It follows that

$$\lim_{n \to \infty} S_{2n+1} \tag{2}$$

exists.

From (1) and (2) and the relations

$$S_{2n+1} = S_{2n} + b_{2n+1}, \qquad \lim_{n \to \infty} b_{2n+1} = 0,$$

we conclude that the limits in (1) and (2) are the same. Denote by S the common limit. Then, for any $\varepsilon > 0$ there exist positive integers n_0 and n_1 such that

$$|S_{2n} - S| < \varepsilon \qquad \text{if } n \geq n_0,$$
$$|S_{2n+1} - S| < \varepsilon \qquad \text{if } n \geq n_1.$$

Consequently,

$$|S_m - S| < \varepsilon \qquad \text{if } m \geq \max(n_0, n_1).$$

This proves that $\lim S_m$ exists.

EXAMPLE.   The series $\sum (-1)^{n-1}/n$ is convergent. Since it is not absolutely convergent, it is conditionally convergent.

Let $\sum a_n$ be any series. We shall denote by $\{p_m\}$ the subsequence of $\{a_n\}$ consisting of all the positive numbers of the sequence. We shall denote by $\{-q_m\}$ the subsequence of $\{a_n\}$ consisting of all the negative numbers of the sequence. Let

$$P_n = \sum_{m=1}^{n} p_m, \qquad Q_n = \sum_{m=1}^{n} q_m.$$

Then, for any n,

$$S_n = \sum_{k=1}^{n} a_k = P_{n'} - Q_{n''}, \tag{3}$$

$$\sum_{k=1}^{n} |a_k| = P_{n'} + Q_{n''} \tag{4}$$

for suitable n', n'', and $n' + n'' \leq n$. Note that if all the numbers a_m are different from 0, then $n' + n'' = n$.

If there are infinitely many positive numbers a_m, then $n' \to \infty$ as $n \to \infty$. Similarly, if there are infinitely many negative numbers a_m, then $n'' \to \infty$ as $n \to \infty$. In what follows, whenever we introduce the series $\sum p_m, \sum q_m$, we shall suppose that $n' \to \infty$ and $n'' \to \infty$ as $n \to \infty$. If one of the series, say $\sum p_m$, consists of only a finite number of terms, then the relation "$n' \to \infty$ as $n \to \infty$" should be replaced by "$n' = \bar{n}$ for all n sufficiently large," where $\bar{n}$ is some fixed positive integer.

If $\sum a_n$ is absolutely convergent then, for all n,

$$\sum_{m=1}^{n'} p_m \leq \sum_{m=1}^{n} |a_m| \leq K, \qquad K \text{ constant.}$$

It follows that $\sum p_m$ is convergent. Similarly, $\sum q_m$ is convergent. These assertions are not valid if $\sum a_n$ is only conditionally convergent:

THEOREM 3.   *If $\sum a_n$ is conditionally convergent, then the series* $\sum p_n, \sum q_n$ *are both divergent.*

Proof.   Suppose $\sum p_m$ is convergent. From (3) we deduce, upon taking $n \to \infty$, that $\lim_{n'' \to \infty} Q_{n''}$ exists. Thus both series

$$\sum p_m, \qquad \sum q_m$$

are convergent. From (4) it then follows that $\sum |a_m|$ is convergent, a contradiction. Similarly, one proves that the series $\sum q_m$ is not convergent.

DEFINITION.   Let $\sum a_n$ be a series, and let $n = n(k)$ be a one-to-one correspondence from the set of all the positive integers k onto the set of all the positive integers n. Thus any positive integer n has the form $n(k)$, where k is uniquely determined by n. Then the series $\sum b_k$, where $b_k = a_{n(k)}$, is called a *rearrangement* of the series $\sum a_n$.

THEOREM 4.   *Let $\sum a_n$ be an absolutely convergent series, and let $\sum b_n$ be any rearrangement of $\sum a_n$. Then $\sum b_n$ is convergent to $\sum_{n=1}^{\infty} a_n$.*

Proof.   Suppose first that $a_n \geq 0$ for all n. Since any term b_k in $\sum b_j$ is a term $a_{n(k)}$, we have, for any positive integer m,

$$\sum_{k=1}^{m} b_k \leq \sum_{n=1}^{h} a_n$$

for any h larger than each of the numbers $n(1),n(2),\cdots,n(m)$. Setting

$$S = \sum_{n=1}^{\infty} a_n,$$

we get

$$\sum_{k=1}^{m} b_k \leq S.$$

It follows that the series $\sum_{k=1}^{\infty} b_k$ is convergent and its sum is $\leq S$. Since $\sum a_n$

can be considered to be a rearrangement of $\sum b_n$, we also get

$$\sum_{k=1}^{m} a_k \leq \sum_{n=1}^{\infty} b_n \qquad \text{for any } m.$$

Hence

$$S \leq \sum_{n=1}^{\infty} b_n.$$

This completes the proof that

$$\sum_{n=1}^{\infty} a_n = \sum_{n=1}^{\infty} b_n.$$

Consider now the general case where the a_n are not necessarily non-negative numbers. Introduce for $\sum a_n$ the series $\sum p_n, \sum q_n$ as before. We also introduce the analogous series $\sum \tilde{p}_n, \sum \tilde{q}_n$ for $\sum b_n$. The series

$$\sum p_n, \qquad \sum q_n$$

are convergent. Furthermore, $\sum \tilde{p}_n$ is a rearrangement of $\sum p_n$, and $\sum \tilde{q}_n$ is a rearrangement of $\sum q_n$. By the special case proved above it follows that $\sum \tilde{p}_n$ and $\sum \tilde{q}_n$ are convergent and

$$\sum \tilde{p}_n = \sum p_n, \qquad \sum \tilde{q}_n = \sum q_n. \tag{5}$$

For any positive integer n,

$$\sum_{m=1}^{n} b_m = \sum_{m=1}^{\tilde{n}'} \tilde{p}_m - \sum_{m=1}^{\tilde{n}''} \tilde{q}_m \tag{6}$$

where $\tilde{n}' \to \infty$, $\tilde{n}'' \to \infty$ as $n \to \infty$. From (5) and (6) it follows that $\sum_{n=1}^{\infty} b_m$ is convergent to

$$\sum_{n=1}^{\infty} p_n - \sum_{n=1}^{\infty} q_n = \sum_{n=1}^{\infty} a_n.$$

This completes the proof.

The converse of Theorem 4 is also true. In fact, we can state a somewhat stronger result:

THEOREM 5.   *If any rearrangement of $\sum a_n$ is convergent, then $\sum a_n$ is absolutely convergent.*

Proof. If $\sum a_n$ is not absolutely convergent, then, by Theorem 3, neither of the series

$$\sum p_m, \quad \sum q_m$$

is convergent. (They are both, necessarily, infinite series.) Since $\sum p_m$ is divergent to ∞, we can find positive integers $m_1, m_2, \cdots$, such that

$$p_1 + p_2 + \cdots + p_{m_1} > 1 + q_1,$$

$$p_{m_1+1} + p_{m_1+2} + \cdots + p_{m_2} > 2 + q_2,$$

$$\cdots$$

$$p_{m_k+1} + p_{m_k+2} + \cdots + p_{m_{k+1}} > k + 1 + q_{k+1},$$

$$\cdots.$$

Suppose all the a_n are different from 0. We then introduce the following rearrangement:

$$p_1, p_2, \ldots, p_{m_1}, -q_1, p_{m_1+1}, \ldots, p_{m_2}, -q_2, \ldots, p_{m_k+1}, \ldots, p_{m_{k+1}}, -q_{k+1}, \ldots . \quad (7)$$

The partial sums $\hat{S}_n$ satisfy

$$\hat{S}_n > j \quad \text{if } n > m_j + j.$$

Thus $\lim \hat{S}_n = \infty$. This contradicts the assumption of the theorem. If some of the a_n are zero, we obtain a rearrangement by inserting in (7) one zero between p_{m_1} and $-q_1$, another zero between p_{m_2} and $-q_2$, and so on. The assertion $\lim \hat{S}_n = \infty$ remains valid.

PROBLEMS

1. Check the following series for absolute and for conditional convergence:

(a) $\sum \dfrac{(-1)^n}{2n}$;

(b) $\sum \dfrac{(-1)^n}{\log n}$;

(c) $\sum \dfrac{(-1)^n}{\sqrt{n + 4/n}}$;

(d) $\sum \left(\dfrac{1}{\sqrt{n}} - \dfrac{1}{n} \right)$;

THEOREM 1.   *The uniform limit of a sequence of continuous functions is a continuous function.*

Proof.   Let $f(x)$ be the uniform limit of a sequence of continuous functions $f_n(x)$ on an interval I. Then for any $\varepsilon > 0$ there is an n_0 such that

$$|f_n(x) - f(x)| < \frac{\varepsilon}{3} \qquad \text{if } n \geq n_0, x \in I. \tag{1}$$

Since $f_{n_0}(x)$ is continuous in I, for any $x_0 \in I$, there is a $\delta > 0$ such that

$$|f_{n_0}(x) - f_{n_0}(x_0)| < \frac{\varepsilon}{3} \qquad \text{if } |x - x_0| < \delta, x \in I.$$

Using this and (1) with $n = n_0$, we get

$$|f(x) - f(x_0)| \leq |f(x) - f_{n_0}(x)| + |f_{n_0}(x) - f_{n_0}(x_0)| + |f_{n_0}(x_0) - f(x_0)|$$

$$< \frac{\varepsilon}{3} + \frac{\varepsilon}{3} + \frac{\varepsilon}{3} = \varepsilon \qquad \text{if } |x - x_0| < \delta, x \in I.$$

This proves the continuity of $f(x)$ at x_0.

COROLLARY.   *If the pointwise limit $f(x)$ of a sequence of continuous functions $f_n(x)$ on an interval I is not continuous on I, then $\{f_n\}$ does not converge to f uniformly on I.*

EXAMPLE.   Let $f_n(x) = 1/(1 + x^{2n})$ on $0 \leq x \leq 1$. Then $\lim f_n(x) = f(x)$, where

$$f(x) = \begin{cases} 1 & \text{if } 0 \leq x < 1, \\ \dfrac{1}{2} & \text{if } x = 1. \end{cases}$$

Since $f(x)$ is discontinuous at $x = 1$, the convergence is not uniform on $[0,1]$.

If a sequence of functions converges uniformly on $[0,1)$, and if it is also convergent at $x = 1$, then it converges uniformly on $[0,1]$. It follows that the sequence $\{f_n(x)\}$ in the above example does not converge uniformly already in the interval $[0,1)$.

We shall give a criterion for uniform convergence. It is called the *Cauchy criterion for uniform convergence.*

THEOREM 2. *A sequence $\{f_n(x)\}$ defined on an interval I is uniformly convergent to some function $f(x)$ on I if, and only if, for any $\varepsilon > 0$ there is a positive integer n_0 such that, for all $x \in I$,*

$$|f_n(x) - f_m(x)| < \varepsilon \qquad if\ n \geq n_0, m \geq n_0. \tag{2}$$

Proof. Suppose first that $\{f_n(x)\}$ is uniformly convergent on I to some function $f(x)$. Then, for any $\varepsilon > 0$ there is an n_0 such that, for all $x \in I$,

$$|f_n(x) - f(x)| < \frac{\varepsilon}{2} \qquad \text{if } n \geq n_0,$$

$$|f_m(x) - f(x)| < \frac{\varepsilon}{2} \qquad \text{if } m \geq n_0.$$

It follows that, for all $x \in I$,

$$|f_n(x) - f_m(x)| \leq |f_n(x) - f(x)| + |f(x) - f_m(x)| < \frac{\varepsilon}{2} + \frac{\varepsilon}{2} = \varepsilon$$

if $n \geq n_0, m \geq n_0$.

Suppose conversely that for any $\varepsilon > 0$ there is an n_0 such that (2) holds for all $x \in I$. Then, for a fixed x, we can apply Theorem 1 of Section 1 to the sequence $\{f_n(x)\}$. It follows that $\lim f_n(x)$ exists. Set $f(x) = \lim f_n(x)$.

Consider the sequence $\{a_m\}$ where

$$a_m = |f_n(x) - f_m(x)|, \qquad x \text{ fixed in } I, n \text{ fixed}, n \geq n_0.$$

Since $f_m(x) \to f(x)$ as $m \to \infty$, $a_m \to |f_n(x) - f(x)|$ as $m \to \infty$. Since, by (2), $a_m < \varepsilon$ if $m \geq n_0$, it follows that $\lim a_m \leq \varepsilon$, that is,

$$|f_n(x) - f(x)| \leq \varepsilon \qquad \text{if } n \geq n_0, x \in I.$$

This shows that $\{f_n(x)\}$ converges to $f(x)$ uniformly on I.

Consider an infinite series $\sum_{n=0}^{\infty} u_n(x)$ of functions $u_n(x)$ defined on an interval I. Let

$$S_n(x) = \sum_{m=0}^{n} u_m(x).$$

If $\{S_n(x)\}$ is uniformly convergent on I to a function $S(x)$, then we say that the series $\sum_{n=0}^{\infty} u_n(x)$ is *uniformly convergent* to $S(x)$. If $\{S_n(x)\}$ is pointwise convergent, then we say that the series $\sum_{n=0}^{\infty} u_n(x)$ is *pointwise convergent*.

From Theorems 1 and 2 we immediately obtain:

THEOREM 3.   *If a series $\sum u_n(x)$ of continuous functions on an interval I is uniformly convergent on I, then its sum $S(x)$ is continuous on I.*

THEOREM 4.   *A series $\sum u_n(x)$ of functions defined on an interval I is uniformly convergent on I, if and only if, for any $\varepsilon > 0$ there is an integer n_0 such that, for all $x \in I$,*

$$|u_n(x) + u_{n+1}(x) + \cdots + u_m(x)| < \varepsilon \qquad \text{whenever } m > n \geq n_0. \qquad (3)$$

If $\sum |u_n(x)|$ is uniformly convergent, then we say that $\sum u_n(x)$ is *absolutely uniformly convergent*, or *normally convergent*. Normal convergence implies uniform convergence.

The following theorem gives a criterion for uniform (or even normal) convergence. It is called the *Weierstrass M-test*.

THEOREM 5.   *Let $\{u_n(x)\}$ be a sequence of functions defined on an interval I. Let $\{M_n\}$ be a sequence of positive constants, such that $|u_n(x)| \leq M_n$ for all $x \in I$. If $\sum M_n$ is convergent, then $\sum u_n(x)$ is normally (and therefore also uniformly) convergent.*

Proof.   Since $\sum M_n$ is convergent for any $\varepsilon > 0$, there is an n_0 such that

$$M_n + M_{n+1} + \cdots + M_m < \varepsilon \qquad \text{if } m > n \geq n_0.$$

It follows that, for all $x \in I$,

$$|u_n(x)| + |u_{n+1}(x)| + \cdots + |u_m(x)| < \varepsilon.$$

Hence, by Theorem 4, $\sum |u_n(x)|$ is uniformly convergent.

PROBLEMS

1. Check for pointwise and uniform convergence each of the following sequences:

(a) $\left\{\dfrac{x^n}{n}\right\}$ in $0 \leq x \leq 1$;   (b) $\left\{\dfrac{nx}{1 + n^2 x^2}\right\}$ in $-1 \leq x \leq 1$;

(c) $\left\{\dfrac{x^n}{1 + x^{2n}}\right\}$ in $0 \leq x < \infty$;   (d) $\left\{\dfrac{1}{n} e^{-n^2 x^2}\right\}$ in $0 \leq x < \infty$;

(e) $\{nxe^{-nx^2}\}$ in $0 \leq x \leq 1$;   (f) $\{nxe^{-nx^2}\}$ in $1 \leq x < \infty$;

(g) $\left\{\dfrac{nx}{1 + nx}\right\}$ in $0 \leq x < \infty$.

2. Prove that the series $\sum(1/n^x)$ is uniformly convergent in any interval $\alpha \leq x < \infty$ where $\alpha > 1$. Is it uniformly convergent in $1 < x < \infty$?

3. Check for pointwise and uniform convergence in each of the following:

(a) $\sum \dfrac{n}{n+1} x^n$ in $0 < x < 1$;

(b) $\sum \dfrac{x^n}{n^2}$ in $0 \leq x \leq 1$;

(c) $\sum \dfrac{x^2}{(1+x^2)^n}$ in $\varepsilon < x < 1$, where $0 < \varepsilon < 1$;

(d) $\sum \dfrac{\sin nx}{n^2}$ in $-\infty < x < \infty$;

(e) $\sum \dfrac{1}{n^2+x^2}$ in $-\infty < x < \infty$;

(f) $\sum e^{-nx}$ in $\varepsilon < x < \infty$, where $\varepsilon > 0$;

(g) $\sum e^{-nx}$ in $0 < x < \infty$.

4. If $\sum u_n(x)$ is uniformly convergent in an interval I, then $\{u_n(x)\}$ is uniformly convergent to 0 on I.

5. If $u_n(x) \geq u_{n+1}(x) \geq 0$ for all x in an interval I and for all $n = 1,2,\cdots$, and if $\lim u_n(x) = 0$ uniformly on I, then $\sum(-1)^n u_n(x)$ is uniformly convergent on I. [*Hint:* Use the proof of Theorem 2, Section 2.]

6. The series $\sum(-1)^n/(n+x^2)$ is uniformly convergent in the interval $0 \leq x < \infty$, but is not absolutely convergent for any $x \geq 0$.

7. If $\{f_n\}$ and $\{g_n\}$ are uniformly convergent on a bounded interval I to bounded functions f and g, respectively, then $\{f_n g_n\}$ is uniformly convergent on I to fg.

8. If $\sum u_n(x)$ is uniformly convergent and if $g(x)$ is a bounded function, then $\sum g(x)u_n(x)$ is uniformly convergent.

5. INTEGRATION AND DIFFERENTIATION OF SERIES

THEOREM 1.   *Let $\{f_n(x)\}$ be a sequence of continuous functions on a closed, bounded interval $a \leq x \leq b$. Assume that $\{f_n(x)\}$ converges to a function $f(x)$ uniformly on the interval. Then $f(x)$ is integrable on $[a,b]$, and*

$$\lim_{n \to \infty} \int_a^b f_n(x)\, dx = \int_a^b f(x)\, dx. \tag{1}$$

Proof. By Theorem 1 of Section 4, $f(x)$ is continuous on $a \leq x \leq b$; hence it is integrable. For any $\varepsilon > 0$, there is an n_0 such that, for all x in $[a,b]$,

$$|f_n(x) - f(x)| < \frac{\varepsilon}{b-a} \qquad \text{if } n \geq n_0.$$

Therefore

$$\left| \int_a^b f_n(x)\, dx - \int_a^b f(x)\, dx \right| = \left| \int_a^b [f_n(x) - f(x)]\, dx \right|$$

$$\leq \int_a^b |f_n(x) - f(x)|\, dx$$

$$\leq \int_a^b \frac{\varepsilon}{b-a}\, dx = \varepsilon \qquad \text{if } n \geq n_0.$$

This proves (1).

If the convergence of $\{f_n(x)\}$ to $f(x)$ is not uniform, then the assertion (1) is generally false. Consider, for example,

$$f_n(x) = \begin{cases} 4n^2 x & \text{if } 0 \leq x \leq \dfrac{1}{2n}, \\[2mm] -4n^2 x + 4n & \text{if } \dfrac{1}{2n} < x \leq \dfrac{1}{n}, \\[2mm] 0 & \text{if } \dfrac{1}{n} < x \leq 1. \end{cases}$$

Here, $f_n(x) \to 0$ for any x in $[0,1]$, but

$$\int_0^1 f_n(x)\, dx = 1 \qquad \text{for any } n.$$

THEOREM 2. *Let* $\displaystyle\sum_{n=0}^{\infty} u_n(x)$ *be an infinite series of continuous functions* $u_n(x)$ *on a closed, bounded interval* $a \leq x \leq b$. *If* $\displaystyle\sum_{n=0}^{\infty} u_n(x)$ *is uniformly convergent to* $S(x)$ *on this interval, then* $S(x)$ *is integrable and*

$$\int_a^b S(x)\, dx = \sum_{n=0}^{\infty} \int_a^b u_n(x)\, dx. \tag{2}$$

Proof. Set $S_m(x) = \displaystyle\sum_{n=0}^{m} u_n(x)$. Applying Theorem 1 to the sequence $\{S_m\}$, we conclude that $S(x)$ is integrable and

$$\lim_{m \to \infty} \int_a^b S_m(x)\, dx = \lim_{m \to \infty} \int_a^b \left[\sum_{n=0}^m u_n(x) \right] dx = \lim_{m \to \infty} \sum_{n=0}^m \int_a^b u_n(x)\, dx$$

exists and equals $\int_a^b S(x)\, dx$. This gives (2).

Assertion (2) can also be written in the form:

$$\int_a^b \left[\sum_{n=0}^\infty u_n(x) \right] dx = \sum_{n=0}^\infty \int_a^b u_n(x)\, dx. \tag{3}$$

This form shows that in order to integrate the sum of a uniformly convergent series one may integrate each term separately and then take the sum of the series of the integrals. This explains why Theorem 2 is often referred to as *the theorem on term-by-term integration*.

EXAMPLE. The series

$$\frac{1}{1-x} = 1 + x + x^2 + \cdots$$

is uniformly convergent if $-t \le x \le t$ where $0 < t < 1$. Integrating term by term we get

$$\log \frac{1+t}{1-t} = 2 \left(t + \frac{t^3}{3} + \frac{t^5}{5} + \frac{t^7}{7} + \cdots \right).$$

We shall now derive the analogs of Theorems 1 and 2 for differentiation.

THEOREM 3. *Let $f_n(x)$ be functions with continuous derivatives in a closed, bounded interval $a \le x \le b$. Assume that the sequence $\{f_n'(x)\}$ is uniformly convergent in $a \le x \le b$, and that the sequence $\{f_n(x)\}$ is pointwise convergent in $a \le x \le b$ to a function $f(x)$. Then $f(x)$ is differentiable, and, for any $a \le x \le b$,*

$$f'(x) = \lim_{n \to \infty} f_n'(x). \tag{4}$$

Proof. Denote by $g(x)$ the uniform limit of $\{f_n'(x)\}$. The function $g(x)$ is continuous and, by Theorem 1,

$$\int_a^x g(t)\, dt = \lim_{n \to \infty} \int_a^x f_n'(t)\, dt.$$

Since

$$\int_a^x f_n'(t)\, dt = f_n(x) - f_n(a),$$

and since, by assumption,

$$\lim_{n} f_n(x) = f(x), \qquad \lim_{n} f_n(a) = f(a),$$

we find that

$$\int_a^x g(t)\, dt = f(x) - f(a).$$

Since $g(t)$ is continuous, the fundamental theorem of calculus implies that $f'(x)$ exists and it is equal to $g(x)$, that is,

$$f'(x) = g(x) = \lim_{n \to \infty} f_n'(x).$$

THEOREM 4. *Let $u_n(x)$ be functions with continuous derivatives in a closed, bounded interval $a \leq x \leq b$. Assume that $\sum\limits_{n=0}^{\infty} u_n'(x)$ is uniformly convergent in $a \leq x \leq b$, and that $\sum\limits_{n=0}^{\infty} u_n(x)$ is pointwise convergent in $a \leq x \leq b$ to a function $S(x)$. Then $S(x)$ is differentiable, and, for any $a \leq x \leq b$,*

$$S'(x) = \sum_{n=0}^{\infty} u_n'(x). \tag{5}$$

The proof is obtained by applying Theorem 3 to the sequence $\{S_m(x)\}$, where $S_m(x) = \sum\limits_{n=0}^{m} u_n(x)$.

Assertion (5) also can be written in the form

$$\frac{d}{dx} \sum_{n=0}^{\infty} u_n(x) = \sum_{n=0}^{\infty} \frac{d}{dx} u_n(x). \tag{6}$$

This formula shows that to differentiate the series $\sum u_n(x)$, we may first differentiate each term separately and then take the sum of the differentiated terms. This explains why Theorem 4 is often referred to as *the theorem on term-by-term differentiation*.

EXAMPLE.   The series

$$\frac{1}{1-x} = 1 + x + x^2 + \cdots + x^n + \cdots$$

is convergent if $-t \leq x \leq t$, where $0 < t < 1$. The series of its term-by-term derivatives

$$1 + 2x + 3x^2 + \cdots + nx^{n-1} + \cdots$$

is uniformly convergent in $-t \le x \le t$, by the Weierstrass M-test, since

$$|nx^{n-1}| \le nt^{n-1} \quad \text{and} \quad \sum nt^{n-1} < \infty.$$

It follows that

$$\frac{1}{(1-x)^2} = \frac{d}{dx}\frac{1}{1-x} = 1 + 2x + 3x^2 + \cdots + nx^{n-1} + \cdots.$$

PROBLEMS

1. If $f(x) = \sum\limits_{n=1}^{\infty} \dfrac{\cos nx}{n^2}$, show that

    $$\int_0^{\pi/2} f(x)\, dx = \sum_{n=0}^{\infty} \frac{(-1)^n}{(2n+1)^3}.$$

2. If $f_n(x) \to f(x)$ pointwise in $a \le x \le b$, and uniformly in every subinterval $a + \delta \le x \le b - \delta, \delta > 0$, and if also $|f_n(x)| \le M$ for all n and for all x in $[a,b]$, then

    $$\int_a^b f_n(x)\, dx \to \int_a^b f(x)\, dx.$$

    [*Hint:* $\int_a^{a+\delta}|f_n(x)|\, dx + \int_{b-\delta}^b |f_n(x)|\, dx + \int_a^{a+\delta} |f(x)|\, dx + \int_{b-\delta}^b |f(x)|\, dx < \varepsilon/2$ if δ is sufficiently small, and $|\int_{a+\delta}^{b-\delta} [f_n(x) - f(x)]\, dx| < \varepsilon/2$ if n is sufficiently large.]

3. Let $f_n(x) = nx/(1 + nx)$ for $0 \le x < \infty$. Show that

    $$\int_0^x f_n(t)\, dt \to x \qquad \text{as } n \to \infty.$$

4. Let $f_n(x) = n^2xe^{-nx}$ for $0 \le x \le 1$. Then $f_n(x) \to 0$ pointwise. Show that, for any $0 < x \le 1$, $\int_0^x f_n(t)\, dt \to 1$ if $n \to \infty$. Explain why this does not contradict Theorem 1.

5. For each of the sequences $\{f_n\}$ given below, find $f(x) = \lim f_n(x)$ in the interval $[a,b]$ indicated, and check whether

    $$\int_a^b f_n(x)\, dx \to \int_a^b f(x)\, dx; \qquad f_n'(x) \to f'(x):$$

    (a) $f_n(x) = \dfrac{nx}{1 + n^2x^4}$ in $[0,b]$;

    (b) $f_n(x) = \dfrac{2x + n}{x + n}$ in $[0,b]$;

    (c) $f_n(x) = \dfrac{nx^2 + 1}{nx + 1}$ in $[0,1]$.

6. Use the relation

$$\frac{1}{1 + x^2} = 1 - x^2 + x^4 - x^6 + \cdots$$

to show that

$$\tan^{-1} x = x - \frac{x^3}{3} + \frac{x^5}{5} - \frac{x^7}{7} + \cdots \qquad \text{if } |x| < 1.$$

7. Show that $\displaystyle\sum_{n=1}^{\infty} \frac{x}{n(x + n)}$ is uniformly convergent for $0 \le x \le b, b > 0$ to some function $S(x)$. Prove that $S'(x) = \displaystyle\sum_{n=1}^{\infty} 1/(x + n)^2, 0 \le x \le b$.

8. Show that

$$\frac{d}{dx} \sum_{n=1}^{\infty} \frac{\sin nx}{n^2} = \sum_{n=1}^{\infty} \frac{\cos nx}{n} \qquad \text{if } 0 < x < 2\pi.$$

Does this relation hold also at the point $x = 0$?

9. Show that

$$\frac{d}{dx} \sum_{n=1}^{\infty} (-1)^{n-1} \frac{x^n}{n} = \frac{1}{1 + x} \qquad \text{if } |x| < 1.$$

Does this relation hold also at the point $x = 1$?

6. POWER SERIES

A series of functions of the form

$$\sum_{n=0}^{\infty} a_n x^n = a_0 + a_1 x + a_2 x^2 + \cdots$$

is called a *power series*. The numbers $a_0, a_1, a_2, \cdots$ are called the *coefficients* of the power series.

THEOREM 1. *If a power series $\sum a_n x^n$ is convergent for $x = x_0$, $x_0 \ne 0$, then it is absolutely convergent for each x with $|x| < |x_0|$. Furthermore, for any $\varepsilon > 0$, the series is normally convergent in the interval $-|x_0| + \varepsilon \le x \le |x_0| - \varepsilon$.*

Proof. Since $\sum a_n x_0^n$ is convergent, $a_n x_0^n \to 0$ if $n \to \infty$. It follows that the sequence $\{a_n x_0^n\}$ is bounded, that is,

$$|a_n| |x_0|^n \le K \qquad \text{for } n = 0, 1, 2, \cdots \qquad (K \text{ constant}).$$

Now let $|x| < |x_0|$. Set $\theta = |x|/|x_0|$. Then $\theta < 1$ and

$$|a_n x^n| = |a_n| \, |x|^n = |a_n| \, |x_0|^n \left(\frac{|x|}{|x_0|}\right)^n \le K \left(\frac{|x|}{|x_0|}\right)^n = K\theta^n.$$

Since the series $\sum K\theta^n$ is convergent, the comparison test shows that $\sum |a_n x^n|$ is convergent. Thus $\sum a_n x^n$ is absolutely convergent.

Now let $|x| \le |x_0| - \varepsilon$ for some $\varepsilon > 0$. Then

$$\theta = \frac{|x|}{|x_0|} \le \frac{|x_0| - \varepsilon}{|x_0|} = \theta_0 < 1$$

where θ_0 is independent of x. Since

$$|a_n x^n| \le K\theta_0^n, \qquad \sum K\theta_0^n \text{ convergent,}$$

the Weierstrass M-test shows that $\sum |a_n x^n|$ is uniformly convergent in the interval $|x| \le |x_0| - \varepsilon$. Thus $\sum a_n x^n$ is normally convergent in this interval.

COROLLARY.   *If a power series $\sum a_n x^n$ is divergent at a point $x = x_0 \ne 0$, then it is also divergent at any point x with $|x| > |x_0|$.*

Proof.   If $\sum a_n x^n$ is convergent for some x, $|x| > |x_0|$, then, by Theorem 1, $\sum a_n x_0^n$ is convergent, a contradiction.

Denote by A the set of all nonnegative numbers x for which $\sum a_n x^n$ is convergent. If A is bounded above, then denote its supremum by R. Let $|x| < R$. Then $\sum a_n x^n$ is convergent. Indeed, if $\sum a_n x^n$ is divergent, then $\sum a_n t^n$ is divergent for all $t > |x|$. Consequently, the supremum of A is $\le |x|$, that is $R \le |x|$, which contradicts our assumption that $|x| < R$. On the other hand, if $|x| > R$, then $\sum a_n x^n$ is divergent. Indeed, otherwise, point $|x| - \varepsilon$ belongs to A for any $0 < \varepsilon < |x|$, and, therefore, $|x| - \varepsilon \le R$ by the definition of R. This gives $|x| \le R$, which contradicts our assumption that $|x| > R$.

If A is not bounded above, then the series $\sum a_n x^n$ is convergent for all $x > 0$; therefore, also for any real x.

We sum up:

THEOREM 2.   *For every power series $\sum a_n x^n$ there are two possibilities:*

(i)   *The series $\sum a_n x^n$ is convergent for every x.*

(ii)   *There is a nonnegative number R such that $\sum a_n x^n$ is convergent if $|x| < R$ and is divergent if $|x| > R$.*

The number R is called the *radius of convergence* of the series. The interval $-R < x < R$ is called the *interval of convergence*. The series may converge or diverge at the endpoints of this interval.

If $\sum a_n x^n$ is convergent for all x, then we say that the *radius of convergence R is equal to* ∞, and write $R = \infty$.

The next result is called the *Cauchy–Hadamard formula*. It gives a formula for computing the radius of convergence R in terms of the coefficients a_n of the power series.

THEOREM 3. *For any power series* $\sum a_n x^n$, *the radius of convergence R is given by*

$$\frac{1}{R} = \overline{\lim_{n \to \infty}} \sqrt[n]{|a_n|}. \tag{1}$$

In (1) it is to be understood that when the right-hand side is 0, then $R = \infty$, and when the right-hand side is ∞, then $R = 0$.

Proof. Let

$$M = \overline{\lim_{n \to \infty}} \sqrt[n]{|a_n|}.$$

Consider first the case where $M \neq \infty$. For any x,

$$\overline{\lim_{n}} \sqrt[n]{|a_n x^n|} = |x|\, M.$$

From Cauchy's root test it follows that the series $\sum a_n x^n$ is convergent if $|x|\, M < 1$ and is divergent if $|x|\, M > 1$. Thus, if $M > 0$, then $R = 1/M$, whereas if $M = 0$, then the series converges for all x, that is, $R = \infty$.

Suppose next that $M = \infty$. Thus, for any $x \neq 0$,

$$\sqrt[n]{|a_n x^n|} = |x| \sqrt[n]{|a_n|} > 1$$

for a subsequence of n's. It follows that $\sum a_n x^n$ is divergent. Thus, $R = 0$.

We shall use the Cauchy–Hadamard formula in order to show that a power series may be integrated and differentiated term by term.

THEOREM 4. *Let* $\displaystyle\sum_{n=0}^{\infty} a_n x^n$ *be a power series with radius of convergence R. Then the power series*

$$\sum_{n=0}^{\infty} \frac{a_n}{n+1} x^{n+1}, \qquad \sum_{n=1}^{\infty} n a_n x^{n-1} \tag{2}$$

also have radius of convergence R.

Proof. The radius of convergence of the first series in (2) is the same as the radius of convergence of the series

$$\sum_{n=0}^{\infty} \frac{a_n}{n+1} x^n. \tag{3}$$

Since $\sqrt[n]{n+1} \to 1$ as $n \to \infty$,

$$\overline{\lim} \sqrt[n]{\frac{|a_n|}{n+1}} = \overline{\lim}\left(\frac{1}{\sqrt[n]{n+1}} \sqrt[n]{|a_n|}\right) = \left(\lim \frac{1}{\sqrt[n]{n+1}}\right)(\overline{\lim} \sqrt[n]{|a_n|}) = \overline{\lim} \sqrt[n]{|a_n|}.$$

The Cauchy–Hadamard formula now shows that the series (3) has the same radius of convergence R as the series $\sum a_n x^n$.

The radius of convergence of the second series in (2) is the same as the radius of convergence of the series

$$\sum_{n=1}^{\infty} na_n x^n. \tag{4}$$

Since

$$\overline{\lim} \sqrt[n]{n|a_n|} = \overline{\lim}(\sqrt[n]{n} \sqrt[n]{|a_n|}) = (\lim \sqrt[n]{n})(\overline{\lim} \sqrt[n]{|a_n|}) = \overline{\lim} \sqrt[n]{|a_n|},$$

the series (4) has the radius of convergence R.

From Theorems 1 and 4 it follows that the series $\sum a_n x^n$ as well as the series in (2), are uniformly convergent in every interval $|x| \le R - \varepsilon, \varepsilon > 0$. Since term-by-term integration and term-by-term differentiation of $\sum a_n x^n$ lead to the first and second series in (2), Theorems 2 and 4 of Section 5 can be applied to yield the following results:

THEOREM 5.   Let $\sum a_n x^n$ be a power series having radius of convergence $R \ne 0$. Set $f(x) = \sum_{n=0}^{\infty} a_n x^n$ for $|x| < R$. Then $f(x)$ is continuous in $-R < x < R$ and

$$\int_\alpha^x f(t) \, dt = \sum_{n=0}^{\infty} \frac{a_n}{n+1} x^{n+1} - \sum_{n=0}^{\infty} \frac{a_n}{n+1} \alpha^{n+1} \qquad (|x| < R, |\alpha| < R). \tag{5}$$

THEOREM 6.   Let $\sum a_n x^n$ be a power series having radius of convergence $R \ne 0$. Set $f(x) = \sum_{n=0}^{\infty} a_n x^n$ for $|x| < R$. Then $f(x)$ has a continuous derivative $f'(x)$ in $-R < x < R$, and

$$f'(x) = \sum_{n=1}^{\infty} na_n x^{n-1}. \tag{6}$$

If we apply Theorem 6 to the series $\sum na_n x^{n-1}$, we find that $f''(x)$ exists and is continuous, and

$$f''(x) = \sum_{n=2}^{\infty} n(n-1)x^{n-2} \qquad \text{if } |x| < R.$$

Continuing in this way step by step, we arrive at the following result:

COROLLARY.   *The function* $f(x) = \sum\limits_{n=0}^{\infty} a_n x^n$ *is infinitely differentiable in* $-R < x < R$, *and, for any positive integer* k,

$$f^{(k)}(x) = \sum_{n=k}^{\infty} n(n-1)\cdots(n-k+1)x^{n-k}. \tag{7}$$

We conclude this section with a theorem on the product of two power series. Let $\sum a_n x^n$ and $\sum b_n x^n$ be given power series with radii of convergence R_1 and R_2, respectively. Their Cauchy product, as defined in Section 3, is then the power series $\sum c_n x^n$ where

$$c_n = a_0 b_n + a_1 b_{n-1} + \cdots + a_{n-1}b_1 + a_n b_0.$$

Since, by Theorem 1, $\sum a_n x^n$ and $\sum b_n x^n$ are absolutely convergent in the interval $|x| < R$, where $R = \min(R_1, R_2)$, Theorem 1 of Section 3 can be applied. We thus obtain:

THEOREM 7.   *Let* $f(x) = \sum\limits_{n=0}^{\infty} a_n x^n$, $g(x) = \sum\limits_{n=0}^{\infty} b_n x^n$ *and assume that both power series converge for all* $|x| < R$. *Denote by* $\sum c_n x^n$ *the Cauchy product of these power series. Then the radius of convergence of* $\sum c_n x^n$ *is* $\geq R$, *and*

$$f(x)g(x) = \sum_{n=0}^{\infty} c_n x^n \qquad \text{if } |x| < R. \tag{8}$$

PROBLEMS

1. Find the radius of convergence of each of the following series:

    (a) $\sum \dfrac{n}{n+1}x^n$;

    (b) $\sum \alpha^n x^n,\ \alpha \neq 0$;

    (c) $\sum \dfrac{(2n)!}{2^n}x^n$;

    (d) $\sum \dfrac{x^n}{n!}$;

    (e) $\sum \dfrac{x^n}{n^\alpha}, \qquad \alpha > 0$;

    (f) $\sum \dfrac{(n+r)!}{n!(n+s)!}x^n$, \qquad r and s positive integers.

2. Find, for each of the following series, whether it converges at the endpoints of the interval of convergence:

(a) $\displaystyle\sum_{n=1}^{\infty} \frac{x^n}{n}$;

(b) $\displaystyle\sum_{n=1}^{\infty} (-1)^n \frac{x^n}{n^2}$;

(c) $\displaystyle\sum_{n=1}^{\infty} (-1)^n \frac{x^n}{\log n}$;

(d) $\displaystyle\sum_{n=0}^{\infty} n^2 x^n$.

3. Find the sum of the series $\displaystyle\sum_{n=1}^{\infty} nx^n$ for $|x| < 1$. [*Hint:* Denote the sum by $f(x)$. Then $\displaystyle\frac{f(x)}{x} = \sum_{n=1}^{\infty} nx^{n-1} = \left(\sum_{n=0}^{\infty} x^n \right)' = \left(\frac{1}{1-x} \right)' = \frac{1}{(1-x)^2}.$]

4. Find the sum of the following series for $|x| < 1$:

(a) $\displaystyle\sum_{n=0}^{\infty} (n+1)(n+2)x^{n+1}$;

(b) $\displaystyle\sum_{n=1}^{\infty} (-1)^n nx^n$;

(c) $\displaystyle\sum_{n=1}^{\infty} n^2 x^n$;

(d) $\displaystyle\sum_{n=1}^{\infty} n^3 x^n$;

(e) $\displaystyle\sum_{n=1}^{\infty} \frac{x^n}{n(n+1)}$;

(f) $\displaystyle\sum_{n=1}^{\infty} \frac{nx^n}{n+1}$;

(g) $\displaystyle\sum_{n=1}^{\infty} \frac{nx^n}{(n+1)(n+2)}$;

(h) $\displaystyle\sum_{n=0}^{\infty} x^{4n+1}$.

5. Given that

$$\frac{1}{1-x} = 1 + x + x^2 + \cdots \qquad \text{if } |x| < 1,$$

find a power series (in $|x| < 1$) for $1/(1-x)^k$ (k positive integer) in two ways: (a) using the corollary to Theorem 6; (b) using Theorem 7.

7. TAYLOR'S SERIES

In Section 6 we considered power series $\sum a_n x^n$. In this section we shall deal also with series of the form $\sum a_n(x - \alpha)^n$. We call such a series a *power series about α* (or *centered at α*), or, briefly, a power series.

All the results of Section 6 apply almost word for word to power series about a point α.

Let $f(x)$ be an infinitely differentiable function in an interval $(a - R, a + R)$, $R \neq 0$. By Taylor's theorem (Section 3.8) we can write, for any positive integer n:

$$f(x) = f(a) + \frac{f'(a)}{1!}(x - a) + \cdots$$

$$+ \frac{f^{(n)}(a)}{n!}(x - a)^n + R_{n+1}, \qquad (|x - a| < R) \qquad (1)$$

where the remainder R_{n+1} has the form

$$R_{n+1} = \frac{f^{(n+1)}(\xi)}{(n + 1)!}(x - a)^{n+1}, \qquad |\xi - a| < |x - a|. \qquad (2)$$

We conclude:

THEOREM 1. *If $f(x)$ is infinitely differentiable in the interval $-R < x - a < R$, and if*

$$R_{n+1} \to 0 \ as \ n \to \infty \qquad for \ each \ x \in (a - R, a + R), \qquad (3)$$

then

$$f(x) = \sum_{n=0}^{\infty} \frac{f^{(n)}(a)}{n!}(x - a)^n \qquad if \ |x - a| < R. \qquad (4)$$

The power series in (4) is called the *Taylor series* (about a, or at a) of the function $f(x)$. When $a = 0$ it sometimes also is called the *Maclaurin series*. Formula (4) is called the *expansion* of $f(x)$ into a Taylor series (about a).

EXAMPLE. We shall expand $f(x) = e^x$ about $x = 0$. First we compute the derivatives of $f(x)$ and find that

$$f^{(n)}(x) = e^x, \qquad f^{(n)}(0) = 1.$$

We see that for any x,

$$R_{n+1} = \frac{e^\xi}{(n + 1)!} x^{n+1} \to 0 \qquad if \ n \to \infty.$$

Hence

$$e^x = 1 + x + \frac{x^2}{2!} + \cdots + \frac{x^n}{n!} + \cdots \qquad for \ all \ x. \qquad (5)$$

Similarly, one finds that

$$\sin x = x - \frac{x^3}{3!} + \frac{x^5}{5!} - \frac{x^7}{7!} + \cdots = \sum_{n=0}^{\infty} \frac{(-1)^n x^{2n+1}}{(2n + 1)!} \qquad for \ all \ x, \qquad (6)$$

$$\cos x = 1 - \frac{x^2}{2!} + \frac{x^4}{4!} - \frac{x^6}{6!} + \cdots = \sum_{n=0}^{\infty} \frac{(-1)^n x^{2n}}{(2n)!} \qquad for \ all \ x. \qquad (7)$$

Let $f(x)$ be a function having the form

$$f(x) = \sum_{n=0}^{\infty} a_n(x - \alpha)^n \qquad \text{if } |x - \alpha| < R. \tag{8}$$

Then we say that $f(x)$ has a *power series expansion* about α. From the corollary to Theorem 6 of Section 6 we have

$$f^{(k)}(x) = \sum_{n=k}^{\infty} n(n - 1)\cdots(n - k + 1)a_n(x - \alpha)^{n-k}.$$

Hence $f^{(k)}(\alpha) = k!\, a_k$. It follows that

$$f(x) = \sum_{n=0}^{\infty} \frac{f^{(n)}(\alpha)}{n!}(x - \alpha)^n. \tag{9}$$

We can therefore state:

THEOREM 2. *If a function $f(x)$ has a power series expansion about a point α, then this power series is the Taylor series of $f(x)$ about α.*

This theorem has an interesting application:

THEOREM 3. *Suppose two power series $\sum a_n(x - \alpha)^n$, $\sum b_n(x - \alpha)^n$ are convergent if $|x - \alpha| < R$ and suppose*

$$\sum_{n=0}^{\infty} a_n(x - \alpha)^n = \sum_{n=0}^{\infty} b_n(x - \alpha)^n \qquad \text{if } |x - \alpha| < R, \tag{10}$$

then $a_n = b_n$ for all n.

This theorem is called the *uniqueness theorem* for power series.

Proof. Denote the sum in (10) by $f(x)$. By Theorem 2,

$$a_n = \frac{f^{(n)}(\alpha)}{n!}, \qquad b_n = \frac{f^{(n)}(\alpha)}{n!}.$$

Hence $a_n = b_n$ for all n.

An infinitely differentiable function may not have a Taylor series expansion, since condition (3) may not be fulfilled. Consider the function

$$f(x) = \begin{cases} e^{-1/x^2} & \text{if } x \neq 0, \\ 0 & \text{if } x = 0. \end{cases} \tag{11}$$

It is infinitely differentiable for all x. In fact, one can prove by induction on n that $f^{(n)}(x)$ exists for all $x \neq 0$ and

$$f^{(n)}(x) = e^{-1/x^2} Q_n\left(\frac{1}{x}\right) \qquad \text{if } x \neq 0,$$

where $Q_n(t)$ is a polynomial in t. This implies that $(1/x)f^{(n)}(x) \to 0$ as $x \to 0$. But then, by induction on n, $f^{(n)}(0)$ exists and equals 0. Thus $f^{(n)}(x)$ exists for all x, and $f^{(n)}(0) = 0$. If $f(x)$ has a Taylor's series about $x = 0$, then

$$f(x) = \sum_{n=0}^{\infty} \frac{f^{(n)}(0)}{n!} x^n = 0$$

for all x in some interval $|x| < R$, which is impossible. Thus, *the function defined in* (11) *has no Taylor's series about* $x = 0$.

PROBLEMS

1. Prove assertions (6) and (7).

2. Derive the Taylor series expansions:

$$\log(1 + x) = x - \frac{x^2}{2} + \frac{x^3}{3} - \frac{x^4}{4} + \cdots \qquad (|x| < 1),$$

$$-\log(1 - x) = x + \frac{x^2}{2} + \frac{x^3}{3} + \frac{x^4}{4} + \cdots \qquad (|x| < 1).$$

3. Prove that for any real α and for any x, $|x| < 1$,

$$(1 + x)^\alpha = 1 + \alpha x + \frac{\alpha(\alpha - 1)}{2!} x^2 + \cdots + \frac{\alpha(\alpha - 1)\cdots(\alpha - n + 1)}{n!} x^n + \cdots.$$

   The series on the right is called the *binomial series*.

4. Prove that

$$\sinh^{-1} x = x + \sum_{n=1}^{\infty} (-1)^n \frac{1 \cdot 3 \cdots (2n - 1)}{2 \cdot 4 \cdots 2n} \frac{x^{2n+1}}{(2n + 1)!}.$$

5. Find the Taylor series for each of the following functions about $x = 0$, and compute the radius of convergence:

   (a) $\displaystyle\int_0^x \frac{\sin t}{t}\, dt;$ \qquad\qquad (b) $\displaystyle\int_0^x e^{-t^2}\, dt;$

   (c) $a^x, a > 0;$ \qquad\qquad (d) $\displaystyle\int_0^x \cos t^2\, dt.$

6. Find the Taylor series of
   (a) e^x about $x = 1$; \quad (b) $\cos x$ about $x = \pi/2$; \quad (c) $\sqrt{x}$ about $x = 1$.

7. Find the first four nonzero terms in Taylor's series for $\tan x$ about $x = 0$.

8. Let $f(x) = \sum a_n x^n$ for $|x| < R$. Prove: (a) If $f(x)$ is even, that is, if $f(-x) = f(x)$, then $a_{2n+1} = 0$ for all n; (b) If $f(x)$ is odd, that is, if $f(-x) = -f(x)$, then $a_{2n} = 0$ for all n.

9. Find the power series $\sum a_n x^n$ for $\dfrac{d}{dx} \dfrac{e^x - 1}{x}$ and deduce that $1 = \displaystyle\sum_{n=1}^{\infty} \dfrac{n}{(n+1)!}$.

10. Find the sum of the series $\displaystyle\sum_{n=0}^{\infty} \dfrac{x^n}{(n+1)^2 n!}$.

11. Find the Taylor series about $x = 0$ for

   (a) $\dfrac{1}{\sqrt{1-x^2}}$;   (b) $\sin^{-1} x$;   (c) $\sin^2 x$.

12. Use the series expansion in (5) for e^x in order to prove that $e^x e^y = e^{x+y}$.

13. Use the series expansions in (6) and (7) for $\sin x$ and $\cos x$ in order to prove that

   (a) $\sin^2 x + \cos^2 x = 1$;   (b) $\sin(x + y) = \sin x \cos y + \cos x \sin y$.

8. CONVERGENCE OF IMPROPER INTEGRALS

Let $f(x)$ be a function defined in a bounded interval $[a,b)$. If the improper integral

$$\int_a^b f(x)\, dx = \lim_{\delta \to 0} \int_a^{b-\delta} f(x)\, dx \qquad (1)$$

exists, then we say that the *integral is convergent*. If the limit in (1) does not exist, then we say that the *integral is divergent*. We shall prove a *criterion of Cauchy* for convergence of improper integrals.

THEOREM 1.   *Let $f(x)$ be integrable in each interval $[a,b - \delta]$, $\delta > 0$. Then the improper integral $\int_a^b f(x)\, dx$ is convergent if, and only if, for any $\varepsilon > 0$ there is a $\delta > 0$ such that*

$$\left| \int_c^d f(x)\, dx \right| < \varepsilon \qquad \text{if } b - \delta < c < d < b. \qquad (2)$$

Proof. Let

$$g(x) = \int_a^x f(t)\, dt.$$

Then $g(x)$ is a continuous function in the interval $a \le x < b$. The condition (2) means that

$$|g(x) - g(y)| < \varepsilon \qquad \text{if } b - \delta < x < b, b - \delta < y < b. \tag{3}$$

The proof of Theorem 2 in Section 2.7 shows that if this last condition is satisfied, then $g(b - 0)$ exists. Therefore, the integral $\int_a^b f(x)\, dx$ converges. Conversely, if the integral is convergent, then $g(b - 0)$ exists. This implies (3). Indeed, for any ε there is a $\delta > 0$ such that

$$|g(x) - g(b - 0)| < \frac{\varepsilon}{2} \qquad \text{if } b - \delta < x < b,$$

$$|g(y) - g(b - 0)| < \frac{\varepsilon}{2} \qquad \text{if } b - \delta < y < b.$$

Hence

$$|g(x) - g(y)| \le |g(x) - g(b - 0)| + |g(b - 0) - g(y)| < \varepsilon,$$

if x and y lie in $(b - \delta, b)$.

If the improper integral

$$\int_a^b |f(x)|\, dx$$

is convergent, then we say that the integral $\int_a^b f(x)\, dx$ is *absolutely convergent*.

THEOREM 2. *If $\int_a^b f(x)\, dx$ is absolutely convergent, then it is also convergent.*

The proof follows from Theorem 1 and the inequality

$$\left| \int_c^d f(x)\, dx \right| \le \int_c^d |f(x)|\, dx.$$

If an integral is convergent but is not absolutely convergent, then it is said to be *conditionally convergent*.

THEOREM 3. *Let $f(x) \ge 0$ for all $a \le x < b$. Then the integral $\int_a^b f(x)\, dx$ is convergent if, and only if, there exists a constant K such that*

$$\int_a^y f(x)\, dx \le K \qquad \text{for all } a < y < b. \tag{4}$$

The proof follows from the results of Section 2.5 on monotone functions, noting that the function

$$g(y) = \int_a^y f(x)\, dx$$

is monotone increasing.

If condition (4) is not satisfied, then $g(y) \to \infty$ as $y \to b - 0$. We then write

$$\int_a^b f(x)\, dx = \infty.$$

The definitions of convergence, absolute convergence, and conditional convergence extend to improper integrals

$$\int_a^b f(x)\, dx = \lim_{\epsilon \to 0+} \int_{a+\epsilon}^b f(x)\, dx, \qquad \int_a^\infty f(x)\, dx, \text{ and so on.}$$

Theorems 1–3 also extend to these integrals. Thus, for instance, if for any ϵ there is an N_0 such that

$$\left| \int_N^M f(x)\, dx \right| < \epsilon \qquad \text{if } M > N > N_0,$$

then the integral $\int_a^\infty f(x)$ is convergent. If there is a constant K such that

$$\int_a^y |f(x)|\, dx < K \qquad \text{for all } a < y < \infty,$$

then the integral $\int_a^\infty f(x)\, dx$ is absolutely convergent.

THEOREM 4.   *Let $f(x)$ be a function defined in the interval $a \le x < \infty$, and integrable on every subinterval $[a,b]$, $b < \infty$. Suppose that there is a monotone increasing sequence of numbers M_n such that $M_0 = a$ and $\lim M_n = \infty$, and such that the following properties hold:*

(i) $f(x) \ge 0$ *in* (M_{2n}, M_{2n+1}) *and* $f(x) \le 0$ *in* (M_{2n+1}, M_{2n+2}), *for all* $n \ge 0$.

(ii) $\left| \int_{M_{n-1}}^{M_n} f(x)\, dx \right| \ge \left| \int_{M_n}^{M_{n+1}} f(x)\, dx \right| \qquad$ *for all* $n \ge 1$.

(iii) $\int_{M_{n-1}}^{M_n} f(x)\, dx \to 0 \qquad$ *if* $n \to \infty$.

Then $\int_a^\infty f(x)$ is convergent.

Proof.   Let

$$b_n = \int_{M_{n-1}}^{M_n} f(x)\, dx.$$

The alternating series test shows that $\sum b_n$ is convergent. Denote its sum by I. Now, for any $a < y < \infty$ there is a positive integer n such that $M_n \leq y < M_{n+1}$. Hence

$$\int_a^y f(x)\,dx = \sum_{k=1}^n b_k + \int_{M_n}^y f(x)\,dx.$$

Since $f(x)$ has a constant sign in (M_n, M_{n+1}),

$$\left| \int_{M_n}^y f(x)\,dx \right| = \int_{M_n}^y |f(x)|\,dx \leq \int_{M_n}^{M_{n+1}} |f(x)|\,dx = \left| \int_{M_n}^{M_{n+1}} f(x)\,dx \right| = |b_{n+1}|.$$

For any $\varepsilon > 0$ we can choose n_0 so large that if $n \geq n_0$, then

$$\left| I - \sum_{k=1}^n b_k \right| < \frac{\varepsilon}{2}, \qquad |b_{n+1}| < \frac{\varepsilon}{2}.$$

It follows that if $y \geq M_{n_0}$, then

$$\left| I - \int_a^y f(x)\,dx \right| < \varepsilon.$$

This proves that $\int_a^\infty f(x)\,dx$ is convergent to I.

PROBLEMS

1. Prove the following *comparison test* for improper integrals: (i) If $f(x) \geq g(x) \geq 0$ and if the improper integral $\int_a^b f(x)\,dx$ is convergent, then the improper integral $\int_a^b g(x)\,dx$ is convergent. (ii) If $f(x) \geq g(x) \geq 0$ and if the improper integral $\int_a^b g(x)\,dx$ diverges, then the improper integral $\int_a^b f(x)\,dx$ diverges.

2. Check the following improper integrals for convergence and for absolute convergence:

(a) $\displaystyle\int_0^\infty \frac{dx}{\sqrt{1 + x^\alpha}}, \, \alpha > 0;$

(b) $\displaystyle\int_0^\infty \frac{\sin x}{x}\,dx;$

(c) $\displaystyle\int_0^\infty \sin x^3\,dx;$

(d) $\displaystyle\int_0^\infty (\cos x)^3\,dx;$

(e) $\displaystyle\int_0^1 \frac{\log x}{\sqrt{x}}\,dx;$

(f) $\displaystyle\int_0^\pi \frac{x}{\sin x}\,dx;$

(g) $\displaystyle\int_0^\infty \frac{e^{-x}}{\sqrt{x}}\,dx;$

(h) $\displaystyle\int_0^\infty \frac{x}{e^x - 1}\,dx;$

(i) $\displaystyle\int_1^\infty \frac{\sin(1/x)}{x}\,dx;$ (j) $\displaystyle\int_0^{\pi/2} \frac{dt}{\sqrt{1-\sin t}};$

(k) $\displaystyle\int_0^\infty \sin^2 x\,dx.$

3. For each of the following integrals, find the values of α for which the integral converges:

(a) $\displaystyle\int_0^\infty x^{\alpha-1}e^{-x}\,dx;$ (b) $\displaystyle\int_1^\infty \frac{x^{\alpha-1}}{1+x}\,dx;$

(c) $\displaystyle\int_0^\infty \frac{1-e^{-x}}{x^\alpha}\,dx;$ (d) $\displaystyle\int_0^{\pi/2} (\tan t)^{1-\alpha}\,dt;$

(e) $\displaystyle\int_a^b \left(\frac{x-a}{b-x}\right)^\alpha \log\frac{b-x}{x-a}\,dx;$ (f) $\displaystyle\int_a^b \frac{dx}{(b-x)^\alpha(x-a)^{1-\alpha}}.$

4. Let $f(x)$ be a positive continuous function in $[a,\infty)$. Show that if $\varlimsup_{x\to\infty} \dfrac{f(x+1)}{f(x)} < 1$, then $\int_a^\infty f(x)\,dx$ is convergent.

part two

FUNCTIONS
OF
SEVERAL VARIABLES

6

SPACE OF SEVERAL
VARIABLES
AND CONTINUOUS
FUNCTIONS ON IT

1. OPEN AND CLOSED SETS IN THE PLANE

Consider the set of all pairs (x,y) of real numbers x and y. We define equality between such pairs: $(x_0,y_0) = (x_1,y_1)$ if, and only if, $x_0 = x_1$ and $y_0 = y_1$. We shall call a pair (x,y) a *point* and the set of all points the *real plane* or, briefly, the *plane*.

In Section 1.4 we have seen how to form a one-to-one correspondence between the real numbers and the points on a straight line. Analogously, one can correspond to each point (x,y) a point on the plane, where the concept "plane" used here is that (Euclidean plane) employed in Euclidean geometry. Since the student surely is familiar with the way this correspondence is given, in Figure 6.1 we shall indicate the position of only a few points.

We define the *distance* between two points (x_0,y_0) and (x_1,y_1) to be the number

$$d = \sqrt{(x_0 - y_0)^2 + (x_1 - y_1)^2}. \tag{1}$$

This definition is suggested by the fact that, in view of the Pythagoras theorem, d is the length of the segment connecting the point (x_0,y_0) to (x_1,y_1). More

183

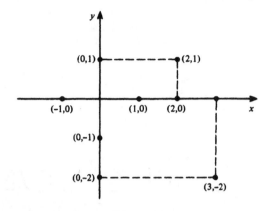

Figure 6.1

precisely, it is the length of the segment connecting the points that we correspond in the Euclidean plane to the pairs (x_0, y_0) and (x_1, y_1).

We shall denote points (x, y) by capital letters also.

We now define *addition* of two points $P = (x_0, y_0)$ and $Q = (x_1, y_1)$ by

$$(x_0, y_0) + (x_1, y_1) = (x_0 + x_1, y_0 + y_1). \tag{2}$$

As seen in Figure 6.2 (and as can easily be verified by Euclidean geometry),
the quadrangle $OPRQ$ is a parallelogram. For this reason, definition (2)
is called the *parallelogram law*. Since the sum of two sides of a triangle is
larger than the third side, we get

$$\overline{OR} \le \overline{OP} + \overline{PR} = \overline{OP} + \overline{OQ}$$

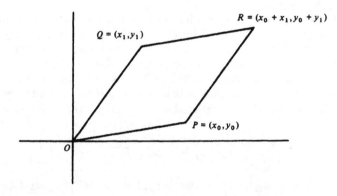

Figure 6.2

where $\overline{AB}$ denotes the length of the segment with endpoints A and B. From the geometric meaning of the definition of the distance we conclude that

$$\sqrt{(x_0 + x_1)^2 + (y_0 + y_1)^2} \leq \sqrt{x_0^2 + y_0^2} + \sqrt{x_1^2 + y_1^2}. \qquad (3)$$

We call this inequality the *triangle inequality*. In deriving it we have used the correspondence between the points (x,y) defined as pairs of real numbers and the points of Euclidean plane, as well as two geometric theorems—the Pythagorean theorem and another theorem on triangles. One also can prove (3) directly, without using any geometry. This will be done, more generally, in Section 3.

We refer to real numbers as *scalars*. We shall now define *multiplication by scalars*. Given a point (x_0,y_0) and a scalar λ, we define the multiplication of (x_0,y_0) by the scalar λ:

$$\lambda(x_0,y_0) = (\lambda x_0, \lambda y_0).$$

It is clear that

$$\sqrt{(\lambda x_0)^2 + (\lambda y_0)^2} = |\lambda|\sqrt{x_0^2 + y_0^2}. \qquad (4)$$

Geometrically, the last relation means that if $P = (x_0,y_0)$, $Q = (\lambda x_0, \lambda y_0)$, $O = (0,0)$, then $\overline{OQ} = |\lambda|\overline{OP}$. The position of Q in the Euclidean plane is given in Figure 6.3.

One introduces for the real plane a terminology that fits in with that employed in geometry. Thus, the *real line determined by two points* (x_0,y_0) and (x_1,y_1) is the set of all points

$$\lambda(x_0,y_0) + (1 - \lambda)(x_1,y_1) \qquad (\lambda \text{ a real number}).$$

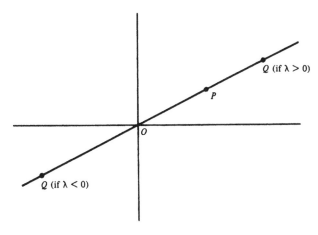

Figure 6.3

The *line segment* connecting (x_0, y_0) to (x_1, y_1) is defined as the set of all points (x, y) of the form

$$t(x_1, y_1) + (1 - t)(x_0, y_0), \qquad 0 \le t \le 1.$$

Point (x_0, y_0) is called the *initial endpoint* of the line segment, and point (x_1, y_1) is called the *terminal endpoint* of the line segment.

The set of all points (x, y) satisfying

$$\sqrt{(x - x_0)^2 + (y - y_0)^2} < r \qquad (r > 0) \tag{5}$$

is called the *open disc* with *center* (x_0, y_0) and *radius* r. It is also called an *r-neighborhood* of (x_0, y_0).

The set of all points (x, y) satisfying

$$\sqrt{(x - x_0)^2 + (y - y_0)^2} \le r$$

is called the *closed disc* with center (x_0, y_0) and radius r.

Let A and B be sets of elements. If every element of A is in B, then we say that the set A is *contained in* the set B and that B contains A; we then write $A \subset B$ or $B \supset A$.

We shall denote the plane by R^2. A set G in R^2 is said to be *open* if for every point P_0 of G there is an open disc with center P_0 that is contained in G.

EXAMPLE. *An open disc is an open set.*

Proof. Let G be an open disc defined by (5), and let (x_1, y_1) belong to G. Then the number

$$\delta = \sqrt{(x_1 - x_0)^2 + (y_1 - y_0)^2}$$

satisfies $\delta < r$. Let G_0 be the open disc defined by

$$\sqrt{(x - x_1)^2 + (y - y_1)^2} < r - \delta.$$

This disc is contained in G, since, by the triangle inequality,

$$\sqrt{(x - x_0)^2 + (y - y_0)^2}$$
$$= \sqrt{[(x - x_1) + (x_1 - x_0)]^2 + [(y - y_1) + (y_1 - y_0)]^2}$$
$$\le \sqrt{(x - x_1)^2 + (y - y_1)^2} + \sqrt{(x_1 - x_0)^2 + (y_1 - y_0)^2}$$
$$< (r - \delta) + \delta = r$$

if (x, y) belongs to G_0. We have thus proved that for any point (x_1, y_1) of G there is an open disc with center (x_1, y_1) contained in G. Hence G is an open set. The geometric description of G and G_0 is given in Figure 6.4.

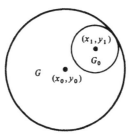

Figure 6.4

The *complement* of a subset G of R^2 is the set of all points in R^2 not contained in G.

Given two sets of elements, A and B, we define the *intersection* $A \cap B$ to be the set consisting of all the elements that belong to both A and B. The *union* $A \cup B$ is defined as the set consisting of all the elements that belong to either A or B.

Similarly, one defines the intersection $A_1 \cap A_2 \cap \ldots \cap A_m$ as the set of all elements that belong to each A_i, $1 \leq i \leq m$, and the union $A_1 \cup A_2 \cup \ldots \cup A_m$ as the set of all the elements that belong to at least one of the sets A_i, $1 \leq i \leq m$.

A set G in R^2 is called a *closed set* if its complement is an open set.

Let G be a subset of R^2. A point P is called an *interior point* of G if there is an open disc with center P that is contained in G. A point P is called an *exterior point* of G if there is an open disc with center P that is contained in the complement of G. Finally, a point P is called a *boundary point* of G if any open disc with center P contains at least one point of G and at least one point of the complement of G. Observe that P is a boundary point of G if, and only if, it is neither an interior point of G nor an exterior point of G.

The set of all interior points of G is called the *interior* of G. Thus, G is an open set if, and only if, the interior of G coincides with G. The set of all the exterior points of G is called the *exterior* of G. Finally, the set of all the boundary points of G is called the *boundary* of G; it is denoted by ∂G.

The set $G \cup \partial G$ is called the *closure* of G.

Let $l_1, \ldots, l_m$ be line segments such that, for each $k = 1, 2, \ldots, m - 1$, the terminal endpoint of l_k is the initial endpoint of l_{k+1}. Then the collection $l = \{l_1, \ldots, l_m\}$ is called a *polygonal line*. The initial endpoint P of l_1 is called the *initial endpoint* of l, and the terminal endpoint Q of l_m is called the *terminal endpoint* of l. We say that l *connects* P to Q.

A subset G of R^2 is said to be *polygonally connected* (or, briefly, *connected*) if for every two points P and Q of G, there is a polygonal line contained in G that connects P to Q.

An open set that is polygonally connected is called a *domain*. If D is a domain, then the closure of D, that is, $D \cup \partial D$, is called a *closed domain*. Any set G satisfying $G \supset D$, $G \subset D \cup \partial D$, where D is a domain, is called a *region*.

A subset G of R^2 is said to be *bounded* if there exists a positive number M such that

$$\sqrt{x^2 + y^2} < M \qquad \text{for all } (x,y) \text{ in } G.$$

This means that G is contained in the open disc with center $(0,0)$ and radius M. If G is not bounded, then it is said to be *unbounded*.

Suppose G is bounded. Consider the set W of all numbers

$$\sqrt{(x_0 - x_1)^2 + (y_0 - y_1)^2},$$

where (x_0,y_0) and (x_1,y_1) vary in G. Since G is bounded, W is a bounded set. It therefore has a supremum. This supremum is called the *diameter* of G.

The notation $S = \{(x,y);\ 2 < x^2 + y^2 < 5\}$ means that S is the set consisting of all points (x,y) having the property $2 < x^2 + y^2 < 5$. Similar notation is used in defining sets by other properties.

PROBLEMS

1. Show that a closed disc is a closed set.

2. Show that the boundary of a set is a closed set.

3. Show that the closure of a set is a closed set.

4. Prove that if G is a closed set, then the closure of G coincides with G, that is, ∂G is contained in G.

5. If $F \supset G$, then the closure of F contains the closure of G.

6. If F is closed and $F \supset G$, then F contains the closure of G.

7. Show that the set defined by $ax + by + c < 0$ (a,b,c are real numbers) is a domain.

8. Let $f(x)$ be a continuous function in an interval $a \le x \le b$. Prove that the graph of f $\{(x,f(x));\ a \le x \le b\}$ is a closed set. Is the set $\{(x,f(x));\ a < x < b\}$ open?

9. Is the set $\{(x,y);\ xy > 0\}$ a domain?

10. Show that the set $\{(x,y);\ 1 < x^2 + y^2 < 2\}$ is a domain.

11. Prove that the union of a finite number of open sets is an open set.

12. Prove that the intersection of a finite number of open sets is an open set.

13. Prove that the union of a finite number of closed sets is a closed set.

14. Prove that the intersection of a finite number of closed sets is a closed set.

15. Is the intersection of two domains again a domain?

16. Is the union of two domains again a domain?

17. Find the diameter of the set $\{(x,y); 0 \le x \le 1, 0 \le y \le 2\}$.

2. CONVERGENCE IN THE PLANE

Consider a sequence of points $P_n = (x_n, y_n)$. Suppose there exists a point $Q = (a,b)$ such that

$$\sqrt{(x_n - a)^2 + (y_n - b)^2} \to 0 \qquad \text{if } n \to \infty. \tag{1}$$

Then we say that $\{P_n\}$ is a *convergent sequence* and that it converges to Q. We write

$$\lim_{n \to \infty} P_n = Q, \qquad \text{or } P_n \to Q \qquad \text{if } n \to \infty.$$

Note that $\{P_n\}$ converges to Q if, and only if, every δ-neighborhood of Q contains all the points P_n with n sufficiently large.

THEOREM 1.   *A sequence $\{(x_n, y_n)\}$ is convergent to (a,b) if, and only if,* $\lim x_n = a, \lim y_n = b$.

THEOREM 2.   *If a sequence $\{(x_n, y_n)\}$ is convergent to (a,b), then every subsequence is convergent to (a,b).*

THEOREM 3.   *If* $\lim_{n} (x_n, y_n) = (a,b), \lim_{n} (\bar{x}_n, \bar{y}_n) = (\bar{a}, \bar{b}),$ *then*

$$\lim_{n}[(x_n, y_n) + (\bar{x}_n, \bar{y}_n)] = (a,b) + (\bar{a}, \bar{b}),$$

$$\lim_{n} \lambda(x_n, y_n) = \lambda(a,b) \qquad (\lambda \text{ scalar}).$$

The proofs of these theorems are left to the student.

DEFINITION.   Let $\{P_n\}$ be a sequence of points. A point Q is called a *limit point* (a *cluster point*, or a *point of accumulation*) of $\{P_n\}$ if there exists a subsequence $\{P_{n_k}\}$ that converges to Q.

DEFINITION.   Let G be a subset of R^2, and let $Q = (a,b)$ be a point in R^2. Suppose that for any $\varepsilon > 0$ there is a point $P = (x,y)$ in G such that $P \neq Q$ and

$$\sqrt{(x-a)^2 + (y-b)^2} < \varepsilon.$$

Then we say that Q is a *limit point* (a *cluster point*, or a *point of accumulation*) of the set G.

THEOREM 4.   *A point Q is a limit point of a set G, if, and only if, there exists a sequence $\{P_n\}$ of mutually distinct points (that is, $P_n \neq P_m$ if $n \neq m$) of G such that* $\lim P_n = Q$.

The proof is similar to the proof of Theorem 1 in Section 1.8, and is left to the student.

The set G' of all the limit points of a set G is called the *derived set* of G.

THEOREM 5.   *For any subset G of R^2,*

$$G \cup G' = G \cup \partial G. \tag{2}$$

Proof. Let $Q \in G'$. By Theorem 4 there exists a sequence of points P_n in G such that $P_n \to Q$ if $n \to \infty$. It follows that every open disc with center Q contains points of G. Hence Q is either an interior point of G or a boundary point of G. We have thus proved that $G \cup G' \subset G \cup \partial G$.

Next let $Q \in \partial G$ and $Q \notin G$. Since $Q \in \partial G$, for any positive integer n there is a point P_n of G lying in the $(1/n)$-neighborhood of Q. $P_n \neq Q$ since $Q \notin G$. It follows that Q belongs to G'. We thus have proved that if $Q \in \partial G$ and $Q \notin G$, then $Q \in G'$. This implies that $G \cup \partial G \subset G \cup G'$.

In Problem 4 of Section 1 it was asserted that if G is closed, then $\partial G \subset G$. On the other hand, if $\partial G \subset G$, then G coincides with its closure and, therefore, by the assertion of Problem 3 of Section 1, G is closed. Combining these remarks with Theorem 5, we can state:

THEOREM 6.   *A set G is closed, if and only if, $\partial G \subset G$, or if, and only if, $G' \subset G$.*

COROLLARY.   *Let G be a closed set in R^2. If $\{P_n\}$ is a sequence of points of G such that $Q = \lim P_n$ exists, then $Q \in G$.*

Indeed, if $Q = P_n$ for some n, then $Q = P_n \in G$. If, on the other hand, $P_n \neq Q$ for all n, then Q is a limit point of the set G, that is, $Q \in G'$, hence $Q \in G$.

THEOREM 7 (Bolzano–Weierstrass Theorem for Sequences). *Every bounded sequence of points in R^2 has at least one limit point.*

Proof. Let $\{P_n\}$ be a bounded sequence of points, and write $P_n = (x_n, y_n)$. Then there is a constant M such that

$$\sqrt{x_n^2 + v_n^2} \leq M \qquad \text{for all } n.$$

It follows that $|x_n| \leq M$. By the Bolzano–Weierstrass theorem for sequences of real numbers, there exists a subsequence $\{x_{n_k}\}$ of $\{x_n\}$ that is convergent to some number a. Since $|y_{n_k}| \leq M$, we again can apply the Bolzano–Weierstrass theorem and conclude that there is a subsequence $\{y_{n_{k'}}\}$ of $\{y_{n_k}\}$ that is convergent to some number b. Since also $\{x_{n_{k'}}\}$ is convergent to a, Theorem 1 implies that

$$(x_{n_{k'}}, y_{n_{k'}}) \to (a, b) \qquad \text{as } k \to \infty.$$

This completes the proof.

THEOREM 8 (Bolzano–Weierstrass Theorem for Sets). *Every bounded, infinite set of points in R^2 has at least one limit point.*

Proof. Let G be a bounded, infinite set of points. We then can find a sequence of mutually distinct points P_n in G. By Theorem 7, $\{P_n\}$ has a limit point Q. But then Q is also a limit point of G.

THEOREM 9. *Let $I_1, I_2, I_3, \ldots$ be a sequence of nonempty closed and bounded sets such that $I_1 \supset I_2 \supset I_3 \supset \ldots$. There then exists at least one point Q contained in all the sets I_n. If, furthermore, the sequence of the diameters $d(I_n)$ of I_n converges to 0, then Q is unique.*

This theorem is called the *nested sequence property*. It extends Theorem 2 of Section 1.8.

Proof. Take any point P_n in I_n. Then $\{P_n\}$ is a bounded sequence and, therefore, by the Bolzano–Weierstrass theorem, there is a subsequence $\{P_{n_k}\}$ that is convergent to some point Q. For fixed m, the points P_{n_k} (which belong to I_{n_k}) belong to I_m if $m < n_k$. Since I_m is a closed set, $Q = \lim P_{n_k}$ also belongs to I_m (by the corollary to Theorem 6). Since this is true for any m, Q belongs to all the sets I_m.

Suppose now that $d(I_m) \to 0$ as $m \to \infty$. If R is another point contained in all the sets I_m, and if we write $Q = (a,b)$, $R = (c,d)$, then

$$\sqrt{(a - c)^2 + (b - d)^2} \leq d(I_m) \to 0 \qquad \text{if } m \to \infty.$$

It follows that $a = c$, $b = d$, that is, $R = Q$.

DEFINITION.   A sequence of points $P_n = (x_n, y_n)$ is called a *Cauchy sequence* if for any $\varepsilon > 0$ there is a positive integer n_0 such that

$$\sqrt{(x_n - x_m)^2 + (y_n - y_m)^2} < \varepsilon \qquad \text{if } n \geq n_0, m \geq n_0. \tag{3}$$

THEOREM 10.   *A sequence $\{P_n\}$ of points in R^2 is convergent if, and only if, it is a Cauchy sequence.*

Proof.   Let $P_n = (x_n, y_n)$. If $\{P_n\}$ is a Cauchy sequence, then $\{x_n\}$ and $\{y_n\}$ are also Cauchy sequences. By Theorem 1 of Section 5.1, the sequences $\{x_n\}$ and $\{y_n\}$ are then convergent. Theorem 1 of Section 1 now shows that $\{P_n\}$ is a convergent sequence. Conversely, if $\lim P_n = Q$ and $Q = (a,b)$, then for any $\varepsilon > 0$ there is an n_0 such that

$$\sqrt{(x_k - a)^2 + (y_k - b)^2} < \frac{\varepsilon}{2} \qquad \text{if } k \geq n_0.$$

Using the triangle inequality we get

$$\sqrt{(x_n - x_m)^2 + (y_n - y_m)^2}$$
$$= \sqrt{[(x_n - a) + (a - x_m)]^2 + [(y_n - b) + (b - y_m)]^2}$$
$$< \frac{\varepsilon}{2} + \frac{\varepsilon}{2} = \varepsilon \qquad \text{if } n \geq n_0, m \geq n_0.$$

Thus $\{P_n\}$ is a Cauchy sequence.

DEFINITION.   Let G be a subset in R^2. Suppose we are given a collection $\{\Omega_\alpha\}$ of subsets of R^2 such that every point of G belongs to at least one of the sets Ω_α. We then say that the collection $\{\Omega_\alpha\}$ forms a *covering* of the set G. If all the sets Ω_α are open sets, we speak of *open covering*.

EXAMPLE.   Let Ω_α be an open disc of radius r_α and center α, where α is a point (x,y) varying over the set G. Then $\{\Omega_\alpha\}$ is an open covering of G.

THEOREM 11 (Heine–Borel Theorem). *Let G be closed, bounded set in R^2 and let $\{\Omega_\alpha\}$ form an open covering of G. There then exists a finite number of sets $\Omega_{\alpha_1},...,\Omega_{\alpha_n}$, which form a covering of G.*

Proof. Suppose the assertion is false, that is, no finite subcollection of $\{\Omega_\alpha\}$ can cover G. Since G is a bounded set, it is contained in a square

$$S_0 = \{(x,y); -R \le x \le R, -R \le y \le R\}.$$

Divide S_0 into four closed squares S_{11}, S_{12}, S_{13}, S_{14} by the straight lines $x = 0$, $y = 0$. Thus

$$S_{11} = \{(x,y); -R \le x \le 0, -R \le y \le 0\},$$

$$S_{12} = \{(x,y); -R \le x \le 0, \quad 0 \le y \le R\},$$

$$S_{13} = \{(x,y); \quad 0 \le x \le R, -R \le y \le 0\},$$

$$S_{14} = \{(x,y); \quad 0 \le x \le R, \quad 0 \le y \le R\}.$$

Not all of the sets $G \cap S_{11}$, $G \cap S_{12}$, $G \cap S_{13}$, $G \cap S_{14}$ can be covered by a finite collection of sets Ω_α; otherwise, G could be covered by a finite collection. Thus, there is at least one set, say, $G \cap S_{11}$, which cannot be covered by a finite number of sets Ω_α. Write $G_1 = G \cap S_{11}$.

We again divide S_{11} into four equal and closed squares, S_{21}, S_{22}, S_{23}, and S_{24}. Arguing as before, we conclude that at least one of the sets $G_1 \cap S_{2i}$ cannot be covered by a finite number of the sets Ω_α. Denote this set by G_2.

Proceeding in this way step by step, we construct a sequence $\{G_j\}$ of closed, bounded sets such that $G_1 \supset G_2 \supset G_3 \supset ...$, and such that

each G_j cannot be covered by a finite number of sets Ω_α.   (4)

The diameter $d(G_j)$ of G_j converges to 0 as $j \to \infty$.

Applying the nested sequence property, we conclude that there exists a unique point Q belonging to all the sets G_j. Since Q then belongs also to G, there is a set Ω_{α_0} containing Q. Since Ω_{α_0} is open, there is a δ-neighborhood of Q that is contained in Ω_{α_0}, that is, if

$$\sqrt{(x - \alpha)^2 + (y - \beta)^2} < \delta,$$

then $(x,y) \in \Omega_{\alpha_0}$, where $Q = (\alpha,\beta)$.

But since every point (x,y) in G_j satisfies

$$\sqrt{(x - \alpha)^2 + (y - \beta)^2} \le d(G_j) < \delta \qquad \text{if } j \ge j_0$$

for some j_0 sufficiently large, we see that $G_j \subset \Omega_{\alpha_0}$ if $j \ge j_0$. This contradicts (4).

Remark. The assertion of the Heine–Borel theorem is generally false for sets G that are not both bounded and closed. Thus, for the bounded set

$G = \{(x,y); 0 < x < 1, 0 < y < 1\}$,   the   sets $\Omega_n = \{(x,y); (1/n) < x < (2/n),$ $0 < y < 1\}$ form an open covering of G having no finite number of subsets that cover G. For the closed set $G = \{(x,y); 1 \le x < \infty, 1 \le y < \infty\}$ the sets $\Omega_n = \{(x,y); x^2 + y^2 < n^2\}$ form an open covering of G having no finite number of subsets that cover G.

PROBLEMS

1. Find the derived sets for each of the following sets:
    (a) $((-1)^n, 1/m)$ where $m,n = 1,2,3,....$
    (b) $(1/m, 1/n)$ where $m,n = 2,4,6,....$
    (c) $\left(\dfrac{(-1)^n}{n}, \dfrac{(-1)^m}{m} \right)$ where $m,n = 1,2,3,....$

2. One defines $\lim P_n = \infty$ if $P_n = (x_n, y_n)$ and $x_n^2 + y_n^2 \to \infty$ as $n \to \infty$. Let $x_n \le x_{n+1}, y_n \le y_{n+1}$ for all n. Prove that either $\lim P_n$ exists, or $\lim P_n = \infty$.

3. Let $P_n = (x_n, y_n)$, $x_n \le x_{n+1}$, $y_n \ge y_{n+1}$. Prove that either $\lim P_n$ exists or $\lim P_n = \infty$.

4. One defines a *countable union* $G_1 \cup G_2 \cup ... \cup G_n \cup ...$, or briefly $\bigcup\limits_{n=1}^{\infty} G_n$, to be the set of all points that belong to at least one of the sets G_n. Prove that a countable union of open sets is open.

5. One defines a *countable intersection* $G_1 \cap G_2 \cap ... \cap G_n \cap ...$, or briefly $\bigcap\limits_{n=1}^{\infty} G_n$, to be the set of all points that belong to all the sets G_n. Give an example in which a countable intersection of open sets is not open.

6. Show that every convergent sequence is bounded.

7. Find the shortest distance from point (x_0, y_0) to the set of points satisfying $ax + by + c = 0$ (a and b are not both zero).

8. Find the shortest distance from a point (x_0, y_0) to the points of the circle $x^2 + y^2 = 1$.

3. THE *n*-DIMENSIONAL SPACE

Let n be a fixed positive integer. We consider the set of all elements x having the form $(x_1, x_2, ..., x_n)$ where $x_1, x_2, ..., x_n$ are any real numbers; we write $x = (x_1, x_2, ..., x_n)$. Two elements $x = (x_1, x_2, ..., x_n)$ and $y = (y_1, y_2, ..., y_n)$ are said to be *equal* if, and only if, $x_1 = y_1, x_2 = y_2, ..., x_n = y_n$. We call these

elements also *n-dimensional points*, or, briefly, *points*. The set of all the *n*-dimensional points is called the *n-dimensional real space*. It is denoted by R^n.
The *zero point* of R^n is the point $(0,0,...,0)$. We denote it also by 0.
By a *scalar* we shall mean any real number.
We define *addition* of two points $x = (x_1, x_2,...,x_n)$ and $y = (y_1, y_2,...,y_n)$ by

$$x + y = (x_1 + y_1, x_2 + y_2,...,x_n + y_n).$$

Multiplication by a scalar is defined by

$$\lambda x = (\lambda x_1, \lambda x_2,...,\lambda x_n) \qquad (\lambda \text{ scalar}).$$

The *distance* between two points x and y is defined by

$$\|x - y\| = \sqrt{\sum_{i=1}^{n} (x_i - y_i)^2}.$$

Note that

$$\|\lambda x\| = |\lambda| \, \|x\|.$$

Notice that the above definitions reduce to definitions given in Section 1 when $n = 2$.

The *scalar product*, or *inner product*, of two points $x = (x_1,...,x_n)$ and $y = (y_1,...,y_n)$ is defined by

$$x \cdot y = \sum_{i=1}^{n} x_i y_i.$$

One often denotes the scalar product of x and y also by (x,y).

The number $\|x\| = \sqrt{(x,x)} = \sqrt{\sum_{i=1}^{n} x_i^2}$ is called the *norm* of x, the *absolute value* of x, or the *length* of x. It is equal to the distance from x to 0. Note that $(x,y) = (y,x)$, $(x' + x'',y) = (x',y) + (x'',y)$, $(x,y' + y'') = (x,y')$ $+ (x,y'')$, and $\lambda(x,y) = (\lambda x,y) = (x,\lambda y)$ (λ scalar).

THEOREM 1.   *The following inequalities hold:*

$$|x \cdot y| \le \|x\| \, \|y\| \qquad \text{(Cauchy's inequality)}; \tag{1}$$

$$\|x + y\| \le \|x\| + \|y\| \qquad \text{(triangle inequality)}. \tag{2}$$

Proof. If $x = 0$, then (1) is obvious. Suppose then that $x \ne 0$. For any scalar λ,

$$0 \le \|\lambda x + y\|^2 = (\lambda x + y, \lambda x + y) = \lambda^2 \|x\|^2 + 2\lambda x \cdot y + \|y\|^2.$$

Taking $\lambda = -x \cdot y/\|x\|^2$, we get

$$0 \le \frac{|x \cdot y|^2}{\|x\|^4} \|x\|^2 - 2\frac{|x \cdot y|^2}{\|x\|^2} + \|y\|^2.$$

Hence

$$-\frac{|x \cdot y|^2}{\|x\|^2} + \|y\|^2 \geq 0.$$

This gives (1).

Next,

$$\|x + y\|^2 = (x + y, x + y) = \|x\|^2 + \|y\|^2 + 2x \cdot y$$

$$\leq \|x\|^2 + \|y\|^2 + 2\|x\| \|y\| = (\|x\| + \|y\|)^2.$$

This gives (2).

The *angle* between two points x and y ($x \neq 0$, $y \neq 0$) is any number θ for which

$$\cos \theta = \frac{x \cdot y}{\|x\| \|y\|}. \tag{3}$$

In view of Cauchy's inequality, the absolute value of the right-hand side in (3) is ≤ 1, so that it can indeed be written as $\cos \theta$ for some θ.

If $x \cdot y = 0$, then we say that x is *orthogonal* to y.

In many problems it is suggestive to use the language of vectors. We therefore introduce the relevant terminology. We shall call a point x in R^n also a *vector* (more precisely, a vector with the initial point 0). The zero point is called the *origin*. Let

$$e_1 = (1,0,0,...,0), \ e_2 = (0,1,0,...,0),..., \ e_n = (0,0,...,0,1).$$

These vectors are orthogonal to each other, and each has length 1. Every vector $x = (x_1, x_2, ..., x_n)$ in R^n can be written in the form

$$x = x_1 e_1 + x_2 e_2 + \cdots + x_n e_n. \tag{4}$$

The number x_i is called the ith *component* of x.

For $n = 3$, one can correspond Figure 6.5 to the decomposition (4). The straight line through P, Q is parallel to the z-axis, and the straight line through Q, R lies in the (x,y)-plane and is parallel to the y-axis. The components of P are x, y, and z.

The set of points x satisfying $\|x - z\| < r$ is called an *open ball* with center z and radius r. It is also called an *r-neighborhood* of z. The *closed ball* with center z and radius r consists of all the points x satisfying $\|x - z\| \leq r$. When $n = 1$, we call an open ball (a closed ball) also an open interval (a closed interval).

A set G in R^n is called an *open* set if for any point x in G there is an open ball with center z that is contained in G. The complement of an open set is called a *closed* set. A *neighborhood* of a point x^0 is any open set containing x^0.

The concepts of the interior of G, the exterior of G, and the boundary ∂G of G are defined precisely as in Section 1, with the word "disc" replaced in each instance by the word "ball."

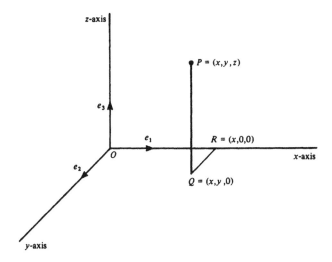

Figure 6.5

The line segment connecting two points x and y is defined to be the set of all points

$$tx + (1 - t)y \qquad \text{where } 0 \le t \le 1.$$

One proceeds to define a polygonal line as in Section 1. The concepts—a polygonally connected set, domain, closed domain, region, bounded set, and diameter of a set—extend almost word for word from R^2 to R^n.

Next we define the concept of convergence. A sequence $\{x^m\}$ of points of R^n is said to be *convergent* if there is a point y in R^n such that

$$\|x^m - y\| \to 0 \qquad \text{if } m \to \infty.$$

We then say that $\{x^m\}$ *converges to* y, and write

$$\lim_{m \to \infty} x^m = y, \qquad \text{or } x^m \to y \qquad \text{if } m \to \infty.$$

It is easily seen that if $x^m \to y$ as $m \to \infty$, then $x^{m_k} \to y$ as $k \to \infty$ for any subsequence $\{x^{m_k}\}$ of $\{x^m\}$. Also, if $x^m \to y$, $z^m \to u$ as $m \to \infty$, then

$$
\begin{aligned}
x^m + z^m &\to y + u & &\text{as } m \to \infty, \\
\lambda x^m &\to \lambda y & &\text{as } m \to \infty \qquad (\lambda \text{ scalar}).
\end{aligned}
\tag{5}
$$

A sequence $\{x^m\}$ of points in R^n is called a *Cauchy sequence* if for any $\varepsilon > 0$ there is a positive integer n_0 such that

$$\|x^m - x^k\| < \varepsilon \qquad \text{if } m \ge n_0, k \ge n_0.$$

We write this condition briefly in the form

$$\lim_{\substack{m \to \infty \\ k \to \infty}} \|x^m - x^k\| = 0.$$

Let $x^m = (x_1{}^m, \ldots, x_n{}^m)$, $y = (y_1, \ldots, y_n)$. Then

$$\|x^m - y\|^2 = (x_1{}^m - y_1)^2 + \cdots + (x_n{}^m - y_n)^2.$$

From this relation we easily deduce the following theorem:

THEOREM 2. $\lim_m x^m = y$ *if, and only if,* $\lim_m x_i{}^m = y_i$ *for* $i = 1, 2, \ldots, n$.

Using Theorem 2 we can proceed analogously to the proof of Theorem 10, Section 2, and establish the following:

THEOREM 3.   *A sequence* $\{x^m\}$ *is convergent if, and only if, it is a Cauchy sequence.*

The concepts of a limit point of a sequence in R^n, and of a limit point of a set in R^n, are the same as in R^2. Theorems 4 and 5 of Section 2 remain valid, with the same proofs. Thus, we have:

A point y is a limit point of a set G in R^n if, and only if, there exists a sequence of mutually distinct points of R^n that is convergent to y.

For any set G in R^n, $G \cup G' = G \cup \partial G$. A set G is closed if, and only if, $\partial G \subset G$, or if, and only if, $G' \subset G$. If G is closed, $x^m \in G$ and $\lim_m x^m = y$, then $y \in G$.

THEOREM 4 (Bolzano–Weierstrass Theorem for Sequences).   *Every bounded sequence in* R^n *has a limit point.*

Proof. Let $x^m = (x_1{}^m, \ldots, x_n{}^m)$ be the points of a bounded sequence. Using the Bolzano–Weierstrass theorem for sequences of real numbers, we first extract from $\{x_1{}^m\}$ a convergent subsequence $\{x_1{}^{m,1}\}$. Next we extract from $\{x_2{}^{m,1}\}$ a convergent subsequence $\{x_2{}^{m,2}\}$, and so on. Finally, we extract from $\{x_n{}^{m,n-1}\}$ a convergent subsequence $\{x_n{}^{m,n}\}$. Each of the sequences $\{x_j{}^{m,n}\}$ ($1 \leq j \leq n$) is convergent. Hence, by Theorem 2, $\{x^{m,n}\}$ also is convergent.

The Bolzano–Weierstrass theorem for bounded, infinite sets in R^n is valid, with the same proof as in R^2. The nested sequence property stated (in R^2) in Theorem 9 of Section 2 is also true in R^n, with the same proof. Finally, the concepts of covering and of open covering of a set in R^2 extend, word for word, to sets in R^n. We also have:

THEOREM 5 (Heine–Borel Theorem).   *Let G be a closed, bounded set in R^n, and let $\{\Omega_\alpha\}$ form an open covering of G. Then there exists a finite number of sets $\Omega_{\alpha_1},...,\Omega_{\alpha_k}$ that form a covering of G.*

The proof is analogous to the proof given in Section 2 in the case of $n = 2$.

PROBLEMS

1. Prove that $\|x - y\|^2 + \|x + y\|^2 = 2\|x\|^2 + 2\|y\|^2$.

2. If $\sum a_n^2$ and $\sum b_n^2$ are convergent series, then the series $\sum a_n b_n$ is absolutely convergent. [*Hint:* Use Cauchy's inequality.]

3. Find when $x \cdot y = \|x\| \, \|y\|$.

4. Find when $\|x + y\| = \|x\| + \|y\|$.

5. Prove the assertions in (5).

6. Find the smallest distance from the point $(1,2,3)$ to the points $(t, 2t + 1, 2 - t)$ where $0 \le t \le 1$.

7. Prove the following *theorem of Dini*: Let $\{f_n(x)\}$ be a monotone increasing sequence of continuous functions on a closed, bounded interval $a \le x \le b$. If the pointwise limit $f(x) = \lim f_n(x)$ exists and is a continuous function in $[a,b]$, then the convergence is uniform. [*Hint:* Let $\varepsilon > 0$. For any x in $[a,b]$ there is an $n_0 = n_0(x)$ such that $0 \le f(x) - f_{n_0}(x) < \varepsilon$. Since f and f_{n_0} are continuous, $0 \le f(y) - f_{n_0}(y) < \varepsilon$ also for all y satisfying $a \le y \le b$, $|y - x| < \delta$ where $\delta = \delta(x)$ is a positive number. The intervals $I_x = (x - \delta, x + \delta)$ form an open covering of $[a,b]$. Use the Heine–Borel theorem in R^1.]

8. Find all the limit points of each of the following sets:

    (a) $\left(\dfrac{1}{m}, \dfrac{1}{n}, \dfrac{1}{k}\right)$, where m, n, k are positive integers.

    (b) $\lambda x + \mu y$ where x,y are fixed points in R^n and λ, μ are any rational numbers.

9. Let K be a closed set in R^n and let G be an open set in R^n. Denote by L the set of all points that belong to K but not to G. Prove that L is a closed set.

4. CONTINUOUS FUNCTIONS

Let S be a subset of R^n. Suppose we are given a rule that assigns to every point $x = (x_1,...,x_n)$ of S a number u. Let us denote this rule by f and write $u = f(x_1,...,x_n)$ or $u = f(x)$. We call f a *function* and the set of pairs $(x,f(x))$ (when x varies in S) the *graph* of the function f. The set S is called the *domain* of f. Finally, the set of all numbers $f(x)$, when x varies in S, is called the *range* of f.

When we speak of *a function $f(x)$ defined on a set S*, we mean that f is a function with domain S.

We write a function f also in the form $f(x)$, or $f(x_1,...,x_n)$. When $n = 2$, we usually write $f(x,y)$ instead of $f(x_1,x_2)$, and when $n = 3$, we often write $f(x,y,z)$ instead of $f(x_1,x_2,x_3)$.

Let $f(x)$ be a function defined on a set S of R^n and let ξ be a point of S. Suppose that for any $\varepsilon > 0$ there exists a $\delta > 0$ such that

$$|f(x) - f(\xi)| < \varepsilon \qquad \text{if } \|x - \xi\| < \delta \text{ and } x \in S. \tag{1}$$

Then we say that $f(x)$ is *continuous* at ξ. If $f(x)$ is continuous at each point of S, then it is said to be continuous on S, or in S.

For $n = 2$ the condition (1) reads:

$$|f(x,y) - f(\xi,\eta)| < \varepsilon \quad \text{if } \sqrt{(x - \xi)^2 + (y - \eta)^2} < \delta \text{ and } (x,y) \in S. \tag{2}$$

THEOREM 1.   *A function $f(x)$ defined on a subset S of R^n is continuous at a point ξ of S if, and only if, for any sequence $\{x^m\}$ of points of S, satisfying $\lim\limits_{m} x^m = \xi$, the limit $\lim\limits_{m} f(x^m)$ exists and is equal to $f(\xi)$.*

This theorem, for the case $n = 1$, was given in Theorems 1 and 2 of Section 2.1. The proof for $n \geq 1$ is very similar to the proof in the case of $n = 1$, and therefore it is omitted.

EXAMPLE.   Consider the function

$$f(x,y) = \begin{cases} \dfrac{xy}{x^2 + y^2} & \text{if } (x,y) \neq (0,0), \\ 0 & \text{if } (x,y) = (0,0). \end{cases}$$

For any real λ,

$$f(x,\lambda x) = \frac{\lambda x^2}{x^2 + \lambda^2 x^2} = \frac{\lambda}{1 + \lambda^2}.$$

Hence, for any sequence $\{(x_m, \lambda x_m)\}$ converging to $(0,0)$, $f(x_m, \lambda x_m) \to \lambda/(1 + \lambda^2)$. Theorem 1 implies that $f(x, y)$ is not continuous at $(0,0)$. Note that the functions $f(x,0)$ and $f(0, y)$ are continuous at 0 (since they are equal to 0 for all x and y, respectively). Thus this example shows that it is possible to have a situation where $f(x_0, y)$ is continuous at $y = y_0$ and $f(x, y_0)$ is continuous at $x = x_0$, but $f(x, y)$ is discontinuous at (x_0, y_0).

THEOREM 2.   *Let $f(x)$ and $g(x)$ be functions defined on a set S of R^n, and continuous at a point ξ of S. Then*

   (i) *$f(x) \pm g(x)$ are continuous at ξ.*

   (ii) *$f(x)g(x)$ is continuous at ξ.*

   (iii) *If $g(x) \neq 0$ for all $x \in S$, then $f(x)/g(x)$ is defined on S and is continuous at ξ.*

For $n = 1$, the assertions of Theorem 2 reduce to Theorems 1–3 of Section 2.2. The proof for $n \geq 1$ is very similar to the proof for $n = 1$.

Let $g(x)$ be a function defined on a set S of R^n. Denote by R the range of g. Let $f(t)$ be a function defined on a set I of R^1 that contains R. Then we can form a new function h with domain S by $h(x) = f(g(x))$. We call it the *composition* of f with g and we write $h = f \circ g$.

THEOREM 3.   *Let $h = f \circ g$ be defined as above. If $g(x)$ is continuous at a point ξ of S, and if $f(t)$ is continuous at the point $g(\xi)$, then $h(x)$ is continuous at the point ξ.*

The proof is similar to the proof of Theorem 4 in Section 2.2, in which the case $n = 1$ was considered.

EXAMPLE 1.   If $f(x)$ is continuous on a set S of R^n, then $|f(x)|$ is also continuous on S.

EXAMPLE 2.   If $f(x)$ is a continuous function on a set S of R^n and if $f(x) \geq 0$ for all $x \in S$, then the function $\sqrt{f(x)}$ is also continuous on S.

A function $f(x)$ defined on a set S of R^n is said to be *bounded above* if there is a real number H such that

$$f(x) \leq H \qquad \text{for all } x \in S.$$

$f(x)$ is said to be *bounded below* if there is a real number K such that

$$f(x) \geq K \qquad \text{for all } x \in S.$$

Finally, $f(x)$ is said to be *bounded* if $|f(x)|$ is bounded above.

Suppose $f(x)$ is bounded above. Then the range R of f is a set (in R^1) bounded above. The supremum M of this set is called the *supremum* of f, or the *least upper bound* of f, and it is denoted by

$$\sup_{x \in S} f(x), \quad \text{or l.u.b.} \ \sup_{x \in S} f(x).$$

M satisfies two properties:

$$f(x) \leq M \quad \text{if } x \in S; \tag{3}$$

$$\text{for any } \varepsilon > 0, \quad f(\bar{x}) > M - \varepsilon \quad \text{for some } \bar{x} \text{ in } S. \tag{4}$$

Suppose now that there exists a point y in S such that $f(x) \leq f(y)$ for all $x \in S$. Then we say that f *has a maximum on* S, and that the maximum on S is assumed at the point y. The number $f(y)$ is called the *maximum of f on S*. It is equal to the supremum of f.

Similarly, one defines the *infimum*, or *greatest lower bound*, of f. It is the number that is the infimum of the range of f. We denote it by

$$\inf_{x \in S} f(x), \quad \text{or g.l.b.} \ \inf_{x \in S} f(x).$$

If there is a point z in S such that $f(x) \geq f(z)$ for all x in S, then we say that f *has a minimum $f(z)$ on* S, and that the minimum is assumed at the point z.

THEOREM 4.   *Let $f(x)$ be a continuous function on a closed, bounded set S of R^n. Then $f(x)$ is bounded.*

Proof.   Suppose $|f(x)|$ is not bounded. Then for any positive integer m there is a point x^m in S such that

$$|f(x^m)| > m. \tag{5}$$

By the Bolzano–Weierstrass theorem, there exists a subsequence $\{x^{m'}\}$ of $\{x^m\}$ that is convergent to some point ξ. Since S is a closed set, ξ belongs to S. From the continuity of $|f(x)|$ at ξ and from Theorem 1 it follows that

$$\lim |f(x^{m'})| = |f(\xi)|.$$

This implies that $\{|f(x^{m'})|\}$ is a bounded sequence, thus contradicting (5).

THEOREM 5.   *Let $f(x)$ be a continuous function on closed, bounded set S of R^n. Then $f(x)$ assumes both its maximum and its minimum.*

Proof.   By Theorem 4, $f(x)$ is bounded above. It therefore has a supremum. Denote the supremum by M. In view of (3) and (4), for any positive integer j there is a point x^j in S such that

$$M - \frac{1}{j} < f(x^j) \leq M. \tag{6}$$

By the Bolzano–Weierstrass theorem there exists a subsequence $\{x^{j'}\}$ of $\{x^j\}$ that is convergent to some point y. Since S is a closed set, $y \in S$. The continuity of $f(x)$ at y then implies that

$$\lim f(x^{j'}) = f(y).$$

Combining this relation with (6), we find that $f(y) = M$. Since $M \geq f(x)$ for all $x \in S$, $f(y)$ is the maximum of $f(x)$ in S.

The proof that $f(x)$ has a minimum in S is similar.

A function $f(x)$, defined on a subset S of R^n, is said to be *uniformly continuous* on S if for any $\varepsilon > 0$ there is a $\delta > 0$ such that

$$|f(x) - f(y)| < \varepsilon \qquad \text{whenever } \|x - y\| < \delta, \, x \in S, \, y \in S. \qquad (7)$$

THEOREM 6.   *Any function f continuous on a closed, bounded set S in R^n is uniformly continuous on S.*

Proof.   Suppose the assertion is false. Then there is a positive number ε for which there exists no $\delta > 0$ such that property (7) is satisfied. Thus, for any $\delta = 1/m$ (m a positive integer) there is a pair of points x^m, y^m for which

$$|f(x^m) - f(y^m)| \geq \varepsilon \quad \text{and} \quad \|x^m - y^m\| < \frac{1}{m}, \, x^m \in S, \, y^m \in S. \qquad (8)$$

By the Bolzano–Weierstrass theorem, there exists a subsequence $\{x^{m'}\}$ of $\{x^m\}$ that converges to some point ξ. Since

$$\|y^{m'} - \xi\| \leq \|y^{m'} - x^{m'}\| + \|x^{m'} - \xi\| < \frac{1}{m'} + \|x^{m'} - \xi\| \to 0$$

as $m' \to \infty$, the sequence $\{y^{m'}\}$ is also convergent to ξ. Recalling that S is closed, we conclude that $\xi \in S$. By the continuity of $f(x)$ at ξ,

$$\lim f(x^{m'}) = f(\xi), \qquad \lim f(y^{m'}) = f(\xi).$$

Therefore

$$|f(x^{m'}) - f(y^{m'})| \leq |f(x^{m'}) - f(\xi)| + |f(\xi) - f(y^{m'})| \to 0$$

if $m' \to \infty$. This contradicts (8).

Suppose $f(x)$ is continuous on a bounded open set G. Can it be extended into the closure $\bar{G}$ of G in such a way that its extension is continuous on $\bar{G}$? If so, then, by Theorem 6, $f(x)$ is necessarily uniformly continuous in G. As will be shown below, this necessary condition is also sufficient.

THEOREM 7.   *Let $f(x)$ be a uniformly continuous function in a bounded, open set G. Then there exists a unique continuous function $F(x)$ defined on the closure $\bar{G}$ of G such that $F(x) = f(x)$ for all x in G.*

F is called a *continuous extension* of f.

Proof. We first prove that F is unique. Suppose H is another continuous extension of f. We have to show that $F(y) = H(y)$ for any $y \in \bar{G}$. If $y \in G$ then, of course, $F(y) = f(y) = H(y)$. If $y \notin G$, then there exists a sequence of points $\{x^m\}$ in G such that $\lim x^m = y$. Since F and H are both continuous at y, and since $F(x^m) = H(x^m)$ for each x^m, we get

$$F(y) = \lim F(x^m) = \lim H(x^m) = H(y).$$

We shall now prove the existence of a continuous extension F. Let $y \in \bar{G}$, $y \notin G$. Take any sequence $\{x^m\}$ of points in G such that $\lim x^m = y$. Then $\{f(x^m)\}$ is a Cauchy sequence. Indeed, for any $\delta > 0$ there is a positive integer n_0 such that

$$\|x^m - x^k\| < \delta \qquad \text{if } m \geq n_0,\ k \geq n_0. \tag{9}$$

Let ε be any positive number and let δ be such that

$$|f(x) - f(y)| < \varepsilon \quad \text{whenever } \|x - y\| < \delta,\ x \in G,\ y \in G. \tag{10}$$

Then (9) implies that

$$|f(x^m) - f(x^k)| < \varepsilon \qquad \text{if } m \geq n_0,\ k \geq n_0.$$

Thus $\{f(x^m)\}$ is a Cauchy sequence. Since any Cauchy sequence is convergent, $\lim f(x^m)$ exists. We define

$$F(y) = \lim f(x^m). \tag{11}$$

To show that this definition is meaningful, one has to prove that if $\{\bar{x}^m\}$ is another sequence in G that converges to y, then

$$\lim f(x^m) = \lim f(\bar{x}^m). \tag{12}$$

Now, for any $\delta > 0$ there is a positive integer m_0 such that

$$\|x^m - \bar{x}^m\| \leq \|x^m - y\| + \|y - \bar{x}^m\| < \frac{\delta}{2} + \frac{\delta}{2} = \delta \qquad \text{if } m \geq m_0. \tag{13}$$

Given $\varepsilon > 0$, take the number δ in (13) to be the number occurring in (10). It follows that

$$|f(x^m) - f(\bar{x}^m)| < \varepsilon \qquad \text{if } m \geq m_0.$$

This proves (12).

So far we have defined the function F at the points of $\bar{G}$ that do not belong to G (that is, at the points of the boundary of G). Now we have to prove the continuity of F at such points y. In view of Theorem 1, it suffices to prove the following:

$$\text{If } y^m \in \bar{G},\ y^m \to y, \qquad \text{then } F(y^m) \to F(y). \tag{14}$$

To prove (14) choose, for each y^m, a point $\bar{x}^m$ in G such that

$$\|\bar{x}^m - y^m\| < \frac{1}{m}, \quad \text{and} \quad |f(\bar{x}^m) - F(y^m)| < \frac{1}{m}.$$

This is possible to do. Indeed, if $y^m \in G$ then take $\bar{x}^m = y^m$. If $y^m \notin G$, then we know (from the definition of $F(y^m)$) that there is a sequence of points $\{x^{m,k}\}$ in G such that

$$x^{m,k} \to y^m, f(x^{m,k}) \to F(y^m) \qquad \text{as } k \to \infty.$$

We therefore can take $\bar{x}^m = x^{m,k}$ for some k sufficiently large.
Since

$$\|\bar{x}^m - y\| \le \|\bar{x}^m - y^m\| + \|y^m - y\| < \frac{1}{m} + \|y^m - y\| \to 0$$

if $m \to \infty$, $\lim \bar{x}^m = y$. But then, by the result of the paragraph following (11),

$$F(y) = \lim_m f(\bar{x}^m).$$

Hence

$$|F(y) - F(y^m)| \le |F(y) - f(\bar{x}^m)| + |f(\bar{x}^m) - F(y^m)|$$

$$< |F(y) - f(\bar{x}^m)| + \frac{1}{m} \to 0$$

if $m \to \infty$. This proves (14).

PROBLEMS

1. Is the function

$$f(x,y) = \begin{cases} \dfrac{x^4}{x^2 + y^2} & \text{if } (x,y) \ne (0,0), \\ 0 & \text{if } (x,y) = (0,0) \end{cases}$$

continuous at $(0,0)$?

2. Is the function

$$f(x,y,z) = \begin{cases} \dfrac{xy - z^2}{x^2 + y^2 + z^2} & \text{if } (x,y,z) \ne (0,0,0), \\ 0 & \text{if } (x,y,z) = (0,0,0) \end{cases}$$

continuous at $(0,0,0)$?

3. Is the function

$$f(x_1,x_2,x_3,x_4) = \begin{cases} \dfrac{x_1^2 - x_2^2 - x_3^2 - x_4^2}{x_1^2 + x_2^2 + x_3^2 + x_4^2} & \text{if } (x_1,x_2,x_3,x_4) \neq 0, \\ 0 & \text{if } (x_1,x_2,x_3,x_4) = (0,0,0,0) \end{cases}$$

continuous at $(0,0,0,0)$?

4. Can the function

$$f(x,y) = e^{-(1/|x-y|)} \qquad \text{when } x \neq y$$

be extended into a continuous function in R^2?

5. A *polynomial* in two variables is a function of the form $\sum a_{ij} x^i y^j$ where the a_{ij} are real numbers and the sum consists of a finite number of terms. A polynomial in n variables is a function of the form

$$\sum a_{i_1 i_2 \dots i_n} x_1^{i_1} x_2^{i_2} \cdots x_n^{i_n}$$

where the $a_{i_1 i_2 \dots i_n}$ are real numbers and the sum consists of a finite number of terms. Prove that any polynomial is a continuous function.

6. Is the function $1/(x^2 + y^2 - 1)$ uniformly continuous in the open *unit disc* (that is, the open disc with center $(0,0)$ and radius 1)?

7. Let $f(x)$ be a continuous function on a closed, bounded set S in R^n. Prove: If $f(x) > 0$ for any $x \in S$, then there is a positive constant m such that $f(x) \geq m$ for all $x \in S$.

8. Is the function

$$f(x,y) = \begin{cases} \dfrac{\sin xy}{x} & \text{if } x \neq 0, \\ y & \text{if } x = 0 \end{cases}$$

continuous in R^2?

9. Show that for any positive α the function

$$f(x,y,z) = \begin{cases} \dfrac{xy\,|z|^\alpha}{x^{2\alpha} + y^2 + z^2} & \text{if } (x,y,z) \neq (0,0,0), \\ 0 & \text{if } (x,y,z) = (0,0,0) \end{cases}$$

is continuous at $(0,0,0)$.

10. Prove Theorem 4 by using the Heine–Borel theorem. [*Hint:* For each $x \in S$, there is a $\delta = \delta(x) > 0$ such that $|f(x) - f(y)| < 1$ if $y \in S$,

$\|y - x\| < \delta$. The collection of open balls with center x and radius $\delta = \delta(x)$ form an open covering of S. Now use the Heine–Borel theorem.]

11.   Prove Theorem 6 using the Heine–Borel theorem.

5.   LIMIT SUPERIOR AND LIMIT INFERIOR

Let G be a region in R^n and let $\xi \in G$. Let $f(x)$ be a function defined at all the points of G that are different from ξ. Suppose there is a number L such that the following property holds: For any $\varepsilon > 0$ there is a $\delta > 0$ such that

$$|f(x) - L| < \varepsilon \qquad \text{if } 0 < \|x - \xi\| < \delta \text{ and } x \in G.$$

We then say that $f(x)$ has a *limit* L as $x \to \xi$, and that $f(x)$ *converges* to L as $x \to \xi$. We write

$$\lim_{x \to \xi} f(x) = L, \qquad \text{or } f(x) \to L \text{ if } x \to \xi.$$

Note that $f(x)$ is continuous at ξ if, and only if, $\lim_{x \to \xi} f(x)$ exists and is equal to $f(\xi)$.

If for any positive number N there is a $\delta > 0$ such that

$$f(x) > N \qquad \text{if } 0 < \|x - \xi\| < \delta, x \in G,$$

then we say that $\lim_{x \to \xi} f(x)$ is equal to ∞ and write

$$\lim_{x \to \xi} f(x) = \infty, \qquad \text{or } f(x) \to \infty \text{ if } x \to \xi.$$

Similarly, one defines $\lim_{x \to \xi} f(x) = -\infty$.

If $\lim_{x \to \xi} f(x)$ does not exist, then we introduce two other limit concepts: $\lim \sup_{x \to \xi} f(x)$ and $\lim \inf_{x \to \xi} f(x)$. These concepts are defined analogously to the case $n = 1$ (see Section 2.6). To define $\lim \sup_{x \to \xi} f(x)$, we first consider the case when $f(x)$ is bounded above (say, by C) in some set $x \in G$, $0 < \|x - \xi\| < \eta$, where $\eta > 0$. We introduce the set U of all real number u obtained in the following manner: There is a sequence $\{x^m\}$ of points in G, $x^m \neq \xi$ for all m, such that

$$\lim_{m \to \infty} x^m = \xi \qquad \text{and} \qquad \lim_{m \to \infty} f(x^m) = u.$$

If U is nonempty, then since it is bounded above by C, it has a supremum A. We call A the *limit superior* of $f(x)$ as $x \to \xi$, and write it in the form

$$\lim \sup_{x \to \xi} f(x) \qquad \text{or } \overline{\lim}_{x \to \xi} f(x).$$

If U is empty then one can prove (as for $n = 1$) that

$$\lim_{x \to \xi} f(x) = -\infty.$$

We write, in this case,

$$\limsup_{x \to \xi} f(x) = -\infty.$$

Finally, if $f(x)$ is not bounded above in any set $x \in G$, $\|x - \xi\| < \eta$ ($\eta > 0$), then one can show that there is a sequence $\{x^m\}$ in G such that $x^m \neq \xi$ for all m and

$$\lim_{m \to \infty} x^m = \xi, \qquad \lim_{m \to \infty} f(x^m) = \infty.$$

We define, in this case,

$$\limsup_{x \to \xi} f(x) = \infty \qquad \text{or} \quad \overline{\lim_{x \to \xi}}\, f(x) = \infty.$$

THEOREM 1.   *If* $A = \limsup_{n \to \xi} f(x)$ *is a real number, then, for any* $\varepsilon > 0$, *there is a* $\delta > 0$ *such that*

$$f(x) < A + \varepsilon \qquad if\ 0 < \|x - \xi\| < \delta,\ x \in G; \tag{1}$$

further, for any $\varepsilon > 0$, $\eta > 0$,

$$f(\bar{x}) > A - \varepsilon \qquad for\ some\ \bar{x} \in G, \qquad 0 < \|\bar{x} - \xi\| < \eta. \tag{2}$$

The proof is similar to the proof of Theorem 1 in Section 2.6, in which the case where G is an interval in R^1 was considered.

The limit inferior is defined similarly. Thus, if $f(x)$ is bounded below in some set $x \in G$, $0 < \|x - \xi\| < \eta$ and if U is nonempty, then the infimum B of the set U is called the *limit inferior* of $f(x)$ as $x \to \xi$. We write B in the form

$$\liminf_{x \to \xi} f(x) \qquad \text{or} \quad \varliminf_{x \to \xi} f(x).$$

If U is empty [when $f(x)$ is bounded below in $x \in G$, $0 < \|x - \xi\| < \eta$], then we have

$$\lim_{x \to \xi} f(x) = \infty.$$

We write, in this case,

$$\liminf_{x \to \xi} f(x) = \infty \qquad \text{or} \quad \varliminf_{x \to \xi} f(x) = \infty.$$

Finally, if $f(x)$ is not bounded below in any set $x \in G$, $0 < \|x - \xi\| < \eta$, then there is a sequence $\{x^m\}$ in G such that $x^m \neq \xi$ for all m, and $x^m \to \xi$, $f(x^m) \to -\infty$ as $m \to \infty$. We then write

$$\liminf_{x \to \xi} f(x) = -\infty \qquad \text{or} \quad \varliminf_{x \to \xi} f(x) = -\infty.$$

The analog of Theorem 1 is valid. That is, if $B = \lim\inf\limits_{x \to \xi} f(x)$ is a real number then, for any $\varepsilon > 0$ there is a $\delta > 0$ such that

$$f(x) > B - \varepsilon \qquad \text{if } 0 < \|x - \xi\| < \delta, \, x \in G,$$

and for any $\varepsilon > 0$, $\eta > 0$ there is a point $\bar{x}$ in G such that

$$f(\bar{x}) < B + \varepsilon \qquad \text{and } 0 < \|\bar{x} - \xi\| < \eta.$$

Theorem 2 of Section 2.6 extends to the present case where x varies in a region of R^n.

We state without proof:

THEOREM 2.   *Let G be a region in R^n and let ξ be a point of G. Let $f(x)$ and $g(x)$ be bounded functions defined at all the points of G that are different from ξ. Then*

$$\overline{\lim_{x \to \xi}} \, [f(x) + g(x)] \leq \overline{\lim_{x \to \xi}} \, f(x) + \overline{\lim_{x \to \xi}} \, g(x); \tag{3}$$

$$\underline{\lim_{x \to \xi}} \, [f(x) + g(x)] \geq \underline{\lim_{x \to \xi}} \, f(x) + \underline{\lim_{x \to \xi}} \, g(x); \tag{4}$$

if $f(x) \geq 0$, $g(x) \geq 0$, then

$$\overline{\lim_{x \to \xi}} \, [f(x)g(x)] \leq \left[\overline{\lim_{x \to \xi}} \, f(x)\right] \cdot \left[\overline{\lim_{x \to \xi}} \, g(x)\right]. \tag{5}$$

PROBLEMS

1. Check whether the following limits exist:

(a) $\lim\limits_{(x,y)\to 0} \dfrac{x^2 y}{x^4 + y^2}$;

(b) $\lim\limits_{(x,y)\to 0} \dfrac{2x^3 - y^3}{x^2 + y^2}$;

(c) $\lim\limits_{(x,y)\to 0} \dfrac{xy^2}{x^2 + y^3}$;

(d) $\lim\limits_{(x,y)\to 0} \dfrac{xy}{|x| + |y|}$;

(e) $\lim\limits_{(x,y)\to 0} \dfrac{x^5 + y^3(x^2 - y^2)}{(x^2 + y^2)^2}$;

(f) $\lim\limits_{(x,y,z)\to 0} \dfrac{\sin(x^2 + y^2 + z^4)}{x^2 + y^2 + z^2}$.

2. Prove the inequality (3).

3. Prove the inequality (4).

4. Prove the inequality (5).

5. Let $f(x)$ be a bounded function in a set G, and let $\xi \in G$. Show that

$$\overline{\lim_{x \to \xi}} \, f(x) = -\underline{\lim_{x \to \xi}} \, [-f(x)].$$

6. VECTOR VALUED FUNCTIONS; CURVES

We shall generalize the concept of a function.

Let S be a set in R^n. Suppose we are given a rule that assigns to each point x of S a point y in some given space R^m. We denote this rule by f and write $y = f(x)$. We call f a *vector valued function* with *domain* S. We also say that $f(x)$ is a function defined on the set S. The set T of all points $f(x)$, when x varies in S, is called the *range* of f. The set of all pairs $(x, f(x))$, when x varies in S, is called the *graph* of f. We also call f a *mapping* or a *transformation* and say that f maps S into R^m, and that f maps S onto T. We write $T = f(S)$ and call it the *image* of S. If $y = f(x)$, then we call y the *image* of x and we call x an *inverse image* of y. A point y may have several inverse images.

Writing $x = (x_1, x_2, \ldots, x_n)$ and $y = (y_1, y_2, \ldots, y_m)$, we can write the relation $y = f(x)$ in the form

$$y_1 = f_1(x_1, x_2, \ldots, x_n),$$
$$y_2 = f_2(x_1, x_2, \ldots, x_n), \tag{1}$$
$$\cdots$$
$$y_m = f_m(x_1, x_2, \ldots, x_n),$$

where the f_i are functions defined as follows: $f_i(x)$ is the ith component of the vector $f(x)$. We thus see that a vector valued function gives rise to m functions $f_1, \ldots, f_m$ defined on the domain S of f. Conversely, to m functions $f_1, \ldots, f_m$ defined on a set S of R^n we can correspond a vector valued function f by $f(x) = (f_1(x), \ldots, f_m(x))$.

Let $f(x)$ be a vector valued function with domain S in R^n and range in R^m. $f(x)$ is said to be *continuous* at a point ξ of S if for any $\varepsilon > 0$ there is a $\delta > 0$ such that

$$\| f(x) - f(\xi) \| < \varepsilon \qquad \text{if } \| x - \xi \| < \delta,\ x \in S. \tag{2}$$

Here

$$\| x - \xi \| = \sqrt{\sum_{i=1}^{n} (x_i - \xi_i)^2}, \qquad \| f(x) - f(\xi) \| = \sqrt{\sum_{i=1}^{m} (f_i(x) - f_i(\xi))^2}.$$

THEOREM 1. *A vector valued function $f(x)$ is continuous at a point ξ if, and only if, all its components $f_1(x), \ldots, f_m(x)$ are continuous at ξ.*

The proof is left to the student.

Theorems 1, part (i) of 2, and 6 and 7 of Section 4 remain true for vector valued functions. The proofs are the same as for functions. Theorem 4 also remains true if we formulate the following definition: $f(x)$ is *bounded* on a set S if there is a constant M such that

$$\|f(x)\| \leq M \qquad \text{for all } x \in M.$$

To extend Theorem 3 of Section 4, we first define the concept of a composite vector valued function. Let g be a vector valued function, mapping a set S in R^n onto a set T in R^m. Let $f(y)$ be a vector valued function mapping a set T', containing T, into R^k. The mapping h defined by $h(x) = f(g(x))$ for $x \in S$ is called the *composite* vector valued function of f and g, and it is denoted by $f \circ g$.

THEOREM 2.   *If g is continuous at a point ξ of S and if f is continuous at the point $g(\xi)$, then the composite vector valued function $h = f \circ g$ is continuous at the point ξ.*

The proof is similar to the proof of Theorem 4 in Section 2.2.

A vector valued function $g(t)$, where t varies in an interval $a \leq t \leq b$, is called a *curve*. If $g(t)$ is also continuous, then we call it a *continuous curve*. The variable t is called the *parameter* of the curve.

Theorem 2 yields the following corollary:

COROLLARY.   *Let $g(t)$ be a continuous curve defined for $a \leq t \leq b$, and let $f(x)$ be a continuous function on a set containing the image of the curve $g(t)$. Then the function $f(g(t))$ is a continuous function for $a \leq t \leq b$.*

Consider a polygonal line $l = \{l_1, l_2, \ldots, l_k\}$ in a space R^n. Denote by z^j the terminal endpoint of the segment l_j, and denote by z^0 the initial endpoint of l_1. The points $z^0, z^1, \ldots, z^k$ are called the *vertices* of l. We define a vector function $g(t)$ for $0 \leq t \leq k$ as follows:

$$g(t) = (1 - t)z^0 + tz^1 \qquad \text{if } 0 \leq t < 1,$$
$$g(t) = (2 - t)z^1 + (t - 1)z^2 \qquad \text{if } 1 \leq t < 2,$$
$$\cdots$$
$$g(t) = (k - t)z^{k-1} + (t - k + 1)z^k \qquad \text{if } k - 1 \leq t \leq k.$$

Looking at the components $g_i(t)$ of $g(t)$, one easily verifies (using Theorem 1) that $g(t)$ is a continuous curve. We shall call $g(t)$ a *polygonal curve* associated with the polygonal line l. Note that the points $g(t)$ belong to l, more precisely, to $l_1 \cup l_2 \cup \cdots \cup l_k$.

We shall use polygonal curves and the corollary to Theorem 2 in order to prove the following *intermediate value theorem*.

THEOREM 3.   *Let $f(x)$ be a continuous function defined in a domain D of R^n. Let y and z be any two points in D and let $f(y) = A, f(z) = B$. If $A < C < B$, then there exists a point ξ in D such that $f(\xi) = C$.*

Proof.   Since D is polygonally connected, there exists a polygonal line l connecting y to z. Let $g(t)$ be a polygonal curve associated with l. Thus $g(t)$ is a continuous vector valued function in some interval $0 \leq t \leq t_0$, $g(0) = y$, $g(t_0) = z$ and $g(t) \in D$ for $0 \leq t \leq t_0$. By the corollary to Theorem 2, the composite function $f(g(t))$ is continuous on $0 \leq t \leq t_0$. Also, $f(g(0)) = f(y) = A$ and $f(g(t_0)) = f(z) = B$. By the intermediate value theorem it follows that there exists a number τ in $(0, t_0)$ such that $f(g(\tau)) = C$. The assertion of the theorem then follows with $\xi = g(\tau)$.

A set S in R^n is said to be *arcwise connected* if for any two points y and z in S there is a continuous curve $g(t)$, $a \leq t \leq b$, such that $g(t) \in S$ for all $a \leq t \leq b$ and $g(a) = y, g(v) = z$.

THEOREM 4.   *If an open set D in R^n is arcwise connected, then it is also polygonally connected (and is therefore a domain).*

Proof.   Let y and z be points of D and $g(t)$, $a \leq t \leq b$, a continuous curve lying in D and connecting them. Since D is an open set, each point $g(\tau)$ has a δ_τ-neighborhood whose closure is contained in D. The set $\{g(t); a \leq t \leq b\}$ is easily seen to be a bounded closed subset of R^n. Hence we can apply the Heine–Borel theorem and conclude that there is a finite number of points $g(\tau_1), \ldots, g(\tau_k)$ such that the corresponding δ_{τ_i}-neighborhoods $B_1, \ldots, B_k$ form a covering of the set $\{g(t); a \leq t \leq b\}$.

Suppose $y = g(a)$ belongs to B_{h_1}. Let s_1 be the supremum of the numbers t such that $g(t) \in \bar{B}_{h_1}$. Since $g(t_i) \in \bar{B}_{h_1}$ for a sequence $\{t_i\}$, $t_i \to s_1 - 0$, and since $\bar{B}_{h_1}$ is closed, we conclude that $g(s_1) \in \bar{B}_{h_1}$. Denote by l_1 the segment connecting y to $g(s_1)$.

Suppose $s_1 < b$. Then the point $g(s_1)$ does not belong to B_{h_1} [otherwise $g(s_1 + \varepsilon) \in B_{h_1}$ for some $\varepsilon > 0$]. Hence it belongs to some ball B_{h_2}, $h_2 \neq h_1$. Let s_2 be the supremum of the numbers t such that $g(t) \in \bar{B}_{h_2}$. Then $s_2 > s_1$ and $g(s_2) \in \bar{B}_{h_2}$. Denote by l_2 the segment connecting $g(s_1)$ to $g(s_2)$. Suppose $s_2 < b$. Then $g(s_2) \notin B_{h_3}$.

$g(s_2)$ belongs to some ball B_{h_2}. $h_3 \neq h_2$ since, otherwise, $g(s_2 + \varepsilon) \in B_{h_2}$ for some $\varepsilon > 0$. Also, $h_3 \neq h_1$ since, otherwise, we get a contradiction to the definition of s_1.

Continuing as before, we construct points $g(s_3), g(s_4)$, and so on, such that $g(s_3)$ is contained in some closed ball $\bar{B}_{h_3}$, but not in B_{h_3}, and $g(t) \notin \bar{B}_{h_3}$ if $t > s_3$; $g(s_4)$ is contained in some closed ball $\bar{B}_{h_4}$, but not in B_{h_4}, and

$g(t) \notin \bar{B}_{h_4}$ if $t > s_4$, and so on. The indices $h_1, h_2, h_3, h_4, \ldots$ are mutually distinct. We denote by l_3 the line segment connecting $g(s_2)$ to $g(s_3)$, by l_4 the line segment connecting $g(s_3)$ to $g(s_4)$, and so forth. The segments l_1, l_2, l_3, l_4, and so on, belong to D. Since the balls B_{h_j} are mutually distinct, there can be at most k of them. Thus, following with the above construction of the points s_j, we arrive after m steps $(m \le k)$ at a point s_m with $g(s_m) = z$. The polygonal line $\{l_1, l_2, \ldots, l_m\}$ is contained in D and connects y to z. See Figure 6.6.

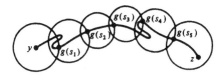

Figure 6.6

PROBLEMS

1. Prove Theorem 2.

2. Let $f_1(t), \ldots, f_m(t)$ be continuous functions for $a \le t \le b$. Let $z^1, \ldots, z^m$ be fixed points in R^m. Prove that the vector function

    $$f_1(t)z^1 + f_2(t)z^2 + \cdots + f_m(t)z^m$$

    is continuous for $a \le t \le b$.

3. Let G be the set of points $\{(t, \sin 1/t); 0 < t \le 1\}$. Prove that the closure $\bar{G}$ of G is not arcwise connected.

4. Let G be the set of points $\{(t, t \sin 1/t); 0 < t \le 1\}$. Prove that the closure of G is arcwise connected.

5. Prove that a continuous vector valued function maps any closed, bounded set onto a closed, bounded set. (This generalizes Theorem 1, Section 2.3.)

6. Let $f(x)$ be a continuous vector valued function whose domain S is an open set in R^n. For any subset T in the range of f, let $f^{-1}(T)$ denote the set of all the inverse images of the points y of T. Prove that if T is open, then $f^{-1}(T)$ is open.

7. Let $f(x)$ be a continuous vector valued function from a set S in R^n into R^n. Assume that the mapping f is one to one and that its inverse is continuous. Prove that S is a domain if, and only if, its range $f(S)$ is a domain.

7

DIFFERENTIATION

1. PARTIAL DERIVATIVES

Let $f(x,y)$ be a function of two variables defined in an open set G, and let (a,b) be a point of G. If

$$\lim_{h \to 0} \frac{f(a + h,b) - f(a,b)}{h}$$

exists, then we say that $f(x,y)$ has a *partial derivative* with respect to x at the point (a,b). We then denote the above limit by

$$f_x(a,b), \quad \frac{\partial f}{\partial x}(a,b), \quad \text{or} \quad \left(\frac{\partial f}{\partial x}\right)_{(a,b)}$$

and call it the partial derivative of f with respect to x at (a,b). Note that $f_x(a,b)$ is the derivative at $x = a$ of the function $f(x,b)$.

If

$$\lim_{k \to 0} \frac{f(a,b + k) - f(a,b)}{k}$$

exists, then we say that $f(x,y)$ has a partial derivative with respect to y at the point (a,b). We denote this limit by

$$f_y(a,b), \qquad \frac{\partial f}{\partial y}(a,b), \qquad \text{or} \left(\frac{\partial f}{\partial y}\right)_{(a,b)}$$

and call it the partial derivative of f with respect to y at (a,b).

If $f_x(a,b)$ exists for all (a,b) in G, then we denote by f_x, or $\partial f/\partial x$, the function whose value at each point $(a,b) \in G$ is $f_x(a,b)$. Similarly, we define the function f_y, or $\partial f/\partial y$.

EXAMPLE.   If $f(x,y) = x^2 + 2y - x \sin y$, then $f_x = 2x - \sin y, f_y = 2 - x \cos y$.

If f_x has a partial derivative with respect to x, then we denote it by

$$f_{xx}, \qquad \text{or} \frac{\partial^2 f}{\partial x^2}.$$

If f_x has a partial derivative with respect to y, then we denote it by

$$f_{xy}, \qquad \text{or} \frac{\partial^2 f}{\partial y\, \partial x},$$

Similarly, we denote by

$$f_{yx}, \qquad \text{or} \frac{\partial^2 f}{\partial x\, \partial y},$$

the partial derivative of f_y with respect to x, and by

$$f_{yy}, \qquad \text{or} \frac{\partial^2 f}{\partial y^2},$$

the partial derivative of f_y with respect to y.

If $f(x,y)$ is as in the above example, then

$$f_{xx} = 2, \qquad f_{xy} = -\cos y, \qquad f_{yx} = -\cos y, \qquad f_{yy} = x \sin y.$$

Notice that f_{xy} and f_{yx} are equal.

THEOREM 1.   *Let f be a function defined in an open set G of R^2. If $f_x, f_y,$ and f_{xy} exist and are continuous in G, then f_{yx} also exists and $f_{xy} = f_{yx}$.*

Proof. We have to show that for any point (a,b) of G, $f_{yx}(a,b)$ exists and is equal to $f_{xy}(a,b)$. There is a δ-neighborhood of (a,b) that is contained in G. We introduce the expression

$$\Delta = f(a + h, b + k) - f(a + h, b) - f(a, b + k) + f(a, b)$$

for $|h|$ and $|k|$ sufficiently small, say $h^2 + k^2 < \delta^2$. Setting

$$\phi(x) = f(x,b + k) - f(x,b),$$

we can write

$$\Delta = \phi(a + h) - \phi(a). \tag{1}$$

The function $\phi(x)$ is differentiable and

$$\phi'(x) = f_x(x,b + k) - f_x(x,b).$$

Applying the mean value theorem, we get from (1):

$$\Delta = h\phi'(a + \theta_1 h) = h[f_x(a + \theta_1 h, b + k) - f_x(a + \theta_1 h, b)] \tag{2}$$

where $0 < \theta_1 < 1$. θ_1 depends on h,k.
    Consider the function

$$g(y) = f_x(a + \theta_1 h, y).$$

It is differentiable, and

$$g'(y) = f_{xy}(a + \theta_1 h, y).$$

Writing (2) in the form

$$\Delta = h[g(b + k) - g(b)],$$

we can apply the mean value theorem and conclude that

$$\Delta = hkg'(b + \theta_2 k) = hkf_{xy}(a + \theta_1 h, b + \theta_2 k) \tag{3}$$

where $0 < \theta_2 < 1$. θ_2 depends on h,k.
    We next write Δ in the form

$$\Delta = [f(a + h, b + k) - f(a + h, b)] - [f(a, b + k) - f(a,b)]$$
$$= kf_y(a + h, b + \theta_3 k) - kf_y(a, b + \theta_4 k)$$

where $0 < \theta_3 < 1, 0 < \theta_4 < 1$; here we again have used the mean value theorem. θ_3 and θ_4 depend on h,k.
    If we compare the last expression for Δ with (3), we get

$$f_{xy}(a + \theta_1 h, b + \theta_2 k) = \frac{f_y(a + h, b + \theta_3 k) - f_y(a, b + \theta_4 k)}{h}. \tag{4}$$

Take a sequence $k = k_m \to 0$. The corresponding $\theta_1 = \theta_1(h, k_m)$ form a bounded sequence. Hence a subsequence $\{\theta_1(h, k_{m'})\}$ is convergent to some $\theta_1^* = \theta_1^*(h)$. Taking $k = k_{m'} \to 0$ in (4), we get

$$f_{xy}(a + \theta_1^* h, b) = \frac{f_y(a + h, b) - f_y(a,b)}{h}.$$

If $h \to 0$, then the term on the left converges to $f_{xy}(a,b)$. Hence

$$\lim_{h \to 0} \frac{f_y(a + h,b) - f_y(a,b)}{h}$$

exists and equals $f_{xy}(a,b)$. Since the limit is, by definition, $f_{yx}(a,b)$, the proof is complete.

So far we have considered the first and second derivatives of f. The third derivatives are defined as follows:

$$\frac{\partial^3 f}{\partial x^3} = \frac{\partial}{\partial x} \frac{\partial^2 f}{\partial x^2}, \qquad \frac{\partial^3 f}{\partial x^2 \, \partial y} = \frac{\partial^3 f}{\partial x \, \partial x \, \partial y} = \frac{\partial}{\partial x} \frac{\partial^2 f}{\partial x \, \partial y}, \qquad \text{and so on.}$$

We also use the notation

$$f_{xxx} = \frac{\partial^3 f}{\partial x^3}, \qquad f_{yxx} = (f_{yx})_x = \frac{\partial}{\partial x} f_{yx} = \frac{\partial^3 f}{\partial x^2 \, \partial y}, \qquad \text{and so on.}$$

Theorem 1 implies that

$$f_{xxy} = f_{xyx} = f_{yxx}, \qquad f_{xyy} = f_{yxy} = f_{yyx} \tag{5}$$

provided f_{xx}, f_{xy}, f_{yy} and all the derivatives in (5) exist and are continuous.

Higher-order partial derivatives are defined similarly. Thus

$$\frac{\partial^{n+1} f}{\partial x^{\alpha+1} \, \partial x^\beta} = \frac{\partial}{\partial x} \frac{\partial^n f}{\partial x^\alpha \, \partial x^\beta}, \qquad \text{and so on.}$$

The derivatives

$$\frac{\partial^m f}{\partial x^m}, \frac{\partial^m f}{\partial y^m} \qquad (m = 1,2,\ldots)$$

are called *pure partial derivatives*. All other derivatives are called *mixed partial derivatives*. Theorem 1 and the relations (5) extend to higher derivatives. Thus, if a function f has partial derivatives of all orders $\leq n$, and if all these derivatives are continuous, then a mixed derivative of order n is independent of the order in which the derivatives with respect to x and the derivatives with respect to y appear. For instance,

$$\frac{\partial^n f}{\partial x^\alpha \, \partial y^\beta \, \partial x^\gamma \, \partial y^\delta} = \frac{\partial^n f}{\partial x^{\alpha+\gamma} \, \partial y^{\beta+\delta}}.$$

Thus, the partial derivatives of order n can be written in the form

$$\frac{\partial^n f}{\partial x^i \, \partial y^{n-i}} \qquad (0 \leq i \leq n).$$

Consider now a function $f(x,y,z)$ of three variables. We define the partial derivative with respect to x by

$$f_x(a,b,c) = \lim_{h \to 0} \frac{f(a + h,b,c) - f(a,b,c)}{h}.$$

The partial derivatives f_y, f_z are defined similarly. The higher-order partial derivatives are defined in the same manner as before. Theorem 1 clearly extends to the present case. Using it successively, we conclude that if all the partial derivatives of $f(x,y,z)$ up to order n exist and are continuous, then, in any partial derivative of order n of f, the order of the differentiations with respect to the variables x, y, and z is immaterial. Consequently, any partial derivative of order n can be written in the form

$$\frac{\partial^n f}{\partial x^i \, \partial y^j \, \partial z^{n-i-j}} \qquad \text{for some } i,j, \text{ where } 0 \leq i + j \leq n.$$

The derivatives

$$\frac{\partial^n f}{\partial x^n}, \frac{\partial^n f}{\partial y^n}, \qquad \text{and} \qquad \frac{\partial^n f}{\partial z^n}$$

are called *pure partial derivatives*; all the other derivatives of order n are called *mixed partial derivatives*.

Consider next a function $f(x) = f(x_1, x_2, \ldots, x_n)$ of n variables. The partial derivative with respect to x_i at a point α is defined to be the limit

$$\frac{\partial f}{\partial x_i}(\alpha) = \lim_{h \to 0} \frac{f(\alpha_1, \ldots, \alpha_{i-1}, \alpha_i + h, \alpha_{i+1}, \ldots, \alpha_n) - f(\alpha_1, \ldots, \alpha_{i-1}, \alpha_i, \alpha_{i+1}, \ldots, \alpha_n)}{h}.$$

Higher-order derivatives are denoted by

$$\frac{\partial^2 f}{\partial x_i \, \partial x_j}, \qquad \frac{\partial^3 f}{\partial x_i \, \partial x_j \, \partial x_k}, \qquad \text{and so on.}$$

Theorem 1 extends to the present case. If we apply it successively, we conclude that if all the partial derivatives of f of orders up to m exist and are continuous, then, in any partial derivative of order m of f, the order in which the differentiations with respect to the variables $x_1, \ldots, x_n$ are performed may be changed without changing the value of the mth order derivative.

When we speak of a function $f(x)$ having first derivatives, we mean first partial derivatives, that is, partial derivatives of the first order.

DEFINITION.   Let $f(x)$ be a function defined in an open set G of R^n. Let α be a point of G and let e be a unit vector, that is, e is a point of R^n with $\|e\| = 1$. If

$$\lim_{h \to 0} \frac{f(\alpha + he) - f(\alpha)}{h} \qquad\qquad (6)$$

exists, then we call it a *directional derivative* of f at α, or the *derivative* of f, at α, *in the direction e*. We denote it by

$$\frac{\partial f}{\partial e}(\alpha).$$

If $e = (0,\ldots,0,1,0,\ldots,0)$ where the digit 1 is in the ith component, then

$$\frac{\partial f}{\partial e}(\alpha) = \frac{\partial f}{\partial x_i}(\alpha).$$

EXAMPLE.   Let $f(x) = x_1{}^2 x_2 x_3$, $\alpha = (1,2,3)$, $e = (1/\sqrt{3}, 1/\sqrt{3}, 1/\sqrt{3})$. Then

$$\alpha + he = \left(1 + \frac{h}{\sqrt{3}}, 2 + \frac{h}{\sqrt{3}}, 3 + \frac{h}{\sqrt{3}}\right)$$

and

$$f(\alpha + he) - f(\alpha) = \left(1 + \frac{h}{\sqrt{3}}\right)^2 \left(2 + \frac{h}{\sqrt{3}}\right)\left(3 + \frac{h}{\sqrt{3}}\right) - 1^2 \cdot 2 \cdot 3$$

$$= \frac{17}{\sqrt{3}} h + ah^2 + bh^3$$

where a, b are some constants. Hence

$$\frac{\partial f}{\partial e}(\alpha) = \frac{17}{\sqrt{3}}.$$

PROBLEMS

1. Let

$$f(x,y) = xy\,\frac{x^2 - y^2}{x^2 + y^2} \qquad \text{if } (x,y) \neq 0,$$

and $f(0,0) = 0$. Prove that $f_x(0,y) = -y, f_y(x,0) = x, f_{xy}(0,0) = -1,$ $f_{yx}(0,0) = 1$.

2. Compute the first and second partial derivatives of the following functions:

(a) $\sqrt{x^2 + y^2 + z^2}$;

(b) $\sin\dfrac{x - y}{x + y}$;

(c) $e^{y\sqrt{1+x^2}}$;

(d) $\tan\dfrac{x}{x^2 + y^2}$;

(e) $x^{y\,\sin z}$;

(f) $x^2 \log(y^2 + z^2)$.

3. Let $f(t)$ have two derivatives, and let $\phi(x,y) = f(x + y) + f(x - y)$. Prove that ϕ has partial derivatives of orders ≤ 2, and $\phi_{xx} - \phi_{yy} = 0$.

4. Let $f(x,y) = \log(1/\sqrt{x^2 + y^2})$. Prove that $f_{xx} + f_{yy} = 0$.

5. Let $f(x,y,z) = 1/\sqrt{x^2 + y^2 + z^2}$. Prove that $f_{xx} + f_{yy} = f_{zz} = 0$.

6. Let $f(x) = \|x\|^{2-n}$ where $\|x\| = \sqrt{x_1^2 + \cdots + x_n^2}$, $n \geq 3$. Prove that

$$\sum_{i=1}^{n} \frac{\partial^2 f}{\partial x_i^2} = 0.$$

7. Show that the function $u(x,y) = e^{(x^2 - y^2)} \sin(2xy)$ satisfies

$$u_{xx} + u_{yy} = 0.$$

8. Compute the derivative of each of the following functions in the direction $(1/\sqrt{3}, \sqrt{2}/\sqrt{3})$, at the points indicated:

(a) $\log(x^2 + y^2)$, at $(1,2)$;   (b) $\dfrac{1 + x^2}{1 + y^2}$, at $(-1,1)$;

(c) e^{x+y}, at $(0,0)$;   (d) $\sqrt{1 + xy}$, at $(1,0)$.

9. Compute the derivative of each of the following functions in the direction $(1/\sqrt{3}, 1/\sqrt{2}, -1/\sqrt{6})$, at the points indicated:

(a) $x^2 + y^2 + z^2$, at $(1,0,-1)$;   (b) $xy\sqrt{1 + z^2}$, at $(1,1,2)$;

(c) $\sin(x + y)\cos z$, at $\left(0,0,\dfrac{\pi}{2}\right)$;   (d) $e^x \cos y + e^z \sin y$, at $\left(0,\dfrac{\pi}{2},0\right)$.

10. Let $f(x,y) = x^{1/3}y^{2/3}$. Determine which of the following derivatives exist and evaluate those that do:

$$\frac{\partial f}{\partial x}(0,0), \frac{\partial f}{\partial x}(0,1), \frac{\partial f}{\partial e}(0,0), \frac{\partial f}{\partial e}(0,1) \quad \text{where } e = \left(\frac{\sqrt{2}}{2}, \frac{\sqrt{2}}{2}\right).$$

2. DIFFERENTIALS; THE CHAIN RULE

Let $f(x,y)$ be a function defined in an open set G of R^2, and let $(a,b) \in G$. Then there is a δ-neighborhood of (a,b) that is contained in G. Consider the expression

$$\Delta f = f(a + h, b + k) - f(a,b)$$

for $h^2 + k^2 < \delta^2$. Suppose there exist numbers A and B such that

$$\Delta f = Ah + Bk + \eta\sqrt{h^2 + k^2} \tag{1}$$

where $\eta = \eta(h,k)$ is a function satisfying

$$\eta(h,k) \to 0 \qquad \text{if } (h,k) \to 0. \tag{2}$$

Then we say that $f(x,y)$ is *differentiable* at (a,b). Taking $k = 0$ in (1), we see that

$$\frac{f(a + h,b) - f(a,b)}{h} = A + \eta\,|h|.$$

It follows that $f_x(a,b)$ exists and is equal to A. Similarly, $f_y(a,b)$ exists and is equal to B. Thus, if $f(x,y)$ is differentiable at (a,b), then

$$f(a + h,b + k) - f(a,b) = hf_x(a,b) + kf_y(a,b) + \eta\sqrt{h^2 + k^2} \tag{3}$$

where η satisfies (2). The linear function [of (h,k)] $f_x(a,b)h + f_y(a,b)k$ is then called the *total differential* (or, briefly, the *differential*) of f at (a,b). It is denoted by df. Thus

$$(df)(h,k) = f_x(a,b)h + f_y(a,b)k. \tag{4}$$

Consider next the case of a function $f(x)$ of n variables, defined in an open set G of R^n. Let $\alpha \in G$ and let $h = (h_1,\ldots,h_n)$ vary in a δ-neighborhood of 0, δ being so small that the ball $\{x;\|x - \alpha\| < \delta\}$ is contained in G. If there exists a vector $A = (A_1,\ldots,A_n)$ such that

$$f(\alpha + h) - f(\alpha) = A \cdot h + \eta\,\|h\| \tag{5}$$

where $\eta = \eta(h) \to 0$ as $\|h\| \to 0$, then we say f is *differentiable* at α. We then easily deduce (as in the case $n = 2$) that $A_i = \partial f(\alpha)/\partial x_i$, so that the condition (5) becomes

$$f(\alpha + h) - f(\alpha) = \sum_{i=1}^{n} h_i \frac{\partial f(\alpha)}{\partial x_i} + \eta\,\|h\|. \tag{6}$$

When f is differentiable at a point α, then we introduce the linear function df defined by

$$(df)(h) = \sum_{i=1}^{n} \frac{\partial f(\alpha)}{\partial x_i}\, h_i. \tag{7}$$

We call this function the *total differential* (or briefly the *differential*) of f at α.

THEOREM 1.   *If $f(x)$ is differentiable at a point α, then $f(x)$ is continuous at α.*

Proof. From (5) and the fact that $\eta(h)$ is a bounded function, it follows that

$$|f(\alpha + h) - f(\alpha)| \leq C \|h\|, \qquad C \text{ constant.}$$

Hence, for any $\varepsilon > 0$,

$$|f(\alpha + h) - f(\alpha)| < C\delta = \varepsilon \qquad \text{if } \|h\| < \delta,$$

provided $\delta = \varepsilon/C$.

EXAMPLE. Let $f(x,y) = \sqrt{|xy|}$. Then $f_x = f_y = 0$ at $(0,0)$. But f is not differentiable at $(0,0)$ since (1) becomes

$$\Delta f = f(h,k) = \eta\sqrt{h^2 + k^2},$$

so that

$$\eta = \frac{\sqrt{|hk|}}{\sqrt{h^2 + k^2}}.$$

This function does not satisfy (2). Indeed, $\eta(h,h) = 1/\sqrt{2}$.

If a function $f(x,y)$ is differentiable at a point (a,b), then one can approximate $f(a + h, b + k)$ for small h,k by

$$f(a,b) + f_x(a,b)h + f_y(a,b)k;$$

the error being $\eta\sqrt{h^2 + k^2}$, where $\eta \to 0$ if $(h,k) \to 0$.

As the previous example shows, the mere existence of $f_x(a,b)$, $f_y(a,b)$ is not sufficient to ensure that f is differentiable at (a,b). In the next theorem we give sufficient conditions for f to be differentiable.

THEOREM 2. *Let $f(x,y)$ be defined in an open set G of R^2, and let f_x, f_y exist and be continuous in G. Then f is differentiable at each point (a,b) of G.*

Proof. Let $(a,b) \in G$. Write

$$\Delta f = f(a + h, b + k) - f(a,b)$$
$$= [f(a + h, b + k) - f(a, b + k)] - [f(a, b + k) - f(a,b)].$$

By the mean value theorem, the expression in the first brackets is equal to

$$hf_x(a + \theta_1 h, b + k) \qquad (0 < \theta_1 < 1)$$

and the expression in the second brackets is equal to

$$kf_y(a, b + \theta_2 k) \qquad (0 < \theta_2 < 1).$$

Hence

$$\Delta f = hf_x(a + \theta_1 h, b + k) + kf_y(a, b + \theta_2 k)$$
$$= hf_x(a,b) + kf_y(a,b) + \tilde{\eta}(h,k)$$

where

$$\tilde{\eta}(h,k) = h[f_x(a + \theta_1 h, b + k) - f_x(a,b)] + k[f_y(a, b + \theta_2 k) - f_y(a,b)].$$

It remains to show that

$$\frac{\tilde{\eta}(h,k)}{\sqrt{h^2 + k^2}} \to 0 \qquad \text{if } (h,k) \to 0. \tag{8}$$

Since f_x and f_y are continuous, for any $\varepsilon > 0$, there is a $\delta > 0$ such that

$$|f_x(x,y) - f_x(a,b)| < \frac{\varepsilon}{2},$$

$$|f_y(x,y) - f_y(a,b)| < \frac{\varepsilon}{2}$$

if $\sqrt{(x - a)^2 + (y - b)^2} < \delta$. Hence, if $\sqrt{h^2 + k^2} < \delta$,

$$|\tilde{\eta}(h,k)| < (|h| + |k|)\frac{\varepsilon}{2} < \sqrt{h^2 + k^2}\,\varepsilon.$$

This proves (8).

The proof of Theorem 2 extends to functions $f(x,y,z)$ and, generally, to functions $f(x)$ of n variables. Thus we have:

THEOREM 3. *Let $f(x)$ be a function defined in an open subset G of R^n. If the first derivatives $\partial f/\partial x_i$ $(1 \le i \le n)$ exist and are continuous in G, then f is differentiable at each point of G.*

Combining Theorems 1 and 3, we get:

COROLLARY. *If $\partial f/\partial x_i$ $(i = 1,...,n)$ exist and are continuous in G (G an open set in R^n), then f is continuous in G.*

We shall use Theorem 2 to prove the following theorem on the partial derivatives of a composite function, called the *chain rule*.

THEOREM 4. *Let $u(x,y)$, $v(x,y)$ be functions defined in an open set G of R^2, having continuous first derivatives in G. Let $f(u,v)$ be a function defined on an open set W of R^2 containing the points $(u(x,y),v(x,y))$, where (x,y) varies* *in G. If f(u,v) has continuous first derivatives in W, then the composite function* $g(x,y) = f(u(x,y),v(x,y))$ *has continuous first derivatives in G, and*

$$\frac{\partial g}{\partial x} = \frac{\partial f}{\partial u}\frac{\partial u}{\partial x} + \frac{\partial f}{\partial v}\frac{\partial v}{\partial x}, \tag{9}$$

$$\frac{\partial g}{\partial y} = \frac{\partial f}{\partial u}\frac{\partial u}{\partial y} + \frac{\partial f}{\partial v}\frac{\partial v}{\partial y}. \tag{10}$$

Proof. Let $(a,b) \in G$ and set $u(a,b) = \alpha$, $v(a,b) = \beta$, $u_x(a,b) = u_x$, $u_y(a,b) = u_y$, $f_x(\alpha,\beta) = f_x$, $f_y(\alpha,\beta) = f_y$. There is a δ-neighborhood of (a,b) that is contained in G. If $\sqrt{h^2 + k^2} < \delta$, then $(a+h, b+k)$ is in G, so that $u(a+h, b+k)$, $v(a+h, b+k)$ and $g(a+h, b+k)$ are well defined. We can then write

$$\Delta u = u(a+h, b+k) - u(a,b) = hu_x + ku_y + \eta_1\sqrt{h^2 + k^2}, \tag{11}$$

$$\Delta v = v(a+h, b+k) - v(a,b) = hv_x + kv_y + \eta_2\sqrt{h^2 + k^2}, \tag{12}$$

$$\begin{aligned}\Delta g &= f(u(a+h, b+k), v(a+h, b+k)) - f(u(a,b), v(a,b)) \\ &= f(\alpha + \Delta u, \beta + \Delta v) - f(\alpha, \beta) \\ &= f_x(\alpha,\beta)\Delta u + f_y(\alpha,\beta)\Delta v + \eta_3\sqrt{(\Delta u)^2 + (\Delta v)^2}.\end{aligned} \tag{13}$$

The functions η_1, η_2, η_3 are functions of (h,k), and $\eta_1 \to 0$, $\eta_2 \to 0$ if $(h,k) \to 0$. As for η_3,

$$\eta_3 \to 0 \quad \text{if } \sqrt{(\Delta u)^2 + (\Delta v)^2} \to 0.$$

Since, however, $\Delta u \to 0$ and $\Delta v \to 0$ if $(h,k) \to 0$, we conclude that $\eta_3 \to 0$ if $(h,k) \to 0$.

Substituting Δu, Δv from (11) and (12) into (13), we get

$$\begin{aligned}\Delta g &= f_x(hu_x + ku_y + \eta_1\sqrt{h^2 + k^2}) \\ &\quad + f_y(hv_x + kv_y + \eta_2\sqrt{h^2 + k^2}) + \eta_3\sqrt{(\Delta u)^2 + (\Delta v)^2} \\ &= h(f_x u_x + f_y v_x) + k(f_x u_y + f_y v_y) + \eta\sqrt{h^2 + k^2}\end{aligned} \tag{14}$$

where

$$\eta = f_x\eta_1 + f_y\eta_2 + \eta_3\frac{\sqrt{(\Delta u)^2 + (\Delta v)^2}}{\sqrt{h^2 + k^2}}.$$

Since, by (11) and (12),

$$|\Delta u| \le |u_x||h| + |u_y||k| + |\eta_1|\sqrt{h^2 + k^2} \le C\sqrt{h^2 + k^2}$$

$$|\Delta v| \le |v_x||h| + |v_y||k| + |\eta_2|\sqrt{h^2 + k^2} \le C\sqrt{h^2 + k^2}$$

where C is a constant, it follows that $\eta \to 0$ if $(h,k) \to 0$. From (14) we then deduce that g is differentiable at (a,b) and (9) and (10) hold.

Having proved (9) and (10) for any point of G, the continuity of g_x and g_y follows from the continuity of the right-hand sides of (9) and (10).

Theorem 4 is a special case of the following *chain rule*:

THEOREM 5.   *Let $u(x) = (u_1(x),...,u_n(x))$ be a vector valued function defined on an open set G of R^m. Let $f(u)$ be a function defined on an open set U of R^n, containing the range of u. If the first derivatives of the functions u_i exist and are continuous in G, and if the first derivatives of f exist and are continuous in U, then the composite function $g(x) = f(u(x))$ has continuous first derivatives in G, given by*

$$\frac{\partial g}{\partial x_i} = \sum_{k=1}^{n} \frac{\partial f}{\partial u_k} \frac{\partial u_k}{\partial x_i}. \tag{15}$$

The proof of Theorem 5 is analogous to the proof of Theorem 4, and therefore is left to the student.

Consider the special case of a curve $u = u(t) = (u_1(t),...,u_n(t))$, where t varies in an interval $a \le t \le b$. If the functions $u_i(t)$ have continuous first derivatives, then we say that $u(t)$ is a *continuously differentiable curve*. If $f(u)$ is defined in an open set containing the range of the curve $u(t)$, then, by Theorem 5, $f(u(t))$ is differentiable for $a < t < b$, and

$$\frac{d}{dt} f(u(t)) = \sum_{k=1}^{n} \frac{\partial f(u(t))}{\partial u_k} \frac{du_k}{dt}. \tag{16}$$

The proof of Theorem 5 shows that (16) holds also at the points $t = a$, $t = b$, with d/dt taken as right and left derivative, respectively.

If $u(t) = \alpha + te$, $e = (e_1,...,e_n)$, $\|e\| = 1$, then

$$\frac{df(u(t))}{dt}\bigg|_{t=0} = \lim_{h \to 0} \frac{f(u(h)) - f(u(0))}{h} = \lim_{h \to 0} \frac{f(\alpha + he) - f(\alpha)}{h}.$$

The right-hand side is the directional derivative $\partial f(\alpha)/\partial e$. Since, by (16), the left-hand side is equal to

$$\sum_{k=1}^{n} \frac{\partial f(0)}{\partial u_k} e_k,$$

we get:

COROLLARY.   *Let $f(x)$ be a function having continuous first derivatives in a neighborhood of a point α of R^n. Let $e = (e_1,...,e_n)$ be a unit vector. Then*

$$\frac{\partial f(\alpha)}{\partial e} = \sum_{i=1}^{n} e_i \frac{\partial f(\alpha)}{\partial x_i}. \tag{17}$$

This formula enables us to compute directional derivatives in terms of the partial derivatives $\partial f/\partial x_i$.

We shall write du/dt, or $u'(t)$, for $(du_1/dt,...,du_n/dt)$.

From (16) we immediately deduce:

THEOREM 6.   *Let* $u(t)$ *and* $v(t)$ *be continuously differentiable curves satisfying* $u(\tau) = v(\tau)$, $u'(\tau) = v'(\tau)$ *for some point* τ. *If* f *has continuous first derivatives in a neighborhood of* $u(\tau)$, *then*

$$\frac{d}{d\tau} f(u(\tau)) = \frac{d}{d\tau} f(v(\tau)). \tag{18}$$

We call $(d/d\tau)f(u(\tau))$ the *derivative of* f *with respect to the curve* $u(t)$ at $t = \tau$. Theorem 6 asserts that this derivative depends only on the vectors $u(\tau)$, $u'(\tau)$.

If $u'(\tau) \neq 0$, then the straight line

$$v(t) = u(\tau) + u'(\tau)(t - \tau) \tag{19}$$

is called the *tangent line* to the curve $u(t)$ at $t = \tau$. It satisfies $v(\tau) = u(\tau)$, $v'(\tau) = u'(\tau)$. Hence the derivative of f with respect to the curve $u(t)$ at $t = \tau$ is equal to the derivative of f with respect to the tangent line of $u(t)$ at $t = \tau$.

When $v(t) = \alpha + et$ ($\alpha \in R^n$, $e \in R^n$, $\|e\| = 1$), then

$$\frac{f(v(\tau + h)) - f(v(\tau))}{h} = \frac{f(v(\tau) + eh) - f(v(\tau))}{h}. \tag{20}$$

Hence the derivative $df(v(\tau))/d\tau$ of f with respect to $v(t)$ at $t = \tau$ is the directional derivative $\partial f(v(\tau))/\partial e$. If $\|e\| \neq 1$, then, setting $\bar{e} = e/\|e\|$ and taking $h = k/\|e\|$ in (20), we get

$$\frac{df(v(\tau))}{d\tau} = \lim_{k \to 0} \frac{f(v(\tau) + \bar{e}k) - f(v(\tau))}{k/\|e\|} = \frac{f(v(\tau))}{\partial \bar{e}} \|e\|.$$

Applying the last remark in the case of the tangent line (19), we conclude:

THEOREM 7.   *Let* $u(t)$ ($a \leq t \leq b$) *be a continuously differentiable curve and let* $u'(\tau) \neq 0$ *for some* $a \leq \tau \leq b$. *If* f *has continuous first derivatives in a neighborhood of* $u(\tau)$, *then*

$$\frac{df(u(\tau))}{d\tau} = \frac{\partial f(u(\tau))}{\partial w} \|u'(\tau)\| \qquad where \; w = \frac{u'(\tau)}{\|u'(\tau)\|}. \tag{21}$$

PROBLEMS

1. Prove Theorem 3.

2. Prove Theorem 5.

3. Let $f(u) = u_1{}^2 + u_2{}^2$, $u_1 = 1 + t$, $u_2 = 1 + t^2$. Compute $(d/dt)f(u(t))$ in two ways: (i) by using Theorem 5; (ii) by first substituting u_1 and u_2 into $f(u)$ and then differentiating the resulting function of t.

4. Let $f(x,y) = \tan^{-1}(y/x)$, $x = \cos t$, $y = \sin t$. Compute $(d/dt)f(x(t),y(t))$ in two ways (as in the preceding problem).

5. Let $f(x,y)$ have continuous first derivatives, and let $x = s \cos \alpha - t \sin \alpha$, $y = s \sin \alpha + t \cos \alpha$, α constant. Calculate

$$\left(\frac{\partial f}{\partial s}\right)^2 + \left(\frac{\partial f}{\partial t}\right)^2.$$

6. Let $f(x,y,z)$ have continuous derivatives and let $x = \alpha_1 u + \beta_1 v$, $y = \alpha_2 u + \beta_2 v$, $z = \alpha_3 u + \beta_3 v$. Compute

$$\left(\frac{\partial f}{\partial u}\right)^2 + \left(\frac{\partial f}{\partial v}\right)^2.$$

7. Compute the derivative of the function $x^2 + 4xy - z^2 + 2yz$ with respect to the curve $x = t$, $y = 2t^2$, $z = 1 - t^3$.

8. Compute the derivative of the function $xyz + x^2 + y^2 - z^2$ with respect to the curve $x = 2t + 1$, $y = \sin t$, $z = e^t$ at the point $t = 0$.

9. A function $f(x) = f(x_1, x_2, \ldots, x_n)$ is said to be *homogeneous of degree k* if

$$f(tx_1, tx_2, \ldots, tx_n) = t^k f(x_1, x_2, \ldots, x_n)$$

for any positive number t. Prove that a function $f(x)$ having continuous first derivatives in R^n is homogeneous of degree k if, and only if, it satisfies

$$\sum_{i=1}^{n} x_i \frac{\partial f}{\partial x_i} - kf = 0 \qquad \text{(Euler's equation)}.$$

[*Hint:* Let $F(x,t) = t^{-k}f(tx)$. Prove that

$$t^{k+1} \frac{\partial F}{\partial t} = \sum_{i=1}^{n} x_i \frac{\partial f}{\partial x_i} - kf.\Big]$$

10. Which of the following functions is homogeneous and of what degree?

(a) $\tan \dfrac{x}{\sqrt{y^2 + z^2}}$;

(b) $\dfrac{x^2}{y} + \dfrac{y^2}{z} + \dfrac{z^2}{x} + \sqrt{x^2 + y^2}$;

(c) $xe^{y/z} + \dfrac{x^2}{y} + 2z$;

(d) $\sqrt{x^2 + 1} + \sqrt{y^2 + 1} - 1$.

11. If $f(x)$ is a homogeneous function of degree k having continuous first and second derivatives, show that $\partial f/\partial x_i$ is homogeneous of degree $k - 1$, for any $i = 1,...,n$.

12. If $w = f(x,y)$, and $y = g(x)$, find dw/dx and d^2w/dx^2.

13. Let $u = f(xy)$. Show that $x(\partial u/\partial x) - y(\partial u/\partial y) = 0$.

14. Use Equation (17) to compute the directional derivatives in Problems 8 and 9 of Section 1.

15. Let $h(x,y) = f(x^2 - y^2, 2xy)$ where $f(u,v)$ is a function having second order continuous derivatives. Compute $h_{xx} + h_{yy}$ in terms of the derivatives of f.

16. Do the same for $h(x,y) = f(e^x \sin y, e^x \cos y)$.

3. THE MEAN VALUE THEOREM AND TAYLOR'S THEOREM

THEOREM 1.   *Let $f(x,y)$ be a function having continuous first derivatives in an open set G of R^2. Let (a,b) and $(a + h, b + k)$ be two points of G such that the line segment connecting them also belongs to G. Then*

$$f(a + h, b + k) - f(a,b) = hf_x(a + \theta h, b + \theta k) + kf_y(a + \theta h, b + \theta k) \quad (1)$$

for some $0 < \theta < 1$.

This theorem is called the *mean value theorem* for a function of two variables.

Proof.   Consider the function

$$\phi(t) = f(a = th, b + tk).$$

By Theorem 5 of Section 2, $\phi'(t)$ exists for $0 \le t \le 1$, and

$$\phi'(t) = hf_x(a + th, b + tk) + kf_y(a + th, b + tk).$$

By the mean value theorem for a function of one variable,

$$\phi(1) - \phi(0) = \phi'(\theta), \qquad 0 < \theta < 1.$$

Hence

$$f(a + h, b + k) - f(a,b) = \phi(1) - \phi(0)$$
$$= hf_x(a + \theta h, b + \theta k) + kf_y(a + \theta h, b + \theta k).$$

We shall apply Theorem 1 to prove the following theorem.

THEOREM 2.   *If $f(x,y)$ is a function having continuous first derivatives in a domain D of R^2, and if $f_x = 0, f_y = 0$ throughout D, then $f = $ constant.*

Proof.   We have to show that for any two points (x_0, y_0) and (x_1, y_1) in D, $f(x_0, y_0) = f(x_1, y_1)$. Let $l = \{l_1, ..., l_k\}$ be a polygonal line with vertices $z^0, z^1, ..., z^k$ that connects (x_0, y_0) to (x_1, y_1) and that is contained in D. We can write

$$z^0 = (a,b), \qquad z^1 = (a + h, b + k).$$

Applying (1) we see that $f(z^0) = f(z^1)$. Similarly, we find that $f(z^1) = f(z^2), ..., f(z^{k-1}) = f(z^k)$. Consequently,

$$f(x_0, y_0) = f(z^0) = f(z^k) = f(x_1, y_1).$$

Let $f(x,y)$ be a function having continuous derivatives of all orders $\leq n + 1$ in an open set G of R^2. Let a line segment connecting two points (a,b) and $(a + h, b + k)$ be contained in G, and set

$$g(t) = f(a + th, b + tk) \qquad \text{for } 0 \leq t \leq 1.$$

Then $g(t)$ is differentiable, and

$$g'(t) = hf_x(a + th, b + tk) + kf_y(a + th, b + tk).$$

Each term on the right is differentiable. Hence $g'(t)$ is differentiable, and

$$g''(t) = h[hf_{xx}(a + th, b + tk) + kf_{xy}(a + th, b + tk)]$$
$$+ k[hf_{yx}(a + th, b + tk) + kf_{yy}(a + th, b + tk)]$$
$$= h^2 f_{xx}(a + th, b + tk) + 2hk f_{xy}(a + th, b + tk)$$
$$+ k^2 f_{yy}(a + th, b + tk).$$

We can proceed in this way step by step. Suppose we have already proved that $g^{(m)}(t)$ exists, and

$$g^{(m)}(t) = h^m \frac{\partial^m f}{\partial x^m}(a+th,b+tk) + \binom{m}{1}h^{m-1}k\frac{\partial^m f}{\partial x^{m-1}\partial y}(a+th,b+tk) + \cdots$$

$$+ \binom{m}{i}h^{m-i}k^i \frac{\partial^m f}{\partial x^{m-i}\partial y^i}(a+th,b+tk) + \cdots \qquad (2)$$

$$+ k^m \frac{\partial^m f}{\partial y^m}(a+th,b+tk).$$

If $m < n+1$, then each term on the right has a continuous derivative. Using the relations

$$\frac{d}{dt}\left[\frac{\partial^m f}{\partial x^{m-i}\partial y^i}(a+th,b+tk)\right] = h\frac{\partial^{m+1}f}{\partial x^{m+1-i}\partial y^i}(a+th,b+tk)$$

$$+ k\frac{\partial^{m+1}f}{\partial x^{m-i}\partial y^{i+1}}(a+th,b+tk),$$

we find that $g^{(m+1)}(t)$ exists and is given by the right-hand side of (2) with m replaced by $(m+1)$.

We shall denote the right-hand side of (2), briefly, by

$$\left(h\frac{\partial}{\partial x} + k\frac{\partial}{\partial y}\right)^m f(x,y)\Bigg|_{(a+th,b+tk)}. \qquad (3)$$

This form is suggested by the analogy with the binomial theorem.

If we substitute $g^{(m)}$ from (2) into Taylor's formula

$$g(1) = g(0) + g'(0) + \cdots + \frac{g^{(n)}\,0)}{n!} + \frac{g^{(n+1)}(\theta)}{(n+1)!} \qquad (0 < \theta < 1)$$

and then use the notation (3), we obtain the following theorem:

THEOREM 3.   *Let $f(x,y)$ be a function having continuous partial derivatives of all orders $\leq n+1$ in an open set G. Let the line segment connecting two points (a,b) and $(a+h,b+k)$ be contained in G. Then*

$$f(a+h,b+k) = f(a,b) + \sum_{m=1}^{n}\frac{1}{m!}\left(h\frac{\partial}{\partial x} + k\frac{\partial}{\partial y}\right)^m f(x,y)\Bigg|_{(a,b)} + R_{n+1} \qquad (4)$$

where

$$R_{n+1} = \frac{1}{(n+1)!}\left(h\frac{\partial}{\partial x} + k\frac{\partial}{\partial y}\right)^{n+1} f(x,y)\Bigg|_{(a+\theta h,b+\theta k)} \qquad (0 < \theta < 1). \qquad (5)$$

Theorem 3 is called *Taylor's theorem* for a function of two variables. Formula (4) is called *Taylor's formula*, and R_{n+1} is called the *remainder*.

COROLLARY. *If $f(x,y)$ has partial derivatives of all orders, and if $R_{n+1} \to 0$ as $n \to \infty$, then*

$$f(a + h, b + k) = f(a,b) + \sum_{m=1}^{\infty} \frac{1}{m!} \left(h \frac{\partial}{\partial x} + k \frac{\partial}{\partial y} \right)^m f(x,y) \Big|_{(a,b)}. \qquad (6)$$

The series on the right is called *Taylor's series* for $f(x,y)$ about (a,b).

Theorems 1–3 can be extended to functions f of any number of variables. The proofs are similar to the above proofs.

PROBLEMS

1. State and prove Theorem 1 for a function $f(x)$ of n variables.

2. State and prove Theorem 2 for a function $f(x)$ of n variables.

3. State and prove Theorem 3 for a function $f(x,y,z)$ of three variables.

4. Find Taylor's series for $\sin x \sin y$ about $(0,0)$.

5. Find Taylor's series for $x^2 y + xy^2 + 1$ about $(1,1)$.

6. Write all the terms of Taylor's series of e^{x^2+y} about $(0,0)$, which involve powers $x^\alpha y^\beta$ up to order 3 (that is, $\alpha + \beta \le 3$).

7. Do the same for the function $1/(x^2 + y^2)$ about $(1,1)$.

4. COMPUTATION OF EXTREMUM

Let $f(x)$ be a function defined in a domain D of R^n, and let $a \in D$. Suppose there exists a δ-neighborhood B_δ of x such that

$$f(x) \le f(a) \qquad \text{for all } x \in B_\delta, \, x \in D.$$

Then we say that f has a *relative maximum* (or a *local maximum*) at a. Similarly, if for some point b of D,

$$f(x) \ge f(b)$$

for all x in D and in some δ_0-neighborhood of b, then we say that f has a *relative minimum* (or a *local minimum*) at b.

If $f(x)$ has either a relative maximum or a relative minimum at a point e of D, then we say that e is a point of *relative extremum*.

THEOREM 1.   *Let $f(x)$ have first derivatives in a domain D of R^n. If $f(x)$ has a relative extremum at a point α of D, then*

$$\frac{\partial f}{\partial x_i}(\alpha) = 0 \qquad for\ i = 1,...,n. \tag{1}$$

Proof.   Write $\alpha = (\alpha_1,...,\alpha_n)$. The function

$$\phi(t) = f(\alpha_1,...,\alpha_{i-1},t,\alpha_{i+1},...,\alpha_n)$$

has a relative extremum at $t = \alpha_i$. Hence $\phi'(\alpha_i) = 0$. This gives (1).

A point α for which (1) holds is called a *critical point* of f. Thus Theorem 1 asserts that points of relative extremum are critical points. We are now led to the problem of determining when a critical point is a point of relative maximum or minimum. We first consider the case $n = 2$.

THEOREM 2.   *Let $f(x,y)$ have continuous first two partial derivatives in a domain D of R^2 and let (a,b) be a critical point of f in D. Set*

$$A = f_{xx}(a,b), \qquad B = f_{xy}(a,b), \qquad C = f_{yy}(a,b).$$

Then:

(a)   *If $B^2 - AC < 0$ and $A > 0$, then f has relative minimum at (a,b).*

(b)   *If $B^2 - AC < 0$ and $A < 0$, then f has relative maximum at (a,b).*

(c)   *If $B^2 - AC > 0$, then f has neither relative maximum nor relative minimum at (a,b).*

If $B^2 - AC = 0$, then no conclusion can be drawn, and one has to compute higher derivatives in order to analyze the nature of the critical point.

Proof.   From Taylor's formula we get

$$f(a + h, b + k) - f(a,b) = Ah^2 + 2Bhk + Ck^2 + \eta\sqrt{h^2 + k^2}$$

where $\eta \to 0$ if $(h,k) \to 0$. Set

$$h = r \cos \phi, \qquad k = r \sin \phi,$$

$$M(\phi) = A \cos^2\phi + 2B \sin \phi \cos \phi + C \sin^2 \phi.$$

Then

$$f(a + h, b + k) - f(a,b) = r^2[M(\phi) + \gamma(r,\phi)] \tag{2}$$

where

$$\sup_{\phi} |\gamma(r,\phi)| \to 0 \qquad if\ r \to 0.$$

Consider the function

$$\psi(t) = At^2 + 2Bt + C.$$

If $B^2 - AC < 0$, then $A \neq 0$, and

$$\psi(t) = A\left(t + \frac{B}{A}\right)^2 + \frac{AC - B^2}{A}.$$

We see that $\psi(t) > 0$ if $A > 0$ and $\psi(t) < 0$ if $A < 0$. Suppose $A > 0$. Then $M(\phi) > 0$ if $0 \le \phi \le 2\pi$. It follows that $M(\phi) \ge c$ for some positive constant c. If $\bar{r}$ is such that $\gamma(r,\phi) \le c/2$ for $r < \bar{r}$, $0 \le \phi \le 2\pi$, then (2) gives

$$f(a + h, b + k) - f(a,b) \ge r^2\left(c - \frac{c}{2}\right) = \frac{c}{2} r^2 \qquad \text{if } r \le \bar{r}, 0 \le \phi \le 2\pi.$$

This shows that f has a relative minimum at (a,b).

Similarly, if $A < 0$, then f has a relative maximum at (a,b). Suppose finally that $B^2 - AC > 0$. If $A \neq 0$ then $\psi(t) = 0$ for two values of t, given by

$$t_1 = \frac{-B + \sqrt{B^2 - AC}}{A}, \qquad t_2 = \frac{-B - \sqrt{B^2 - AC}}{A}.$$

It is easily verified that

$$\psi(t) = A(t - t_1)(t - t_2).$$

Hence $\psi(t) < 0$ if $t_2 < t < t_1$ and $\psi(t) > 0$ if $t > t_1$ or if $t < t_2$. (We have assumed here, for definiteness, that $t_2 < t_1$.) It follows that there are values ϕ_1 and ϕ_2 such that

$$M(\phi_1) > 0, \qquad M(\phi_2) < 0. \tag{3}$$

If $B^2 - AC > 0$ and $A = 0$, then $\psi(t) = 2Bt + C$, and the assertion (3) is again valid.

Taking r_0 to be a sufficiently small positive number, we get

$$M(\phi_1) + \gamma(r,\phi_1) > 0, \qquad M(\phi_2) + \gamma(r,\phi_2) < 0$$

if $0 < r \le r_0$. From (2) it then follows that $f(x,y)$ cannot have relative maximum or relative minimum at (a,b).

We shall extend Theorem 2 to functions of n variables.

Let

$$B = (B_{ij}) = \begin{pmatrix} B_{11} B_{12} \cdots B_{1n} \\ B_{21} B_{22} \cdots B_{2n} \\ \cdot \quad \cdot \quad \cdots \quad \cdot \\ B_{n1} B_{n2} \cdots B_{nn} \end{pmatrix}$$

be a matrix with n rows and n columns. We denote by det B the determinant of B. We say that B is *symmetric* if $B_{ij} = B_{ji}$. For a real variable λ, consider the determinant

$$\det(B - \lambda I)$$

where I is the identity matrix, that is, $I = (I_{ij})$ where $I_{ij} = 0$ if $i \neq j$ and $I_{jj} = 1$. The above determinant is a polynomial in λ, called the *characteristic polynomial* of B. If B is symmetric, then all the *characteristic roots* (that is, the roots of the characteristic polynomial) are real numbers.

A symmetric matrix B is said to be *positive definite* if

$$\sum_{i,j=1}^{n} B_{ij} x_i x_j > 0 \qquad \text{whenever } \|x\| \neq 0.$$

It is said to be *negative definite* if

$$\sum_{i,j=1}^{n} B_{ij} x_i x_j < 0 \qquad \text{whenever } \|x\| \neq 0.$$

It is said to be *indefinite* if there are vectors x and y such that

$$\sum_{i,j=1}^{n} B_{ij} x_i x_j > 0$$

and

$$\sum_{i,j=1}^{n} B_{ij} y_i y_j < 0.$$

It is known from linear algebra that if the characteristic roots of B are all positive, then B is positive definite, whereas if the characteristic roots of B are all negative, then B is negative definite. Finally, if there is at least one positive root and at least one negative root, then B is indefinite.

From Taylor's formula for a function $f(x)$ of n variables we get: If α is a critical point of f, then

$$f(\alpha + h) - f(\alpha) = \sum_{i,j=1}^{n} \frac{\partial^2 f(\alpha)}{\partial x_i \, \partial x_j} h_i h_j + \eta \|h\|^2 \quad (h = (h_1, \cdots, h_n)) \qquad (4)$$

where $\eta = \eta(h) \to 0$ if $h \to 0$. Set

$$A_{ij} = \frac{\partial^2 f(\alpha)}{\partial x_i \, \partial x_j}. \qquad (5)$$

Note that the matrix $A = (A_{ij})$ is symmetric. We now state the extension of Theorem 2 to functions of n variables:

THEOREM 3.   *Let $f(x)$ have continuous first two derivatives in a domain D of R^n, and let α be a critical point of f in D. Define A_{ij} by (5). Then:*

(a) *If (A_{ij}) is positive definite, then f has relative minimum at α.*

(b) *If (A_{ij}) is negative definite, then f has relative maximum at α.*

(c) *If (A_{ij}) is indefinite, then f has neither relative minimum nor relative maximum at α.*

Proof. Suppose (a) holds. The function $\sum A_{ij}x_i x_j$ is positive on the set $\{x; \|x\| = 1\}$. Thus it attains a positive minimum c. It follows that

$$\sum A_{ij}x_i x_j \geq c\,\|x\|^2 \qquad \text{for all } x \neq 0.$$

Using this in (4) we get

$$f(\alpha + h) - f(\alpha) \geq c\,\|h\|^2 + \eta\,\|h\|^2.$$

Since $\eta(h) \to 0$ if $h \to 0$, it follows that $|\eta(h)| < c/2$ if $\|h\| \leq \delta$ for some $\delta > 0$. Hence

$$f(\alpha + h) - f(\alpha) > \frac{c}{2}\,\|h\|^2 > 0 \qquad \text{if } \|h\| < \delta.$$

Thus f has a relative minimum at α.

The proof of (b) is similar. The proof of (c) is left to the student.

PROBLEMS

1. Prove the assertion (c) of Theorem 3.

2. Find the critical points of each of the following functions, and use Theorem 3 to test the nature of the critical points:

(a) $x^2 - 8x^2 y + 3y^2$;

(b) $x^4 + y^4 - 2x^2 + 4xy - 2y^2$;

(c) $\dfrac{1}{xy} - \dfrac{4}{x^2 y} - \dfrac{8}{xy^2}$;

(d) $x^3 + 12xy + y^3 - 1$;

(e) $\dfrac{xy}{5} + \dfrac{1}{x} - \dfrac{2}{y}$;

(f) $(x + y)e^{-x^2 - y^2}$;

(g) $6x \sin y - 2x^2 \sin y + \dfrac{1}{2}x^2 \sin y \cos y$;

(h) $x^3 y^2 (2 - x - y)$.

3. Find the critical points and discuss their nature, for each of the following functions:

(a) $\dfrac{a}{x} + \dfrac{b}{y} + xy \qquad (a > 0, b > 0)$;

(b) $(ax^2 + by^2)e^{-x^2 - y^2} \qquad (b > a > 0)$.

4. Find the shortest distance from the point $(1,-1,1)$ to the set of points given by $z = xy$.

5. Find the shortest distance from the point $(0,0,a)$ to the set of points given by $z = xy$, where $a > 0$.

6. Find the critical points of

$$f(x,y,z) = (ax^2 + by^2 + cz^2)e^{-x^2-y^2-z^2} \qquad (a > b > c > 0)$$

and analyze their nature.

5. IMPLICIT FUNCTION THEOREMS

Let $F(x,y)$ be a function defined in an open set G of R^2. We are concerned with the set of points (x,y) in G satisfying $F(x,y) = 0$. Can we write the points of this set in the form $y = g(x)$? If so, then we say that the equation $F(x,y) = 0$ defines the function $y = g(x)$ *implicitly*, and we call the equation

$$F(x,y) = 0 \qquad (1)$$

the *implicit equation* for $y = g(x)$.

If $F(x,y) = x^2 + y^2$ in R^2, then the only solution of (1) is $(0,0)$. Thus there is no function $y = g(x)$ that is defined on some interval of positive length.

If $F(x,y) = x^2 + y^2 - 1$ in R^2, then there are two functions $y = g(x)$ for which $(x,g(x))$ satisfies (1) for $-1 \le x \le 1$, namely,

$$y = g(x) = \pm\sqrt{1 - x^2}.$$

Note, however, that if we restrict (x,y) to belong to a small neighborhood of a solution (x_0,y_0) of (1) with $|x_0| < 1$, then there is a unique solution $y = g(x)$ satisfying $g(x_0) = y_0$.

We shall now prove a general theorem asserting that the solutions (x,y) of (1) in a small neighborhood of a point (x_0,y_0) have the form $y = g(x)$. This theorem is called the *implicit function theorem* for a function of two variables.

THEOREM 1. *Let $F(x,y)$ be a function defined in an open set G of R^2, having continuous first derivatives and let (x_0,y_0) be a point of G for which*

$$F(x_0,y_0) = 0, \qquad (2)$$

$$F_y(x_0,y_0) \ne 0. \qquad (3)$$

Then there exists a rectangle R in G defined by

$$|x - x_0| < \alpha, \qquad |y - y_0| < \beta, \qquad (4)$$

such that the points (x,y) in R that satisfy (1) *have the form* $(x,g(x))$ *where* $g(x)$ *is a function having a continuous derivative in* $|x - x_0| < \alpha$. *Furthermore,*

$$g'(x) = -\frac{F_x(x,g(x))}{F_y(x,g(x))} \qquad if \ |x - x_0| < \alpha. \tag{5}$$

The condition (3) cannot be omitted, as shown by the example of $F(x,y) = x^2 + y^2$, $(x_0,y_0) = (0,0)$.

Proof. We may suppose that $F_y(x_0,y_0) > 0$. By continuity, $F_y(x,y) > 0$ if $|x - x_0| \le d$, $|y - y_0| \le d$ where d is a small positive number. Consider the function $\phi(y) = F(x_0,y)$. It is strictly monotone increasing since $\phi'(y) = F_y(x_0,y) > 0$. Since $\phi(y_0) = F(x_0,y_0) = 0$, it follows that

$$F(x_0,y_0 - d) = \phi(y_0 - d) < 0 < \phi(y_0 + d) = F(x_0,y_0 + d).$$

Using the continuity of $F(x,y_0 - d)$ and of $F(x,y_0 + d)$ we deduce that

$$F(x,y_0 - d) < 0, \qquad F(x,y_0 + d) > 0 \tag{6}$$

if $|x - x_0|$ is sufficiently small, say, if $|x - x_0| \le d_1$. See Figure 7.1.

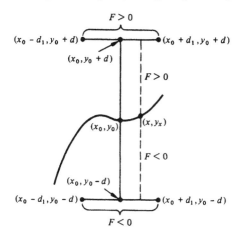

Figure 7.1.

Consider, in the interval $|y - y_0| \le d$, the continuous function $\psi(y) = F(x,y)$ for x fixed, $|x - x_0| \le d_2$, $d_2 = \min(d,d_1)$. It is strictly monotone increasing since $\psi_y = F_y(x,y) > 0$. Also, by (6), $\psi(y_0 - d) < 0$, $\psi(y_0 + d) > 0$. By the intermediate value theorem, there is a point y_x in the interval $(y_0 - d, y_0 + d)$ satisfying $\psi(y_x) = 0$. Since ψ is strictly monotone, y_x is unique. Writing $y_x = g(x)$, we have proved that the solutions of Equation (1) for $|x - x_0| < d_2$, $|y - y_0| < d$ have the form $(x,g(x))$.

We shall next prove that $g(x)$ is continuous. For any $\varepsilon > 0$, $\varepsilon \leq d$, $F(x_0, y_0 + \varepsilon) > 0 > F(x_0, y_0 - \varepsilon)$. Repeating the argument given above with $y_0 \pm d$ replaced by $y_0 \pm \varepsilon$, we see that there exists a number $d_2(\varepsilon)$ such that if $|x - x_0| < d_2(\varepsilon)$, then there is a unique $y = \tilde{y}_x$ in $(y_0 - \varepsilon, y_0 + \varepsilon)$ satisfying $F(x, \tilde{y}_x) = 0$. By uniqueness of the solution y of $F(x, y) = 0$ in $(y_0 - d, y_0 + d)$, it follows that $\tilde{y}_x = g(x)$. Hence

$$|g(x) - y_0| < \varepsilon \qquad \text{if } |x - x_0| \leq d_2(\varepsilon).$$

This proves the continuity of $g(x)$ at x_0.

Now let x_1 be any point in $|x - x_0| < d_2$. Then $F(x_1, y_1) = 0$ where $y_1 = g(x_1)$, and $F_y(x_1, y_1) > 0$. We therefore can apply the proof of the continuity of $g(x)$ at x_0 and deduce the continuity of $g(x)$ at x_1.

We proceed to prove that $g(x)$ is differentiable. We begin with the relation

$$F(x_0 + h, g(x_0 + h)) - F(x_0, g(x_0)) = 0$$

where $|h| < d_2$. Writing $g(x_0 + h) = g(x_0) + \Delta g$ and using the differentiability of F, we get

$$hF_x(x_0, y_0) + \Delta g \cdot F_y(x_0, y_0) + \eta \sqrt{h^2 + (\Delta g)^2} = 0 \qquad (7)$$

where $\eta = \eta(h, \Delta g) \to 0$ if $(h, \Delta g) \to 0$. Since $g(x)$ is continuous, $\Delta g \to 0$ if $h \to 0$. Hence $\eta \to 0$ if $h \to 0$.

Writing $F_x(x_0, y_0) = F_x$, $F_y(x_0, y_0) = F_y$, and dividing both sides of (7) by hF_y, we find that

$$\frac{\Delta g}{h} = -\frac{F_x}{F_y} + \frac{\eta}{hF_y} \sqrt{h^2 + (\Delta g)^2}. \qquad (8)$$

If $|h|$ is sufficiently small, $|\eta/F_y| < 1/2$. Hence

$$\frac{|\Delta g|}{|h|} \leq \frac{|F_x|}{|F_y|} + \frac{1}{2}\left(1 + \frac{|\Delta g|}{|h|}\right).$$

It follows that

$$\frac{|\Delta g|}{|h|} \leq C, \qquad C \text{ constant.}$$

Using this in (8) and taking $h \to 0$, we conclude that

$$\lim_{h \to 0} \frac{\Delta g}{h}$$

exists and is equal to $-F_x/F_y$.

We have thus proved that $g'(x)$ exists at x_0 and that (5) holds at x_0. The same argument can be applied at any point x in $|x - x_0| < d_2$. Thus $g'(x)$

exists and it satisfies (5). Since the right-hand side of (5) is continuous, the same is true of $g'(x)$. This completes the proof of the theorem.

If instead of (3) we assume that $F_x(x_0,y_0) \neq 0$, then we can prove an analog of Theorem 1 with the roles of x and y interchanged.

The proof of Theorem 1 extends to the case where $x = (x_1,\dots,x_n)$. The result is the following *implicit function theorem* for a function of several variables:

THEOREM 2. *Let* $F(x,y) = F(x_1,\dots,x_n,y)$ *be a function defined in an open set* G *of* R^{n+1} *and let* $(x^0,y^0) = (x_1{}^0,\dots,x_n{}^0,y^0)$ *be a point of* G. *Assume that*

$$F(x^0,y^0) = 0, \qquad F_y(x^0,y^0) \neq 0. \qquad (9)$$

Then there exists an $(n + 1)$*-dimensional rectangle* R *defined by*

$$|x_i - x_i{}^0| < \alpha \,(1 \leq i \leq n), \qquad |y - y^0| < \beta \qquad (10)$$

such that for any x *as in* (10) *there exists a unique solution* $y = y_x$ *of* $F(x,y) = 0$ *in the interval* $|y - y^0| < \beta$. *Writing* $y_x = g(x)$, *the function* $g(x)$ *is continuously differentiable, and*

$$\frac{\partial}{\partial x_i} g(x) = - \frac{(\partial F/\partial x_i)(x,g(x))}{(\partial F/\partial y)(x,g(x))} \qquad (1 \leq i \leq n). \qquad (11)$$

Consider next a more complicated situation in which we want to solve two equations simultaneously:

$$F(x,y,z,u,v) = 0, \qquad (12)$$

$$G(x,y,z,u,v) = 0. \qquad (13)$$

We introduce the determinant

$$J = \frac{\partial(F,G)}{\partial(u,v)} = \begin{vmatrix} F_u & F_v \\ G_u & G_v \end{vmatrix}$$

called the *Jacobian* of F, G with respect to u, v.

THEOREM 3. *Let* F *and* G *have continuous first derivatives in an open set* D *of* R^5 *containing a point* $P_0 = (x_0,y_0,z_0,u_0,v_0)$. *Assume that*

$$F(x_0,y_0,z_0,u_0,v_0) = 0, \qquad G(x_0,y_0,z_0,u_0,v_0) = 0 \qquad (14)$$

and

$$\left. \frac{\partial(F,G)}{\partial(u,v)} \right|_{(x_0,y_0,z_0,u_0,v_0)} \neq 0. \qquad (15)$$

Then there exists a cube

$$R: |x - x_0| < \alpha, \qquad |y - y_0| < \alpha, \qquad |z - z_0| < \alpha$$

and a rectangle

$$S: |u - u_0| < \beta_1, \qquad |v - v_0| < \beta_2$$

such that for any (x,y,z) in R there is a unique pair (u,v) in S for which (12) *and* (13) *hold. Writing*

$$u = f(x,y,z), \qquad v = g(x,y,z),$$

the functions f and g have continuous first derivatives in R, and

$$f_x = -\frac{1}{J}\frac{\partial(F,G)}{\partial(x,v)} = -\frac{1}{J}\begin{vmatrix} F_x & F_v \\ G_x & G_v \end{vmatrix}, \tag{16}$$

$$g_x = -\frac{1}{J}\frac{\partial(F,G)}{\partial(u,x)} = -\frac{1}{J}\begin{vmatrix} F_u & F_x \\ G_u & G_x \end{vmatrix}. \tag{17}$$

Similar formulas hold for f_y, f_z, g_y, g_z.

Proof. Since $J \neq 0$ at P_0, either $F_v \neq 0$ or $G_v \neq 0$ at P_0. Suppose $F_v \neq 0$ at P_0. By Theorem 2, if (x,y,z,u) lies in a small rectangle T with center (x_0,y_0,z_0,u_0), then there exists a unique solution $v = \phi(x,y,z,u)$ of (12) in some small interval $|v - v_0| < \beta_2$. Let

$$H(x,y,z,u) = G(x,y,z,u,\phi(x,y,z,u)).$$

Then (u,v) is a solution of (12) and (13) (when $(x,y,z,u) \in T$, $|v - v_0| < \beta_2$) if, and only if, $v = \phi(x,y,z,u)$ and

$$H(x,y,z,u) = 0. \tag{18}$$

ϕ has continuous first derivatives and $\phi_u = -F_u/F_v$. Hence

$$H_u = G_u + G_v\phi_u = G_u - \frac{G_v F_u}{F_v} = \frac{G_u F_v - G_v F_u}{F_v} = -\frac{J}{F_v} \neq 0$$

at P_0. We therefore can apply Theorem 2. We conclude that for any (x,y,z) in a small cube R with center (x_0,y_0,z_0) there is a unique solution u of (18) in some interval $|u - u_0| < \beta_1$; the points (x,y,z,u) belong to T. Furthermore, this solution u has the form

$$u = g(x,y,z)$$

where g has continuous first derivatives. It follows that

$$v = \phi(x,y,z,g(x,y,z))$$

also has continuous first derivatives.

It remains to prove (16) and (17). To do this we differentiate the equations

$$F(x,y,z,f(x,y,z),g(x,y,z)) = 0$$
$$G(x,y,z,f(x,y,z),g(x,y,z)) = 0$$

with respect to x, and get

$$F_x + F_u f_x + F_v g_x = 0, \qquad G_x + G_u f_x + G_v g_x = 0.$$

Solving for f_x, g_x, we get (16) and (17).

We conclude this section with a statement of the most general implicit function theorem for a system of functions. Let

$$F_i(x,u) = F_i(x_1,...,x_n,u_1,...,u_r) \qquad (1 \le i \le r)$$

be functions having continuous first derivatives in an open set containing a point (x^0,u^0). The matrix

$$\begin{pmatrix} \dfrac{\partial F_1}{\partial u_1} & \dfrac{\partial F_1}{\partial u_2} & \cdots & \dfrac{\partial F_1}{\partial u_r} \\[2mm] \dfrac{\partial F_2}{\partial u_1} & \dfrac{\partial F_2}{\partial u_2} & \cdots & \dfrac{\partial F_2}{\partial u_r} \\[2mm] \cdot & \cdot & \cdots & \cdot \\[2mm] \dfrac{\partial F_r}{\partial u_1} & \dfrac{\partial F_r}{\partial u_2} & \cdots & \dfrac{\partial F_r}{\partial u_r} \end{pmatrix}$$

or briefly, $(\partial F_i/\partial u_j)$ is called the *Jacobian matrix* of $(F_1,...,F_r)$ with respect to $(u_1,...,u_r)$. The determinant of this matrix is called the *Jacobian* of $(F_1,...,F_r)$ with respect to $(u_1,...,u_r)$ and is denoted by

$$J = \frac{\partial(F_1,...,F_r)}{\partial(u_1,...,u_r)}.$$

THEOREM 4.   *Let $F_1,...,F_r$ have continuous first derivatives in a neighborhood of a point (x^0,u^0). Assume that*

$$F_i(x^0,u^0) = 0 \,(1 \le i \le r), \quad \frac{\partial(F_1,...,F_r)}{\partial(u_1,...,u_r)} \ne 0 \qquad at \ (x^0,u^0). \tag{19}$$

Then there is a δ-neighborhood R of x^0 and a γ-neighborhood S of u^0 such that for any x in R there is a unique solution u of

$$F_i(x,u) = 0 \qquad (1 \le i \le r) \tag{20}$$

in S. The vector valued function $u(x) = (u_1(x),...,u_r(x))$ thus defined has continuous first derivatives in R.

In order to compute $\partial u_i/\partial x_j$ $(1 \leq i \leq r)$ for a fixed j, we differentiate the equations

$$F_1(x,u(x)) = 0,\ldots, F_r(x,u(x)) = 0$$

with respect to x_j. We obtain the system of linear equations for $\partial u_i/\partial x_j$:

$$\frac{\partial F_k}{\partial x_j} + \sum_{i=1}^{r} \frac{\partial F_k}{\partial u_i} \frac{\partial u_i}{\partial x_j} = 0 \qquad (1 \leq k \leq r). \tag{21}$$

The system of linear equations (21) in the unknowns $\partial u_i/\partial x_j$ can be uniquely solved, since the determinant of the coefficients matrix, which is precisely the Jacobian $\partial(F_1,\ldots,F_r)/\partial(u_1,\ldots,u_r)$, is different from 0.

We briefly give the proof of Theorem 4. It is based upon induction on r. Without loss of generality we may assume that

$$\frac{\partial(F_1,\ldots,F_{r-1})}{\partial(u_1,\ldots,u_{r-1})} \neq 0.$$

Therefore, by the inductive assumption, the solution of

$$F_i(x,u) = 0, \qquad 1 \leq i \leq r - 1$$

in a neighborhood of (x^0,u^0) is given by $u_i = \phi_i(x,u_r)$ $(1 \leq i \leq r - 1)$. Let

$$G(x,u_r) = F_r(x,\phi_1(x,u_r),\ldots,\phi_{r-1}(x,u_r),u_r). \tag{22}$$

If we show that

$$\frac{\partial G}{\partial u_r} \neq 0 \qquad \text{at } (x^0,u^0), \tag{23}$$

then we can use Theorem 2 to solve the equation $G(x,u_r) = 0$. To prove (23), differentiate the equations

$$F_i(x,\phi_1(x,u_r),\ldots,\phi_{r-1}(x,u_r),u_r) = 0 \qquad (1 \leq i \leq r - 1)$$

with respect to u_r to obtain

$$\sum_{j=1}^{r-1} \frac{\partial F_i}{\partial u_j} \frac{\partial \phi_j}{\partial u_r} + \frac{\partial F_i}{\partial u_r} = 0 \qquad (1 \leq i \leq r - 1). \tag{24}$$

Differentiate also (22) with respect to u_r to obtain:

$$\sum_{j=1}^{r-1} \frac{\partial F_r}{\partial u_j} \frac{\partial \phi_j}{\partial u_r} - \frac{\partial G}{\partial u_r} + \frac{\partial F_r}{\partial u_r} = 0. \tag{25}$$

Solving the linear system (24) and (25) (in the unknowns $\partial \phi_j/\partial u_r$, $\partial G/\partial u_r$) for $\partial G/\partial u_r$, we obtain

$$\frac{\partial G}{\partial u_r} = \frac{\partial(F_1,\ldots,F_r)/\partial(u_1,\ldots,u_r)}{\partial(F_1,\ldots,F_{r-1})/\partial(u_1,\ldots,u_{r-1})}.$$

This gives (23).

PROBLEMS

1. Can the equation $\sqrt{x^2 + y^2 + z^2} - \cos z = 0$ be solved uniquely for y in terms of x,z in a neighborhood of $(0,1,0)$? Can it be solved uniquely for z in terms of x,y in such a neighborhood?

2. Can the equation $x + y + z - \sin xyz = 0$ be solved uniquely in the form $z = f(x,y)$ in a neighborhood of $(0,0,0)$?

3. Let $y = y(x)$ be a twice continuously differentiable function satisfying $F(x,y) = 0$, where $F(x,y)$ is a function having first two continuous derivatives. Prove that if $F_y \neq 0$, then

$$F_y{}^3 y'' = \begin{vmatrix} F_{xx} & F_{xy} & F_x \\ F_{xy} & F_{yy} & F_y \\ F_x & F_y & 0 \end{vmatrix}.$$

4. Find the solution $y = f(x)$ of $x^2 + y^2 - x^3 = 0$ near each of the following points: (a) $(5,10)$; (b) $(10, -30)$.

5. Can the equation $xy - z \log y + e^{xz} = 1$ be solved uniquely in the form $z = f(x,y)$ in a neighborhood of $(0,1,1)$? Can it be solved uniquely in the form $y = g(x,z)$ in such a neighborhood?

6. If in Theorem 1, $F(x,y)$ has continuous derivatives of all orders $\leq k$, then $g(x)$ also has continuous derivatives of all orders $\leq k$.

7. If in Theorem 3, F and G have second continuous derivatives, then the same is true of f and g.

8. Can the system of equations

$$u^2 + v^2 - x^2 - y = 0, \qquad u + v - x^2 + y = 0$$

be uniquely solved for u,v in terms of x,y in a neighborhood of

$$(x_0,y_0,u_0,v_0) = (2,1,1,2)?$$

9. Can the system of equations

$$xy^2 + xzu + yv^2 = 3, \qquad u^3yz - 2xv + u^2v^2 = 0$$

be uniquely solved for u,v in terms of x,y,z, in a neighborhood of $(x_0,y_0,z_0,u_0,v_0) = (1,1,1,1,1)$?

10. Show that the assertion of Theorem 3 remains true when either R is a ball with center (x_0,y_0,z_0), or S is a disc with center (u_0,v_0), or both.

6. INVERSE MAPPINGS

A vector valued function also is called a mapping or a transformation.

Let $f(x)$ be a mapping with domain D and range T. We say that f is one to one if for any y in T there exists a unique x in D such that $f(x) = y$. We then define the *inverse mapping* g by $g(y) = x$. It is clear that

$$g(f(x)) = x \quad \text{if } x \in D,$$
$$f(g(y)) = y \quad \text{if } y \in T.$$

EXAMPLE.   The inverse mapping of

$$u = 2x - 3y, \qquad v = -x + 2y$$

is given by

$$x = 2u + 3v, \qquad y = u + 2v.$$

Let $y = f(x)$ be a mapping from a set in R^n into R^n. Write this mapping in the more explicit form

$$
\begin{aligned}
y_1 &= f_n(x_1, \dots, x_n), \\
&\quad \dots \\
y_n &= f_n(x_1 \dots, x_n).
\end{aligned}
\tag{1}
$$

We say that f is *continuously differentiable* if all the functions f_i have continuous first derivatives.

The *Jacobian of the transformation* (1) is the determinant

$$J = \frac{\partial(f_1, \dots, f_n)}{\partial(x_1, \dots, x_n)} = \det\left(\frac{\partial f_i}{\partial x_j}\right).$$

THEOREM 1.   *Let $y = f(x)$ be a continuously differentiable mapping from an open set D of R^n into R^n, and let $x^0 \in D$, $y^0 = f(x^0)$. If $J(x^0) \neq 0$, then there exist neighborhoods*

$$R = \{x; \|x - x^0\| < \alpha\}, \qquad S = \{y; \|y - y^0\| < \beta\} \tag{2}$$

such that for each $y \in S$ there is a unique $x = g(y)$ such that $f(x) = y$. Furthermore, the mapping g is continuously differentiable in S.

Theorem 1 is called an *inverse mapping theorem*. It asserts the existence of an inverse g to f. The existence of the inverse g is asserted only *locally*, that is, in some neighborhood of a given point. If one can establish the existence of an inverse in the whole range of f, then one speaks of a *global* inverse.

Proof. Let

$$F_i(x,y) = f_i(x) - y_i \qquad (1 \le i \le n).$$

Then $y = f(x)$ if, and only if,

$$F_i(x,y) = 0 \qquad (1 \le i \le n). \tag{3}$$

By assumption, $F_i(x^0,y^0) = 0$ for $1 \le i \le n$. Furthermore, since $\partial F_i/\partial x_j = \partial f_i/\partial x_j$,

$$\frac{\partial(F_1,\ldots,F_n)}{\partial(x_1,\ldots,x_n)} = J \ne 0 \qquad \text{at } (x^0,y^0).$$

We therefore can use the implicit function theorem (Theorem 4 of Section 5) and conclude that there exist neighborhoods R and S as in (2) such that for any $y \in S$ there is a unique $x = g(y)$ in R for which (3) holds. Furthermore, $g(y)$ is continuously differentiable in S. This completes the proof.

Consider the transformation

$$x = r \cos \phi, \tag{4}$$
$$y = r \sin \phi,$$

from the set of all points (r,ϕ) with $0 \le r < \infty$, $-\infty < \phi < \infty$, into R^2. This is a differentiable transformation with Jacobian

$$\frac{\partial(x,y)}{\partial(r,\phi)} = \begin{vmatrix} \cos \phi & -r \sin \phi \\ \sin \phi & r \cos \phi \end{vmatrix} = r.$$

Thus if $r > 0$, then (4) has a local, continuously differentiable inverse. In fact, this inverse is given by

$$r = \sqrt{x^2 + y^2}, \tag{5}$$
$$\phi = \tan^{-1} \frac{y}{x}$$

where the "branch" of $\tan^{-1}$ taken will depend on the points (r_0,ϕ_0), (x_0,y_0) around which one wants to set up the local inverse.

PROBLEMS

1. Consider the transformation

$$x = r \sin \theta \cos \phi,$$
$$y = r \sin \theta \sin \phi, \tag{6}$$
$$z = r \cos \theta$$

where $r > 0, 0 \le \phi < 2\pi, 0 \le \theta \le \pi$. Show that the Jacobian is equal to $r^2 \sin \theta$. Show also that the inverse is given by

$$
\begin{aligned}
r &= \sqrt{x^2 + y^2 + z^2}, \\
\theta &= \tan^{-1} \frac{\sqrt{x^2 + y^2}}{z}, \\
\phi &= \tan^{-1} \frac{y}{x}.
\end{aligned}
\tag{7}
$$

2. Let $u(x,y)$ be a function having two continuous derivatives. Denote by $v(r,\phi)$ the composite function $u(x(r,\phi),y(r,\phi))$ where $x(r,\phi)$, $y(r,\phi)$ are given by (4). Prove that

$$
\frac{\partial^2 u}{\partial x^2} + \frac{\partial^2 u}{\partial y^2} = \frac{\partial^2 v}{\partial r^2} + \frac{1}{r} \frac{\partial v}{\partial r} + \frac{1}{r^2} \frac{\partial^2 v}{\partial \phi^2}.
$$

3. Use the result of the preceding problem to prove that $u = \log r$ satisfies $u_{xx} + u_{yy} = 0$.

4. Let $u(x,y,z)$ be a function having two continuous derivatives. Denote by $v(r,\phi,\theta)$ the composition of $u(x,y,z)$ with the mapping (6). Prove that

$$
\frac{\partial^2 u}{\partial x^2} + \frac{\partial^2 u}{\partial y^2} + \frac{\partial^2 u}{\partial z^2} = \frac{1}{r^2} \frac{\partial}{\partial r} \left(r^2 \frac{\partial v}{\partial r} \right) + \frac{1}{r^2 \sin^2 \theta} \frac{\partial^2 v}{\partial \phi^2} + \frac{1}{r^2 \sin \theta} \frac{\partial}{\partial \theta} \left(\sin \theta \frac{\partial v}{\partial \theta} \right).
$$

5. Use the result of the preceding problem to compute $u_{xx} + u_{yy} + u_{zz}$ when $u = r^\alpha$. For which α is $u_{xx} + u_{yy} + u_{zz} = 0$?

6. A function $u(x,y,z)$ is called *harmonic* in an open set G if it satisfies in G $\Delta u = u_{xx} + u_{yy} + u_{zz} = 0$. It is called *biharmonic* if it satisfies $\Delta^2 u = \Delta(\Delta u) = 0$. Prove that if u and v are harmonic functions in G, then $w = (x^2 + y^2 + z^2)u + v$ is biharmonic in G.

7. JACOBIANS

THEOREM 1.   *Let* $y = g(x)$ *be a continuous differentiable mapping with domain* R *and range* S, *and let* $z = f(y)$ *be a continuous differentiable mapping with domain* S' *and range* T, *where* R, S, S', T *are sets in* R^n, R *and* S' *are open, and* $S' \supset S$. *Then the composite mapping* $z = f(g(x))$ *is continuously differentiable and*

$$
\frac{\partial(z_1,...,z_n)}{\partial(x_1,...,x_n)} = \frac{\partial(z_1,...,z_n)}{\partial(y_1,...,y_n)} \frac{\partial(y_1,...,y_n)}{\partial(x_1,...,\partial x_n)}.
\tag{1}
$$

Here we have used the notation

$$\frac{\partial(y_1,...,y_n)}{\partial(x_1,...,\partial x_n)} = \frac{\partial(g_1,...,g_n)}{\partial(x_1,...,x_n)}, \quad \text{and so on.}$$

Proof. The only assertion that needs to be proved is assertion (1). This will follow from the chain rule. We write $z = f(g(x))$ in terms of components

$$z_i = f_i(g_1(x),...,g_n(x)) \quad (1 \le i \le n).$$

By the chain rule,

$$\frac{\partial z_i}{\partial x_j} = \sum_{k=1}^{n} \frac{\partial f_i}{\partial y_k} \frac{\partial g_k}{\partial x_j}. \tag{2}$$

Introduce the Jacobian matrices

$$Z = \left(\frac{\partial z_i}{\partial x_j}\right), \quad Y = \left(\frac{\partial f_i}{\partial y_j}\right), \quad G = \left(\frac{\partial g_i}{\partial x_j}\right). \tag{3}$$

If $A = (A_{ij})$, $B = (B_{ij})$, $C = (C_{ij})$ are $n \times n$ matrices, and if

$$C_{ij} = \sum_{j=1}^{n} A_{ik} B_{kj},$$

then, by definition of product of matrices, $C = AB$. From (2) and (3) we then see that

$$Z = YG.$$

If we now use the fact that the determinant of a product of two $n \times n$ matrices is equal to the product of the determinants of the two matrices, we obtain (1).

In the special case that $y = f(x)$, and $x = g(y)$ is the inverse of f, the composite mapping is $y = y$. The Jacobian of this mapping is 1. Hence:

COROLLARY. *If $y = f(x)$ is a continuously differentiable mapping from an open set D of R^n into R^n, and if $x^0 \in D$ and $\partial(f_1,...,f_n)/\partial(x_1,...,x_n) \neq 0$ at x^0, then the inverse mapping $x = g(y)$ exists in a neighborhood of $f(x^0)$, it is continuously differentiable, and*

$$\frac{\partial(g_1,...,g_n)}{\partial(y_1,...,y_n)} = \frac{1}{\partial(f_1,...,f_n)/\partial(x_1,...,x_n)}. \tag{4}$$

We shall next consider mappings whose Jacobian vanishes identically. Consider functions

$$f_1(x),...,f_m(x)$$

defined in an open set D of R^n. Denote by R the range of the mapping $f = (f_1,...,f_m)$. Suppose there exists a function $F(u) = F(u_1,...,u_m)$ defined in an open set R' containing R such that $F(u)$ is not identically zero in any open subset of R', and such that

$$F(f_1(x),...,f_m(x)) = 0 \qquad \text{for all } x \text{ in } D. \tag{5}$$

Then we say that the functions $f_1,...,f_m$ are *functionally dependent* in D (by means of F). If $f_1,...,f_m$ are not functionally dependent in D, then they are said to be *functionally independent* in D.

We shall consider below only the case where $m = n$.

THEOREM 2. *Let $f_1,...,f_n$ be functionally dependent by means of a function F as above. Assume that $f_1,...,f_n$ and F are continuously differentiable. Then*

$$\frac{\partial(f_1,...,f_n)}{\partial(x_1,...,x_n)} = 0 \qquad \text{for all } x \in D. \tag{6}$$

Proof. Differentiating (5) (with $m = n$) with respect to x_i, we get, by the chain rule,

$$\sum_{k=1}^{n} \frac{\partial F}{\partial u_k} \frac{\partial f_k}{\partial x_i} = 0 \qquad (1 \le i \le n). \tag{7}$$

If (6) is not satisfied at a point x^0 of D, then, by continuity, it is also not satisfied in some neighborhood X of x^0. Consider the system (7) as a homogeneous linear system of equations in the unknowns $\partial F/\partial u_k$. Since the determinant of the coefficients matrix is not zero, we conclude that

$$\frac{\partial F}{\partial u_k} (f(x)) = 0 \qquad (1 \le k \le n) \qquad \text{if } x \in X. \tag{8}$$

Since $\partial(f_1,...,f_n)/\partial(x_1,...,x_n) \ne 0$ at x^0, there is a δ-neighborhood U of $u^0 = f(x^0)$ and a neighborhood X_0 of x ($X_0 \subset X$) such that for every $u \in U$ there is a unique x in X_0 for which $f(x) = u$. Therefore, (8) implies that

$$\frac{\partial F}{\partial u_k} (u) = 0 \qquad (1 \le k \le n) \qquad \text{for all } u \in U. \tag{9}$$

But then $F(u) = \text{const.}$ in U. Since $F(u^0) = F(f(x^0)) = 0$, the constant is zero, that is, $F(u) = 0$ for all $u \in U$. This contradicts our assumptions on F.

Theorem 2 has a converse. For simplicity we shall state and prove it only in the case $n = 3$.

THEOREM 3. *Let*

$$u = f(x,y,z), \qquad v = g(x,y,z), \qquad w = h(x,y,z)$$

be functions having continuous derivatives in an open set G of R^3. Assume that

$$\frac{\partial(f,g,h)}{\partial(x,y,z)} = 0 \qquad \text{for all } (x,y,z) \text{ in } G.$$

Assume also that $\partial(f,g)/\partial(x,y) \neq 0$ at some point (x_0,y_0,z_0) of G. Then there exists a neighborhood G_0 of (x_0,y_0,z_0) and a continuously differentiable function $H(u,v)$ defined in a neighborhood of $(u_0,v_0) = (f(x_0,y_0,z_0), g(x_0,y_0,z_0))$ such that

$$h(x,y,z) = H(f(x,y,z),g(x,y,z)) \qquad \text{for all } (x,y,z) \text{ in } G_0. \tag{10}$$

The relation (10) implies functional dependence of f, g, h in G_0 by means of the function

$$F(u,v,w) = w - H(u,v).$$

Proof. Since $\partial(f,g)/\partial(x,y) \neq 0$ at (x_0,y_0), the implicit function theorem can be used to solve the equations $u = f(x,y,z)$, $v = g(x,y,z)$ locally in the form

$$x = \phi(u,v,z), \qquad y = \psi(x,y,z). \tag{11}$$

The relations

$$u = f(\phi(u,v,z),\psi(u,v,z),z),$$
$$v = g(\phi(u,v,z),\psi(u,v,z),z), \tag{12}$$

hold for all values of u, v, z in a neighborhood of (u_0,v_0,z_0). Differentiating with respect to z, we get

$$f_x\phi_z + f_y\psi_z + f_z = 0,$$
$$g_x\phi_z + g_y\psi_z + g_z = 0. \tag{13}$$

Let

$$h(\phi(u,v,z),\psi(u,v,z),z) = H(u,v,z). \tag{14}$$

Then

$$h_x\phi_z + h_y\psi_z + h_z = H_z. \tag{15}$$

If we multiply the first column of the Jacobian

$$J = \begin{vmatrix} f_x & f_y & f_z \\ g_x & g_y & g_z \\ h_x & h_y & h_z \end{vmatrix}$$

by ϕ_z and the second column by ψ_z and add to the third column, then we obtain, after using (13) and (15),

$$J = \begin{vmatrix} f_x & f_y \\ g_x & g_y \end{vmatrix} H_z.$$

Since $J = 0$, $H_z = 0$ in some δ-neighborhood of (u_0,v_0,z_0). It follows that $H(u,v,z)$ is independent of z. Writing $H(u,v) = H(u,v,z)$, the relation (14) gives the assertion (10).

PROBLEMS

1. Show that the following sets of functions are functionally dependent, and find explicitly the dependence:

    (a) $u = \log x - \log y$, $v = \dfrac{x^2 + y^2}{xy}$;

    (b) $u = x + y + z$, $v = xy + xz$, $w = x^2 + y^2 + z^2 + 2yz$;

    (c) $u = \log(x + y)$, $v = x^2 + y^2 + 2xy + 1$;

    (d) $u = \dfrac{1 - xy}{x + y}$, $v = \dfrac{(x + y)^2}{(1 + x^2)(1 + y^2)}$.

2. Find whether the following sets of functions are functionally dependent, and if so find the dependence:

    (a) $u = \dfrac{x + y}{x}$, $v = \dfrac{x + y}{y}$;

    (b) $u = \tan^{-1} x - \tan^{-1} y$, $v = \dfrac{x - y}{1 + xy}$;

    (c) $u = x + y$, $v = x^2 + y^2$.

3. Let x,y,z be functions of (u,v). Set $J(x,y) = \dfrac{\partial(x,y)}{\partial(u,v)}$, $J(y,z) = \dfrac{\partial(y,z)}{\partial(u,v)}$, $J(z,x) = \dfrac{\partial(z,x)}{\partial(u,v)}$. Prove that

    $$x_u J(y,z) + y_u J(z,x) + z_u J(x,y) = 0,$$
    $$x_v J(y,z) + y_v J(z,x) + z_v J(x,y) = 0.$$

4. Let x,y,z be as in the preceding problem. Prove that

    $$J(x,J(y,z)) + J(y,J(z,x)) + J(z,J(x,y)) = 0.$$

5. Under what conditions on a,b,c,d will the linear functions $ax + by$, $cx + dy$ be functionally dependent?

6. Find the inverse of the transformation

$$u = \frac{x}{r^2}, \qquad v = \frac{y}{r^2}, \qquad w = \frac{z}{r^2} \qquad \text{where } r^2 = x^2 + y^2 + z^2.$$

7. Let $F(u,v,w)$ be a twice continuously differentiable function, and let u,v,w be the functions defined in the preceding problem. Denote by $G(x,y,z)$ the composite function. Prove that $F_{uu} + F_{vv} + F_{ww} = 0$ if, and only if, $G_{xx} + G_{yy} + G_{zz} = 0$.

8. VECTORS

We shall consider vectors in R^3. We write them in the form

$$\mathbf{u} = u_1\mathbf{i} + u_2\mathbf{j} + u_3\mathbf{k}$$

where $\mathbf{i} = (1,0,0)$, $\mathbf{j} = (0,1,0)$, $\mathbf{k} = (0,0,1)$. If

$$\mathbf{v} = v_1\mathbf{i} + v_2\mathbf{j} + v_3\mathbf{k},$$

then $\mathbf{u} + \mathbf{v}$, $\lambda\mathbf{u}$ (λ scalar) and $\mathbf{u} \cdot \mathbf{v}$ (scalar product) are given by

$$\mathbf{u} + \mathbf{v} = (u_1 + v_1)\mathbf{i} + (u_2 + v_2)\mathbf{j} + (u_3 + v_3)\mathbf{k},$$

$$\lambda\mathbf{u} = \lambda u_1\mathbf{i} + \lambda u_2\mathbf{j} + \lambda u_3\mathbf{k},$$

$$\mathbf{u} \cdot \mathbf{v} = u_1v_1 + u_2v_2 + u_3v_3.$$

The length of $\mathbf{u}$ is given by

$$\|\mathbf{u}\| = \sqrt{u_1{}^2 + u_2{}^2 + u_3{}^2}.$$

The *vector product* $\mathbf{u} \times \mathbf{v}$ is defined by

$$\mathbf{u} \times \mathbf{v} = (u_2v_3 - u_3v_2)\mathbf{i} + (u_3v_1 - u_1v_3)\mathbf{j} + (u_1v_2 - u_2v_1)\mathbf{k}. \tag{1}$$

It is suggestive to write $\mathbf{u} \times \mathbf{v}$ in the symbolic form

$$\mathbf{u} \times \mathbf{v} = \begin{vmatrix} \mathbf{i} & \mathbf{j} & \mathbf{k} \\ u_1 & u_2 & u_3 \\ v_1 & v_2 & v_3 \end{vmatrix}.$$

Thus if we pretend that $\mathbf{i}$, $\mathbf{j}$, $\mathbf{k}$ in the last determinant are numbers, and expand this determinant, then we obtain the right-hand side of (1).

We state without proof:

$$\mathbf{u} \times \mathbf{v} = -\mathbf{v} \times \mathbf{u}; \tag{2}$$

$$\mathbf{u} \cdot (\mathbf{u} \times \mathbf{v}) = 0, \qquad \mathbf{v} \cdot (\mathbf{u} \times \mathbf{v}) = 0; \tag{3}$$

$$\|\mathbf{u} \times \mathbf{v}\|^2 = \|\mathbf{u}\|^2 \|\mathbf{v}\|^2 - (\mathbf{u} \cdot \mathbf{v})^2; \tag{4}$$

$$\|\mathbf{u} \times \mathbf{v}\| = \|\mathbf{u}\| \|\mathbf{v}\| |\sin(\mathbf{u},\mathbf{v})|, \text{ where } \cos(\mathbf{u},\mathbf{v}) = \frac{\mathbf{u} \cdot \mathbf{v}}{\|\mathbf{u}\| \|\mathbf{v}\|}; \tag{5}$$

$$(\mathbf{u} \times \mathbf{v}) \cdot \mathbf{w} = \begin{vmatrix} u_1 & u_2 & u_3 \\ v_1 & v_2 & v_3 \\ w_1 & w_2 & w_3 \end{vmatrix} = (\mathbf{v} \times \mathbf{w}) \cdot \mathbf{u} = (\mathbf{w} \times \mathbf{u}) \cdot \mathbf{v}; \tag{6}$$

$$(\mathbf{u} \times \mathbf{v}) \times \mathbf{w} = (\mathbf{u} \cdot \mathbf{w})\mathbf{v} - (\mathbf{v} \cdot \mathbf{w})\mathbf{u}. \tag{7}$$

We can write a curve $u(t) = (u_1(t), u_2(t), u_3(t))$ also in the form

$$\mathbf{u}(t) = u_1(t)\mathbf{i} + u_2(t)\mathbf{j} + u_3(t)\mathbf{k}.$$

If the $u_i(t)$ are differentiable functions, then we say that $\mathbf{u}(t)$ is differentiable, and we write

$$\frac{d\mathbf{u}(t)}{dt} = \frac{du_1(t)}{dt}\mathbf{i} + \frac{du_2(t)}{dt}\mathbf{j} + \frac{du_3(t)}{dt}\mathbf{k}.$$

If $\mathbf{v}(t)$ is another curve and if $f(t)$ is a scalar function, then the following rules can easily be verified:

$$\frac{d}{dt}(\mathbf{u} + \mathbf{v}) = \frac{d\mathbf{u}}{dt} + \frac{d\mathbf{v}}{dt}; \tag{8}$$

$$\frac{d}{dt}(f\mathbf{u}) = f\frac{d\mathbf{u}}{dt} + \frac{df}{dt}\mathbf{u}; \tag{9}$$

$$\frac{d}{dt}(\mathbf{u} \cdot \mathbf{v}) = \mathbf{u} \cdot \frac{d\mathbf{v}}{dt} + \frac{d\mathbf{u}}{dt} \cdot \mathbf{v}; \tag{10}$$

$$\frac{d}{dt}(\mathbf{u} \times \mathbf{v}) = u \times \frac{d\mathbf{v}}{dt} + \frac{d\mathbf{u}}{dt} \times \mathbf{v}. \tag{11}$$

The vector $d\mathbf{u}(\tau)/dt$ is called the *tangent vector* to $\mathbf{u}(t)$ at $t = \tau$. If $\|\mathbf{u}(t)\| = 1$ for all t, then

$$0 = \frac{d}{dt}(\mathbf{u} \cdot \mathbf{u}) = 2\mathbf{u} \cdot \frac{d\mathbf{u}}{dt}.$$

Thus, the tangent vector at each point t is orthogonal to $\mathbf{u}(t)$.

Now let u_1, u_2, u_3 be functions defined in a set D of R^3. The vector

$$\mathbf{u}(x,y,z) = u_1(x,y,z)\mathbf{i} + u_2(x,y,z)\mathbf{j} + u_3(x,y,z)\mathbf{k}$$

is then called a *vector field* on D. If D is an open set and if all the components u_i are differentiable, then we say that $\mathbf{u}$ is differentiable in D.

Given a differentiable function f in a domain D of R^3, we define the *gradient* of f to be the vector field

$$\operatorname{grad} f = \frac{\partial f}{\partial x}\mathbf{i} + \frac{\partial f}{\partial y}\mathbf{j} + \frac{\partial f}{\partial z}\mathbf{k}.$$

Given a differentiable vector field

$$\mathbf{F} = f\mathbf{i} + g\mathbf{j} + h\mathbf{k}, \tag{12}$$

we define a scalar function, called the *divergence* of $\mathbf{F}$, by

$$\operatorname{div} \mathbf{F} = \frac{\partial f}{\partial x} + \frac{\partial g}{\partial y} + \frac{\partial h}{\partial z}.$$

It is convenient to introduce the symbol ∇, called *nabla*,

$$\nabla = \frac{\partial}{\partial x}\mathbf{i} + \frac{\partial}{\partial y}\mathbf{j} + \frac{\partial}{\partial z}\mathbf{k}.$$

In terms of ∇, we can write grad f and div $\mathbf{F}$ in a more suggestive form:

$$\operatorname{grad} f = \nabla f, \quad \text{or} \quad \operatorname{div} \mathbf{F} = \nabla \cdot \mathbf{F}.$$

The *curl* of a vector field $\mathbf{F}$ given by (12) is the vector field

$$\operatorname{curl} \mathbf{F} = \left(\frac{\partial h}{\partial y} - \frac{\partial g}{\partial z}\right)\mathbf{i} + \left(\frac{\partial f}{\partial z} - \frac{\partial h}{\partial x}\right)\mathbf{j} + \left(\frac{\partial g}{\partial x} - \frac{\partial f}{\partial y}\right)\mathbf{k}.$$

It is suggestive to write it in the form

$$\operatorname{curl} \mathbf{F} = \nabla \times \mathbf{F} = \begin{vmatrix} \mathbf{i} & \mathbf{j} & \mathbf{k} \\ \frac{\partial}{\partial x} & \frac{\partial}{\partial y} & \frac{\partial}{\partial z} \\ f & g & h \end{vmatrix}.$$

Some authors write rot $\mathbf{F}$ instead of curl $\mathbf{F}$.

PROBLEMS

1. Prove the relations (2)–(7).
2. Prove the relations (8)–(11).

3. Show that div grad $f = f_{xx} + f_{yy} + f_{zz}$.

4. Find ∇f for each of the following functions:

   (a) $\dfrac{x}{x^2 + y^2} + \dfrac{z}{y^2 + z^2}$;

   (b) $\dfrac{x}{x^2 + y^2 + z^2}$;

   (c) $e^{x+y} \sin z$;

   (d) $\dfrac{z}{(x^2 + y^2 + z^2)^{1/2}}$.

5. Prove the following relations:

   $\mathrm{div}(\lambda \mathbf{F}) = \mathrm{grad}\,\lambda \cdot \mathbf{F} + \lambda\,\mathrm{div}\,\mathbf{F}$; $\mathrm{curl}\,(\lambda \mathbf{F}) = (\mathrm{grad}\,\lambda) \times \mathbf{F} + \lambda\,\mathrm{curl}\,\mathbf{F}$.

6. The Newtonian gravity field due to a point mass m situated at the origin is the vector field

$$F(x,y,z) = -\frac{km}{r}\,\mathbf{r}, \qquad k \text{ positive constant}$$

   where $\mathbf{r} = x\mathbf{i} + y\mathbf{j} + z\mathbf{j}$, $r = \|\mathbf{r}\|$. Prove that div grad $\mathbf{F} = 0$.

7. Compute curl $\mathbf{F}$ for the following vector fields:

   (a) $\dfrac{xz}{\sqrt{x^2 + y^2}}\,\mathbf{i} - \dfrac{yz}{\sqrt{x^2 + y^2}}\,\mathbf{j} + \mathbf{k}$;

   (b) $\dfrac{x\mathbf{i} + y\mathbf{j} + z\mathbf{k}}{x^2 + y^2 + z^2}$;

   (c) $\dfrac{x\mathbf{i} + y\mathbf{j}}{x^2 + y^2}$;

   (d) $\sin(yz)\mathbf{i} + \sin(xz)\mathbf{j} + \sin(xy)\mathbf{k}$.

8. Let $\mathbf{r} = x\mathbf{i} + y\mathbf{j} + z\mathbf{k}$, $r = \|\mathbf{r}\|$, and $\mathbf{b}$ a constant vector. Prove:

   (a) $\mathrm{div}(r^n\mathbf{b}) = nr^{n-2}\mathbf{r} \cdot \mathbf{b}$;

   (b) $\mathrm{curl}(r^n\mathbf{b}) = nr^{n-2}\mathbf{r} \times \mathbf{b}$;

   (c) $\mathrm{div}(r^n\mathbf{b} \times \mathbf{r}) = 0$;

   (d) $\mathrm{grad}\,\dfrac{\mathbf{b} \cdot \mathbf{r}}{r^3} = -\mathrm{curl}\,\dfrac{\mathbf{b} \times \mathbf{r}}{r^3}$.

9. Show that curl(grad ϕ) = 0 for any twice continuously differentiable function ϕ.

10. Show that div(curl $\mathbf{F}$) = 0 for any twice continuously differentiable vector field $\mathbf{F}$.

8

MULTIPLE INTEGRALS

1. SETS OF CONTENT ZERO

By an *n-dimensional rectangle* we shall mean a set R of the form

$$\{x = (x_1,\ldots,x_n);\ a_i < x_i < b_i,\ i = 1,\ldots,n\}.$$

The *volume* of R is the number

$$V(R) = (b_1 - a_1)(b_2 - a_2)\cdots(b_n - a_n).$$

Let S be a subset of R^n. Suppose that for any $\varepsilon > 0$ there is a finite number of n-dimensional rectangles $R_1, R_2,\ldots,R_k$ that form a covering of S, and such that

$$V(R_1) + V(R_2) + \cdots + V(R_k) < \varepsilon.$$

Then we say that S has *Jordan content zero* (or briefly, *content zero*) relative to R^n (or, in R^n).

Note that a union of a finite number of sets of content zero (in R^n) is again of content zero (in R^n).

The object of the present section is to find a large class of regions whose boundary has content zero. In the following sections we shall define integrals over such regions.

Let $f(t) = (f_1(t),...,f_n(t))$ be a curve in R^n, defined on an interval $a \le t \le b$, and let $a < t_0 < b$. If $f(t)$ is discontinuous at t_0, but

$$f(t_0 + 0) = \lim_{t \to t_0 + 0} f(t) \quad \text{and} \quad f(t_0 - 0) = \lim_{t \to t_0 - 0} f(t)$$

exist, then we say that $f(t)$ has a *discontinuity of the first kind at* t_0.

Suppose now that for some partition of $[a,b]$,

$$\tau_0 = a < \tau_1 < \tau_2 < \cdots < \tau_{n-1} < \tau_n = b,$$

the curve $f(t)$ is uniformly continuous in each interval (τ_{i-1}, τ_i). Then we say that $f(t)$ is *piecewise continuous* in $[a,b]$. Note that $f(\tau_i + 0)$ and $f(\tau_i - 0)$ exist for each $a < \tau_i < b$, so that the discontinuity of $f(t)$ at τ_i is of the first kind.

A curve $f(t)$ is said to be *piecewise continuously differentiable* if it is continuous and if $f'(t)$ is piecewise continuous.

THEOREM 1. *Let $f(t)$ $(a \le t \le b)$ be a piecewise continuously differentiable curve in R^n, $n \ge 2$. Then the range C of f has content zero in R^n.*

Proof. Denote by $\tau_1,...,\tau_n$ the points of discontinuity of $f'(t)$, arranged in increasing order. Denote by M an upper bound on the function $\|f'(t)\|$. We claim that for any two points t', t'' in $[a,b]$,

$$|f_m(t'') - f_m(t')| \le M|t'' - t'| \qquad (m = 1,2,...,n). \tag{1}$$

To prove (1) we may assume that $t'' > t'$. If t', t'' lie in an interval $[\tau_i, \tau_{i+1}]$, then (1) follows from the mean value theorem. Suppose then that there exist points τ_i, τ_j such that

$$\tau_{i-1} \le t' < \tau_i < \cdots < \tau_j < t'' \le \tau_{j+1}.$$

Using the mean value theorem, we get

$$|f_m(t'') - f_m(t')| \le |f_m(t'') - f_m(\tau_j)| + |f(\tau_j) - f(\tau_{j-1})| + \cdots + |f(\tau_i) - f(t')|$$

$$\le M(t'' - \tau_j) + M(\tau_j - \tau_{j-1}) + \cdots + M(\tau_i - t') = M(t'' - t').$$

Let ε be any positive number, and let p be a positive integer depending on ε, to be determined later on. We introduce the points

$$s_r = a + \frac{b-a}{p} r \qquad \text{for } r = 0,1,2,...,p.$$

If $s_r \le t \le s_{r+1}$, then, by (1),

$$|f_m(t) - f_m(s_r)| \le M|t - s_r| \le M \frac{b-a}{p}. \tag{2}$$

Set $M_p = M(b-a)/p$, and denote by T_r the rectangle defined by

$$f_m(s_r) - 2M_p < x_m < f_m(s_r) + 2M_p \qquad (1 \le m \le n).$$

From (2) it follows that the rectangles $T_0, T_1, \ldots, T_{p-1}$ form a covering of the range of f. But

$$V(T_0) + V(T_1) + \cdots + V(T_{p-1}) = p(4M_p)^n = [4M(b-a)]^n \, \frac{1}{p^{n-1}} < \varepsilon$$

if p is sufficiently large, since $n - 1 > 0$. This completes the proof.

A *projectable hypersurface* in R^n $(n \ge 2)$ is a function having the form

$$x_i = F(x_1, \ldots, x_{i-1}, x_{i+1}, \ldots, x_n) \tag{3}$$

for some i, where F is defined and continuous in some bounded, closed domain D. Denote by S this projectable hypersurface and denote by C its graph. The points x of C corresponding to $(x_1, \ldots, x_{i-1}, x_{i+1} \ldots x_n)$ in the interior of D are called *interior points* of S (or C). A point of C that is not an interior point of S is called a *boundary point* of S (or C).

In R^3 a projectable hypersurface is called a *projectable surface*. It has one of the following forms:

$$z = f(x,y), \qquad y = g(x,z), \qquad x = h(y,z).$$

In R^2, a projectable hypersurface is called a *projectable curve*. It can be written either in the form $y = f(x)$ $(\alpha \le x \le \beta)$, or in the form $x = g(y)$ $(\gamma \le y \le \delta)$.

THEOREM 2.   *The graph of a projectable hypersurface S in R^n has content zero in R^n.*

Proof.   For clarity we give the proof in case $n = 3$; the proof for any n is similar. We may suppose that S is given by

$$z = f(x,y), \qquad (x,y) \in D.$$

We denote by C the graph of S. Since $f(x,y)$ is continuous on the closed, bounded set D, it is uniformly continuous. Hence, for $\varepsilon > 0$ there is a $\delta > 0$ such that

$$|f(x,y) - f(x',y')| < \varepsilon \quad \text{if} \quad \sqrt{(x-x')^2 + (y-y')^2} < \delta. \tag{4}$$

Let p be a positive integer satisfying

$$\frac{2}{p} < \frac{\delta}{\sqrt{2}}. \tag{5}$$

We divide the (x,y)-plane into squares of sides $1/p$ by introducing all the straight lines of the form

$$x = k + \frac{r}{p} \quad (r = 0,1,...,p-1; k = 0,\pm1,\pm2,...)$$

$$y = h + \frac{s}{p} \quad (s = 0,1,...,p-1; h = 0,\pm1,\pm2,...).$$

The squares that intersect D will be denoted by T_i. Denote by $\hat{T}_i$ the square having the same center as T_i but twice its sides. The squares $\hat{T}_i$ form an open covering of D. D is contained in a rectangle defined by $|x| < \lambda$, $|y| < \lambda$. Denote by K the area of the rectangle defined by $|x| < \lambda + 2$, $|y| < \lambda +2$. It is clear that

$$\sum V(T_i) < K.$$

Since $V(\hat{T}_i) = 4V(T_i)$,

$$\sum V(\hat{T}_i) < 4K.$$

Denote by (ξ_i,η_i) some point of $\hat{T}_i \cap D$. From (4) and (5) it follows that

$$|f(x,y) - f(\xi_i,\eta_i)| < \varepsilon \quad \text{if } (x,y) \in \hat{T}_i \cap D.$$

Therefore, the three-dimensional rectangles R_i, defined by

$$R_i : (x,y) \in \hat{T}_i, \quad |z - f(\xi_i,\eta_i)| < \varepsilon,$$

form a covering of the graph C of the surface S. Clearly $V(R_i) = 2\varepsilon V(\hat{T}_i)$, so that

$$\sum V(R_i) = 2\varepsilon \sum V(\hat{T}_i) < 8\varepsilon K.$$

Since K is a constant (independent of ε) and ε is arbitrary, we conclude that C has content zero in R^3.

DEFINITION. Let G be a bounded region in R^n. Suppose that the boundary of G has the form $C_1 \cup C_2 \cup \cdots \cup C_k$, such that the following holds:

(a)   If $n = 2$, each C_j is the graph of a curve l_j, which is either piecewise continuously differentiable or projectable, and any two sets C_i, C_j $(i \neq j)$ have no points in common except, perhaps, boundary points of both l_i and l_j.

(b)   If $n \geq 3$, each C_j is the graph of a projectable hypersurface S_j, and any two sets C_i, C_j $(i \neq j)$ have no points in common except, perhaps, boundary points of both S_i and S_j.

We then say that G is a *normal region*. If G is also a closed domain, then we say that G is a *closed normal domain*.

From Theorems 1 and 2 we obtain:

COROLLARY 1. *The boundary of a normal region in R^n $(n \geq 2)$ has content zero in R^n.*

Let $f(x)$ be a function defined on a region D in R^n $(n \geq 2)$. Suppose we can write

$$D = D_1 \cup D_2 \cup \cdots \cup D_k \cup C_1 \cup C_2 \cup \cdots \cup C_m$$

such that

(i)   The D_i are domains, and $D_i \cap D_j$ is empty if $i \neq j$.

(ii) If $n = 2$, each C_j is the graph of either a projectable curve or of a piecewise continuously differentiable curve.

(iii) If $n \geq 3$, each C_j is the graph of a projectable hypersurface.

(iv) $f(x)$ is uniformly continuous in each domain D_i.

Then we say that $f(x)$ is *piecewise continuous* in D.

From Theorems 1 and 2 we obtain:

COROLLARY 2. *The set of discontinuities of a piecewise continuous function $f(x) = f(x_1,\ldots,x_n)$ has content zero in R^n $(n \geq 2)$.*

PROBLEMS

1. Prove that a piecewise continuously differentiable curve $f(t)$, $(a \leq t \leq b)$, in R^2 satisfying $f'(t + 0) \neq 0$ if $a \leq t < b$, $f'(t - 0) \neq 0$ if $a < t \leq b$, is a finite union of projectable curves.

2. Prove that the ball $x^2 + y^2 + z^2 < 1$ is a normal domain.

3. Prove that the ball $\|x\| < 1$ in R^n $(n \geq 2)$ is a normal domain.

4. Prove that the cube $0 \leq x \leq 1, 0 \leq y \leq 1, 0 \leq z \leq 1$ is a normal domain.

5. Is the intersection of two normal domains a normal domain?

2. DOUBLE INTEGRALS

Let $f(x,y)$ be a bounded function in a closed rectangle G defined by

$$G : a \leq x \leq b, \qquad c \leq y \leq d.$$

A *partition* of G consists of a pair of partitions

$$x_0 = a < x_1 < x_2 < \cdots < x_{n-1} < x_n = b,$$
$$y_0 = c < y_1 < y_2 < \cdots < y_{m-1} < y_m = d \tag{1}$$

of $[a,b]$ and $[c,d]$. Denote R_{ij} the closed rectangle

$$x_{i-1} \leq x \leq x_i, \qquad y_{j-1} \leq y \leq y_j,$$

and define

$$M_{ij} = \sup_{(x,y) \in R_{ij}} f(x,y),$$

$$m_{ij} = \inf_{(x,y) \in R_{ij}} f(x,y),$$

$$\Delta_{ij} = \text{area of } R_{ij} = (x_i - x_{i-1})(y_j - y_{j-1}).$$

The number

$$\max_{i,j} \sqrt{(x_i - x_{i-1})^2 + (y_j - y_{j-1})^2}$$

is called the *mesh* of the partition (1). We denote it briefly by

$$\max_{i,j} |R_{ij}| .$$

We now form the *upper Darboux sum*

$$S = \sum_{i=1}^{n} \sum_{j=1}^{m} M_{ij} \Delta_{ij}$$

and the *lower Darboux sum*

$$s = \sum_{i=1}^{n} \sum_{j=1}^{m} m_{ij} \Delta_{ij}$$

corresponding to the partition (1). We denote by J the infimum of the set of numbers S and by I the supremum of the set of numbers s.

A partition of G is said to be *more refined* than the partition of G given by (1), if the corresponding partitions of $[a,b]$ and $[c,d]$ are, respectively, more refined than the partitions of $[a,b]$ and $[c,d]$ in (1).

We can now proceed similarly to Section 4.1, and establish the following theorems:

THEOREM 1.   *Let S' and s' be upper and lower Darboux sums corresponding to a partition of G that is more refined than the partition of* (1). *Then S' ≤ S and s' ≥ s.*

THEOREM 2.   *For any ε > 0 there is a number δ > 0 such that for any partition* (1) *of G with mesh* <δ,

$$0 \le S - J < \varepsilon, \qquad -\varepsilon < s - I \le 0.$$

THEOREM 3.   *J ≥ I.*

If $J = I$, then we say that f is *Darboux integrable* over (on, or in) G. The number J is then called the *Darboux integral* of f over G and is denoted by

$$(D) \iint\limits_{G} f(x,y)\, dA.$$

THEOREM 4.   *A bounded function f is Darboux integrable if, and only if, for any ε > 0 there is a partition* (1) *for which S − s < ε.*

We turn to the definition of the Riemann integral. For any partition (1) of G, and for any points (ξ_i, η_j) in R_{ij}, we form the *Riemann sum*

$$T = \sum_{i=1}^{n} \sum_{j=1}^{m} f(\xi_i, \eta_j)\, \Delta_{ij}.$$

If there exists a number K such that

$$|T - K| < \varepsilon \qquad \text{whenever} \max_{i,j} |R_{ij}| < \delta,$$

and for any choice of (ξ_i, η_j) in R_{ij}, then we say that f is *Riemann integrable* over (on, or in) G. The number K is then called the *Riemann integral* of f over G, and is denoted by

$$(R) \iint\limits_{G} f(x,y)\, dA.$$

The proof of Theorem 1 in Section 4.2 extends to the present case, and thus yields the following result:

THEOREM 5. *Let $f(x,y)$ be a bounded function in a closed rectangle G. Then f is Riemann integrable if, and only if, f is Darboux integrable, and in this case*

$$(R) \iint_G f(x,y)\, dA = (D) \iint_G f(x,y)\, dA.$$

From now on we shall refer to both the Darboux integral and the Riemann integral briefly as the integral. We shall denote the integral of f over G by

$$\iint_G f(x,y)\, dA. \qquad (2)$$

In order to distinguish it from the integral $\int_a^b f(x)\, dx$ of a function of one variable, we call the integral in (2) a *double integral*.

THEOREM 6. *Let $f(x,y)$ be a bounded function in the closed rectangle R in R^2, and suppose that the set of its discontinuities has content 0 in R^2. Then f is integrable over R.*

Proof. Let $M = \underset{R}{\text{lub}}|f|$. Given any $\varepsilon > 0$, there is a finite number of (open) rectangles $R_1,\ldots,R_h$ such that their union

$$R_0 = R_1 \cup \cdots \cup R_h$$

covers the set of discontinuities of f, and such that

$$V(R_1) + \cdots + V(R_h) < \frac{\varepsilon}{4M} \qquad (V(R_j) = \text{area of } R_j). \qquad (3)$$

Denote by R^* the set of points that belong to R but not to R_0. Since R_0 is an open set and R is a closed set, the set R^* is closed. Therefore, the function f, being continuous in R^*, is also uniformly continuous. Thus, for any $\varepsilon > 0$ there is a $\delta > 0$ such that

$$|f(x,y) - f(x',y')| < \frac{\varepsilon}{2V(R)} \qquad \text{if } \sqrt{(x-x')^2 + (y-y')^2} < \delta \qquad (4)$$

and $(x,y) \in R^*$, $(x',y') \in R^*$; here $V(R) = \text{area of } R$.

The vertices of the rectangles $R_1,\ldots,R_h$ lie on a finite number of lines $x = \hat{x}_i$, $y = \hat{y}_j$. Form a partition (1) of R with mesh $< \delta$, such that the points $\hat{x}_i$ appear in the corresponding partition of $[a,b]$ and such that the points

$\hat{y}_j$ appear in the corresponding partition of $[c,d]$. We shall prove that the corresponding Darboux sums satisfy

$$S - s = \sum_{i=1}^{n} \sum_{j=1}^{m} (M_{ij} - m_{ij})\Delta_{ij} < \varepsilon. \tag{5}$$

In view of Theorem 4, the proof of Theorem 6 is then complete.

The partition that we now have determines also a partition of the closure $\bar{R}_k$ of each open rectangle R_k, since R_k has the form

$$\{(x,y); \ \hat{x}_i < x < \hat{x}_j, \ \hat{y}_r < y < \hat{y}_s\}.$$

Consequently, each $\bar{R}_k$ can be written as a finite union of rectangles R_{ij}; denote them by $R_\lambda^{(k)}$ ($\lambda = 1,...,\alpha_k$). Then

$$\sum_\lambda V(R_\lambda^k) = V(R_k).$$

The closure $\bar{R}_0$ of R_0 is the union of the rectangles $R_\lambda^{(k)}$ ($\lambda = 1,...,\alpha_k$, $k = 1,...,h$). The rectangles $R_\lambda^{(k)}$ are not necessarily different from each other. In fact, any rectangle that belongs to r of the rectangles $\bar{R}_m$, will appear r times among the rectangles $R_\lambda^{(k)}$.

Denote by $\sum'$ the partial sum of $\sum_{i=1}^{n} \sum_{j=1}^{m}$ corresponding to all indices i, j for which R_{ij} lies in $\bar{R}_0$. Denote by $\sum''$ the partial sum corresponding to the remaining indices. Then

$$\sum' \Delta_{ij} = \sum' V(R_{ij}) \le \sum_{k=1}^{h} \sum_{\lambda=1}^{\alpha_k} V(R_\lambda^k) = \sum_{k=1}^{h} V(R_k) < \frac{\varepsilon}{4M}.$$

Therefore

$$\sum' (M_{ij} - m_{ij})\Delta_{ij} \le 2M \sum' \Delta_{ij} < 2M \frac{\varepsilon}{4M} = \frac{\varepsilon}{2}. \tag{6}$$

Each rectangle R_{ij} whose indices i, j appear in the summation $\sum''$, must be contained in R^*. Using (4), we obtain

$$\sum'' (M_{ij} - m_{ij})\Delta_{ij} \le \frac{\varepsilon}{2V(R)} \sum'' \Delta_{ij} \le \frac{\varepsilon}{2V(R)} V(R) = \frac{\varepsilon}{2}. \tag{7}$$

Combining (6) and (7), the inequality in (5) follows, and the proof of the theorem is thereby completed.

Let D be any bounded set in R^2 and let $f(x,y)$ be a bounded function defined in D. We define a function

$$g(x,y) = \begin{cases} f(x,y) & \text{if } (x,y) \in D, \\ 0 & \text{if } (x,y) \notin D, \end{cases}$$

called the *extension of f by* 0.

Let R be any closed rectangle in R^2 that contains D. If g is integrable on R, then we say that f is *integrable* on D (over D, or in D). The *integral* of f over D is defined as the number

$$\iint_R g(x,y)\, dA.$$

We denote it by

$$\iint_D f(x,y)\, dA.$$

THEOREM 7. *The above definition is independent of R, that is, if R^* is another rectangle containing D, then g is integrable over R if, and only if, it is integrable over R^*, and in this case*

$$\iint_R g(x,y)\, dA = \iint_{R^*} g(x,y)\, dA.$$

The proof is left to the student; see Problems 6 and 7.

THEOREM 8. *Let $f(x,y)$ be an integrable function on a bounded set D in R^2. Let C be a subset of D of content 0 in R^2, and denote by $D-C$ the subset of D consisting of all the points that do not belong to C. Let g be a bounded function in D such that $f = g$ at all the points of $D-C$. Then g is integrable on D, and*

$$\iint_D g(x,y)\, dA = \iint_D f(x,y)\, dA.$$

Proof. Suppose first that D is a closed rectangle. For any partition of D, we consider the upper and lower Darboux sums for f

$$S = \sum\sum M_{ij}\Delta_{ij}, \qquad s = \sum\sum m_{ij}\Delta_{ij},$$

and for g

$$\tilde{S} = \sum\sum \tilde{M}_{ij}\Delta_{ij}, \qquad \tilde{s} = \sum\sum \tilde{m}_{ij}\Delta_{ij}.$$

For any $\varepsilon > 0$, let $R_1,...,R_h$ be rectangles that cover C and that satisfy $V(R_1) + \cdots + V(R_h) < \varepsilon/4M$, M being a number lager than l.u.b.$|f|$ and l.u.b.$|g|$. Using partitions as in the proof of Theorem 6 (with $D = R$), we get

$$\left|\sum\nolimits' M_{ij}\Delta_{ij} - \sum\nolimits' \tilde{M}_{ij}\Delta_{ij}\right| < \frac{\varepsilon}{2}, \tag{8}$$

$$\left|\sum\nolimits' \tilde{M}_{ij}\Delta_{ij} - \sum\nolimits' \tilde{m}_{ij}\Delta_{ij}\right| < \frac{\varepsilon}{2}, \tag{9}$$

$$\sum\nolimits'' M_{ij}\Delta_{ij} = \sum\nolimits'' \tilde{M}_{ij}\Delta_{ij}, \qquad \sum\nolimits'' m_{ij}\Delta_{ij} = \sum\nolimits'' \tilde{m}_{ij}\Delta_{ij}. \tag{10}$$

Since f is integrable, there is a $\delta > 0$ such that if $\max|R_{ij}| < \delta$, then

$$\sum (M_{ij} - m_{ij})\Delta_{ij} < \varepsilon, \tag{11}$$

$$\left| \sum M_{ij}\Delta_{ij} - \iint\limits_{D} f(x,y)\, dA \right| < \varepsilon. \tag{12}$$

From (9), (10), and (11) it follows that

$$\sum (\tilde{M}_{ij} - \tilde{m}_{ij})\Delta_{ij} = \sum' (\tilde{M}_{ij} - \tilde{m}_{ij})\Delta_{ij} + \sum'' (M_{ij} - m_{ij})\Delta_{ij} < \frac{3}{2}\,\varepsilon.$$

Since ε is arbitrary, Theorem 4 shows that g is integrable on D.

Next, from (8), (10) and (12) we easily get

$$\left| \sum \tilde{M}_{ij}\Delta_{ij} - \iint\limits_{D} f(x,y)\, dA \right| < \frac{3}{2}\,\varepsilon.$$

It follows that the integral of g over D is equal to $\iint\limits_{D} f(x,y)\, dA$. This completes the proof of the theorem in case D is a rectangle.

When D is an arbitrary bounded set, we extend both f and g by 0, and apply the previous result in some rectangle containing D.

Let $f(x,y)$ be a bounded function defined in a bounded region D. Denote by S the set of discontinuities of f in D. Denote by ∂D the boundary of D. Then the set of discontinuities of the extension of f by zero is contained in $S \cup \partial D$. Applying Theorem 6, we conclude:

THEOREM 9.   *Let $f(x,y)$ be a bounded function in a bounded region D. If both ∂D and the set S of discontinuities of f have content zero in R^2, then $f(x,y)$ is integrable on D.*

From Corollaries 1 and 2 to Theorem 2 of Section 1, we then have:

THEOREM 10.   *A piecewise continuous function on a normal region D in R^2 is integrable.*

PROBLEMS

1.   Prove Theorem 1.

2.   Prove Theorem 2.

3.   Prove Theorem 3.

4. Prove Theorem 4.

5. Prove Theorem 5.

6. Let R and R^* be closed rectangles containing a bounded set D in R^2. Let f be a bounded function in D and let g be its extension by 0. Prove:
   (a) If $R^* \supset R$ and $\iint_R g \, dA$ exists, then $\iint_{R^*} g \, dA$ exists;
   (b) If $R^* \supset R$ and $\iint_{R^*} g \, dA$ exists, then $\iint_R g \, dA$ exists.

7. Prove Theorem 7. [*Hint:* Compare R and R^* with a rectangle R^{**} containing both, and use the preceding problem.]

8. Using the definition, compute the following integrals when $R = \{(x,y); 0 \leq x \leq 1, 0 \leq y \leq 1\}$:
   (a) $\iint_R 1 \cdot dA$;   (b) $\iint_R y \, dA$;
   (c) $\iint_R xy \, dA$;   (d) $\iint_R x^2 dA$.

9. If f is integrable on a bounded set D and $f \geq 0$, then $\iint_D f \, dA \geq 0$.

10. If $f \geq g$ and f, g are integrable over a bounded set D, then $\iint_D f \, dA \geq \iint_D g \, dA$.

11. Let f and g be integrable functions over a bounded set D and let λ, μ be constants. Then $\lambda f + \mu g$ is integrable over D, and

$$\iint_D (\lambda f + \mu g) \, dA = \lambda \iint_D f \, dA + \mu \iint_D g \, dA.$$

12. Let D be a normal domain in R^2 and let $\bar{D}$ be the closure of D. Let S and T be any sets satisfying $D \subset S \subset T \subset \bar{D}$. Prove that if a function f is integrable on T, then it is integrable on S, and $\iint_S f \, dA = \iint_T f \, dA$.

13. Let S be a bounded set in R^2 and let χ be a function defined in S by $\chi(x,y) = 1$ for any $(x,y) \in S$. Prove that S has content zero if, and only if, χ is integrable on S and $\int_S \chi(x,y) \, dA = 0$.

3. ITERATED INTEGRALS

It is very difficult to compute double integrals directly from the definition given in Section 2. We therefore shall develop now tools for computing double integrals.

We first consider integrals over a rectangle $R = \{(x,y); a \leq x \leq b, c \leq y \leq d\}$.

THEOREM 1. *Let $f(x,y)$ be integrable in a rectangle R, and suppose that for each fixed x in $[a,b]$ the function $f(x,y)$ is integrable in y on the interval $c \le y \le d$. Then the function*

$$g(x) = \int_c^d f(x,y)\, dy \tag{1}$$

is integrable in the interval $a \le x \le b$, and

$$\iint_R f(x,y)\, dA = \int_a^b g(x)\, dx. \tag{2}$$

If we substitute $g(x)$ from (1) into (2), we get

$$\iint_R f(x,y)\, dA = \int_a^b \left(\int_c^d f(x,y)\, dy \right) dx. \tag{3}$$

The right-hand side of (3) also is written in the form

$$\int_a^b dx \int_c^d f(x,y)\, dy.$$

The integral on the right-hand side of (3) is called an *iterated integral*. Theorem 1 asserts that a double integral is equal to an iterated integral. It thus reduces the problem of computing a double integral to the problem of computing two integrals on intervals of the real line.

Proof. Consider any partition

$$\begin{aligned} x_0 &= a < x_1 < x_2 < \cdots < x_{n-1} < x_n = b, \\ y_0 &= c < y_1 < y_2 < \cdots < y_{m-1} < y_m = d \end{aligned} \tag{4}$$

of R. Let $x_{i-1} \le \xi_i \le x_i$. By the mean value theorem for integrals,

$$\int_{y_{j-1}}^{y_j} f(\xi_i, y)\, dy = \mu_{ij}(y_j - y_{j-1})$$

where

$$m_{ij} \le \mu_{ij} \le M_{ij},$$

and m_{ij}, M_{ij} are defined as in Section 2. Denoting by S and s the upper and lower Darboux sums corresponding to the partition (4), we have

$$s \le \sum\sum \mu_{ij}\Delta_{ij} \le S.$$

Since f is integrable over R, for any $\varepsilon > 0$ there is a $\delta > 0$ such that

$$\left| S - \iint_R f\, dA \right| < \varepsilon, \qquad \left| s - \iint_R f\, dA \right| < \varepsilon \qquad \text{if } \max |R_{ij}| < \delta.$$

We express this briefly, by writing

$$\lim_{\max|R_{ij}|\to 0} S = \lim_{\max|R_{ij}|} s = \iint_R f \, dA.$$

Therefore

$$\sum\sum \mu_{ij}\Delta_{ij} \to \iint_R f \, dA \qquad \text{as } \max|R_{ij}| \to 0. \tag{5}$$

On the other hand,

$$\sum\sum \mu_{ij}\Delta_{ij} = \sum\sum (x_i - x_{i-1}) \int_{y_{j-1}}^{y_j} f(\xi_i, y) \, dy$$

$$= \sum_i \left[\int_c^d f(\xi_i, y) \, dy \right] (x_i - x_{i-1}) = \sum g(\xi_i)(x_i - x_{i-1}). \tag{6}$$

From (5) and (6) it follows that as $\max(x_i - x_{i-1}) \to 0$, $\max(y_j - y_{j-1}) \to 0$,

$$\sum g(\xi_i)(x_i - x_{i-1}) \to \iint_R f \, dA.$$

Consequently, $g(x)$ is Riemann integrable on $[a,b]$ and (2) holds.

COROLLARY.   *Let $f(x,y)$ be integrable over R, and suppose that for each ξ in $[a,b]$ and η in $[c,d]$, the functions $f(\xi,y)$ and $f(x,\eta)$ are integrable over $[c,d]$ and $[a,b]$, respectively. Then*

$$\int_a^b dx \int_c^d f(x,y) \, dy = \int_c^d dy \int_a^b f(x,y) \, dx.$$

Thus, the order of integration in the iterated integral is irrelevant. The double integral $\iint_R f(x,y) \, dA$ is often written in the form

$$\iint_R f(x,y) \, dx \, dy.$$

By Theorem 1, this integral coincides with

$$\int_c^d \left(\int_a^b f(x,y) \, dx \right) dy$$

provided, for each y, $f(x,y)$ is integrable with respect to x, $a \le x \le b$.

Let $y = \gamma_1(x)$, $y = \gamma_2(x)$ be two continuous functions defined in an interval $\alpha \le x \le \beta$, such that

$$\gamma_1(x) < \gamma_2(x) \qquad \text{if } \alpha < x < \beta.$$

Let D be the domain consisting of all the points (x,y) with

$$\alpha < x < \beta, \qquad \gamma_1(x) < y < \gamma_2(x). \tag{7}$$

THEOREM 2.   *If D is defined by (7) and if f is continuous in the closure of D, then*

$$\iint\limits_D f(x,y)\, dA = \int_\alpha^\beta dx \int_{\gamma_1(x)}^{\gamma_2(x)} f(x,y)\, dy \tag{8}$$

where the integral on the right stands for

$$\int_\alpha^\beta \left(\int_{\gamma_1(x)}^{\gamma_2(x)} f(x,y)\, dy \right) dx.$$

Proof.   Let

$$R = \{(x,y); \quad \alpha \le x \le \beta, \ c \le y \le d\}$$

where $c < \gamma_1(x)$, $\gamma_2(x) < d$ for all x in $[\alpha,\beta]$. Denote by g the extension of f by 0. By Theorem 1,

$$\iint\limits_D f\, dA = \iint\limits_R g\, dA = \int_\alpha^\beta \left(\int_c^d g(x,y)\, dy \right) dx. \tag{9}$$

Write

$$\int_c^d g(x,y)\, dy = \int_c^{\gamma_1(x)} g(x,y)\, dy + \int_{\gamma_1(x)}^{\gamma_2(x)} g(x,y)\, dy + \int_{\gamma_2(x)}^d g(x,y)\, dy.$$

Since $g(x,y) = 0$ if $y < \gamma_1(x)$ or if $y > \gamma_2(x)$, the first and third integrals on the right are equal to zero. It follows that the right-hand side of (9) is equal to

$$\int_\alpha^\beta \left(\int_{\gamma_1(x)}^{\gamma_2(x)} g(x,y)\, dy \right) dx = \int_\alpha^\beta dx \int_{\gamma_1(x)}^{\gamma_2(x)} f(x,y)\, dy.$$

This completes the proof of (8).

The same proof yields also the following result:

COROLLARY.   *Theorem 2 remains true if f is piecewise continuous in D, provided the discontinuities lie on a finite number of continuous curves of the form $y = \delta_j(x)$.*

Let a_{mn} be nonnegative real numbers for $m,n = 1,2,\dots$. Define a function $f(x,y)$ in $0 \le x \le 1, 0 \le y \le 1$ such that

$$f(x,y) = a_{mn} \qquad \text{if } \frac{1}{2^m} < x \le \frac{1}{2^{m-1}}, \quad \frac{1}{2^n} < y \le \frac{1}{2^{n-1}}.$$

If $f(x,y)$ satisfies the conditions of the corollary to Theorem 1, then the assertion of the corollary gives

$$\sum_{m=1}^{\infty} \left(\sum_{n=1}^{\infty} a_{mn} \right) = \sum_{n=1}^{\infty} \left(\sum_{m=1}^{\infty} a_{mn} \right). \tag{10}$$

We shall prove this directly:

THEOREM 3. *If* $\sum_{n=1}^{\infty} a_{mn} < \infty$ *for any m and if the series on the left-hand side of* (10) *is convergent, then* $\sum_{m=1}^{\infty} a_{mn} < \infty$ *for any n and* (10) *holds.*

Proof. Since

$$\sum_{m=1}^{k} a_{mn} \le \sum_{m=1}^{k} \left(\sum_{j=1}^{\infty} a_{mj} \right) \le \sum_{m=1}^{\infty} \left(\sum_{j=1}^{\infty} a_{mj} \right) < \infty,$$

the series $\sum_{m=1}^{\infty} a_{mn}$ is convergent. Next,

$$\sum_{n=1}^{p} \left(\sum_{m=1}^{\infty} a_{mn} \right) = \sum_{m=1}^{\infty} \left(\sum_{n=1}^{p} a_{mn} \right) \le \sum_{m=1}^{\infty} \left(\sum_{n=1}^{\infty} a_{mn} \right) < \infty,$$

for any p. Consequently,

$$\sum_{n=1}^{\infty} \left(\sum_{m=1}^{\infty} a_{mn} \right) \text{ is finite and } \le \sum_{m=1}^{\infty} \left(\sum_{n=1}^{\infty} a_{mn} \right). \tag{11}$$

From the inequalities

$$\sum_{m=1}^{q} \left(\sum_{n=1}^{\infty} a_{mn} \right) = \sum_{n=1}^{\infty} \left(\sum_{m=1}^{q} a_{mn} \right) \le \sum_{n=1}^{\infty} \left(\sum_{m=1}^{\infty} a_{mn} \right)$$

it also follows that

$$\sum_{m=1}^{\infty} \left(\sum_{n=1}^{\infty} a_{mn} \right) \le \sum_{n=1}^{\infty} \left(\sum_{m=1}^{\infty} a_{mn} \right).$$

Combining this with (11), the assertion (10) follows.

DEFINITION. The *area* of a bounded region D is the integral $\iint_{D} dA$. The *centroid* of D is the point $(\bar{x}, \bar{y})$ given by

$$\bar{x} = \frac{\iint x \, dA}{\iint dA}, \qquad \bar{y} = \frac{\iint y \, dA}{\iint dA}.$$

PROBLEMS

1. If $f(x)$ is integrable over $a \le x \le b$ and $g(y)$ is integrable over $c \le y \le d$, then $f(x)g(y)$ is integrable over the rectangle $a \le x \le b$, $c \le y \le d$.

2. Find the centroid of each of the following regions:
    (a) The triangle with vertices $(0,0)$, $(a,0)$, (a,b);
    (b) The semicircular region $x^2 + y^2 \le a^2$, $x \ge 0$;
    (c) The region $x^2/a^2 + y^2/b^2 < 1$, $x \ge 0$, $y \ge 0$;
    (d) The region restricted by $x^n \le y \le 1$, $0 \le x \le 1$;
    (e) The region bounded by $bx^2 = a^2 y$ and the line $ay = bx$, where $a > 0$, $b > 0$;
    (f) The region bounded by the parabolas $y = x^2 + x, y = 2x^2 - 2$.

3. Compute $\iint (x^2 + y^2) \, dx \, dy$ over the triangle with vertices $(0,0),(1,0),(1,1)$.

4. Let $f(x,y)$ be a continuous function in a closed rectangle $a \le x \le b$, $c \le y \le d$, and let

$$g(x,y) = \int_c^y \int_a^x f(s,t) \, ds \, dt.$$

    Show that $g_{xy} = f$.

5. Compute the integrals

    (a) $\displaystyle\int_0^1 dx \int_x^1 e^{x/y} \, dy$;

    (b) $\displaystyle\int_0^1 dy \int_y^1 e^{-x^2} \, dx$.

6. Let $K(x,y)$ be a continuous function for $0 \le x \le 1, 0 \le y \le 1$. Prove that

$$\int_0^1 dx \int_0^x K(x,y) \, dy = \int_0^1 dy \int_y^1 K(x,y) \, dx.$$

4. MULTIPLE INTEGRALS

We shall extend the results of Sections 2 and 3 to functions of n variables. We begin with the case $n = 3$. We shall denote by $x = (x_1, x_2, x_3)$ a point in R^3.

Let R be a three-dimensional closed rectangle given by

$$a_i \le x_i \le b_i \qquad (i = 1,2,3).$$

A partition of R is given by partitions

$$x_{i,0} = a_1 < x_{i,1} < x_{i,2} < \cdots < x_{i,m_i-1} < x_{i,m_i} = b_i \quad (i = 1,2,3). \qquad \textbf{(1)}$$

Denote by $R_{j_1 j_2 j_3}$ the closed rectangle defined by

$$x_{i,j_i-1} \leq x_i \leq x_{i,j_i} \qquad (i = 1,2,3).$$

We denote the volume of $R_{j_1 j_2 j_3}$ by $\Delta_{j_1 j_2 j_3}$. The *mesh* of the partition (1) is the number

$$\max_{j_1,j_2,j_3} \sqrt{(x_{1,j_1} - x_{1,j_1-1})^2 + (x_{2,j_2} - x_{2,j_2-1})^2 + (x_{3,j_3} - x_{3,j_3-1})^2}.$$

We denote it by $\max |R_{j_1 j_2 j_3}|$.

Let $f(x)$ be a bounded function defined on R, and denote by $M_{j_1 j_2 j_3}$ and $m_{j_1 j_2 j_3}$ the supremum and the infimum of f in $R_{j_1 j_2 j_3}$. We form the upper and lower Darboux sums

$$S = \sum_{j_1} \sum_{j_2} \sum_{j_3} M_{j_1 j_2 j_3} \Delta_{j_1 j_2 j_3},$$

$$s = \sum_{j_1} \sum_{j_2} \sum_{j_3} m_{j_1 j_2 j_3} \Delta_{j_1 j_2 j_3}.$$

If $J = \inf S$ is equal to $I = \sup s$, then we say that f is *Darboux integrable* over R, and the number J is then called the *Darboux integral* of f over R. Theorems 1–4 of Section 2 extend to the present case.

We form the Riemann sums

$$T = \sum_{j_1} \sum_{j_2} \sum_{j_3} f(\xi_{j_1},\xi_{j_2},\xi_{j_3}) \Delta_{j_1 j_2 j_3}$$

where $(\xi_{j_1},\xi_{j_2},\xi_{j_3})$ is any point in $R_{j_1 j_2 j_3}$. If there exists a number K such that for any $\varepsilon > 0$ there is a $\delta > 0$ such that

$$|T - K| < \varepsilon \qquad \text{whenever } \max |R_{j_1 j_2 j_3}| < \delta,$$

then we say that f is *Riemann integrable* over R, and the number K is then called the *Riemann integral* of f over R.

Theorem 5 of Section 2 extends to the present case. Thus, f is Riemann integrable over R if, and only if, f is Darboux integrable over R, and in this case the two integrals are equal. We denote the integral of f over R by

$$\iiint_R f(x_1,x_2,x_3)\, dV.$$

THEOREM 1.   *Let $f(x) = f(x_1,x_2,x_3)$ be a bounded function in a closed three-dimensional rectangle R, and suppose that the set of its discontinuities has content 0 in R^3. Then f is integrable over R.*

The proof is the same as the proof of Theorem 6 in Section 2.

Now let D be any bounded set in R^3, and let f be a bounded function defined in D. The function

$$g(x) = \begin{cases} f(x) & \text{if } x \in D, \\ 0 & \text{if } x \notin D \end{cases}$$

is called the *extension* of f by 0.

Let R be any closed rectangle containing D. If g is integrable over R, then we say that f is *integrable* over D, and, in this case, the number

$$\iiint_R g(x)\, dV$$

is called the *integral* of f over D. We denote it by

$$\iiint_D f(x)\, dV, \quad \text{or} \quad \iiint_D f(x_1, x_2, x_3)\, dx_1\, dx_2\, dx_3. \tag{2}$$

Theorem 7 of Section 2 and its proof extend to the present case. Thus the definition (2) is independent of R. Theorem 8 and its proof extend to the present case of sets in R^3. Finally, the analogs of Theorems 9 and 10 are valid, namely:

THEOREM 2. *If $f(x)$ is a bounded function in a bounded region D in R^3, and if both ∂D and the set of discontinuities of f are sets of content zero in R^3, then $f(x)$ is integrable on D.*

THEOREM 3. *A piecewise continuous function in a normal region D in R^3 is integrable.*

All the results developed so far extend to functions $f(x) = f(x_1,...,x_n)$ of n variables. The proofs are essentially unchanged.

The integral (2) is called a *triple integral*. For any n, we write the integral in the form

$$\iint \cdots \int_D f(x)\, dV, \quad \text{or} \quad \iint \cdots \int_D f(x_1,...,x_n)\, dx_1 \cdots dx_n,$$

and call it a *multiple integral*, or *n-tuple integral*.

We state without proof an analog of Theorem 1 of Section 3:

THEOREM 4. *Let $f(x,y,z)$ be an integrable function on the three-dimensional closed rectangle*

$$R: a_1 \le x \le b_1, \qquad a_2 \le y \le b_2, \qquad a_3 \le z \le b_3.$$

Denote by R_{yz} the closed rectangle $a_2 \le y \le b_2$, $a_3 \le z \le b_3$, and assume that the integral

$$g(x) = \int_{R_{yz}} f(x,y,z)\, dA \tag{3}$$

exists. Then $g(x)$ is integrable on $a_1 \le x \le b_1$, and

$$\iiint_R f(x,y,z)\, dV = \int_{a_1}^{b_1} g(x)\, dx. \tag{4}$$

If we substitute $g(x)$ from (3) into (4), we get

$$\iiint_R f(x,y,z)\, dV = \int_{a_1}^{b_1} \left(\iint_{R_{yz}} f(x,y,z)\, dA \right) dx. \tag{5}$$

The integral on the right is called an *iterated integral*. Thus, the relation (5) asserts that a triple integral is equal to an iterated integral.

The next theorem gives another way of iterating a triple integral.

THEOREM 5. *Let $f(x,y,z)$ be an integrable function on the three-dimensional closed rectangle*

$$R : a_1 \le x \le b_1, \qquad a_2 \le y \le b_2, \qquad a_3 \le z \le b_3.$$

Denote by R_{yz} the rectangle $a_2 \le y \le b_2$, $a_3 \le z \le b_3$, and assume that for each (y,z) in R_{yz}, the integral

$$h(y,z) = \int_{a_1}^{b_1} f(x,y,z)\, dx \tag{6}$$

exists. Then $h(y,z)$ is integrable over R_{yz} and

$$\iiint_R f(x,y,z)\, dV = \iint_{R_{yz}} h(y,z)\, dA. \tag{7}$$

If we substitute h from (6) into (7), we get

$$\iiint_R f(x,y,z)\, dV = \iint_{R_{yz}} \left(\int_{a_1}^{b_1} f(x,y,z)\, dx \right) dA. \tag{8}$$

The integral on the right is called an *iterated integral*.

The assertions of Theorems 4 and 5 also can be written in the form

$$\iiint_R f(x,y,z)\, dx\, dy\, dz = \int_{a_1}^{b_1} \left(\iint_{R_{yz}} f(x,y,z)\, dy\, dz \right) dx$$

$$= \iint_{R_{yz}} \left(\int_{a_1}^{b_1} f(x,y,z)\, dx \right) dy\, dz. \tag{9}$$

Consider now the case of a domain D defined by

$$D = \{(x,y,z); \quad (x,y) \in G, \quad \gamma_1(x,y) < z < \gamma_2(x,y)\} \tag{10}$$

where G is a bounded domain in R^2 and $\gamma_1(x,y)$, $\gamma_2(x,y)$ are continuous in the closure of G.

THEOREM 6. *Let D be as in* (10), *and let $f(x,y,z)$ be a continuous function in the closure of D. Then f is integrable on D and*

$$\iiint_D f(x,y,z)\, dV = \iint_G \left(\int_{\gamma_1(x,y)}^{\gamma_2(x,y)} f(x,y,z)\, dz \right) dA. \tag{11}$$

The proof is similar to the proof of Theorem 2 in Section 3.
Theorems 4, 5, and 6 extend to multiple integrals in R^n.
The *volume* of a bounded region D in R^n is the integral $\iint \cdots \int_D dx_1 \cdots dx_n$.

PROBLEMS

1. Prove Theorem 4.

2. Prove Theorem 5.

3. Prove Theorem 6.

4. Compute the volume of the hemiball $x^2 + y^2 + z^2 \leq a^2$, $z \geq 0$.

5. Compute the volume bounded by the ellipsoid

$$\frac{x^2}{a^2} + \frac{y^2}{b^2} + \frac{z^2}{c^2} = 1.$$

6. Compute the volume bounded by the paraboloid $a^2 z = b(a^2 - x^2 - y^2)$ and the (x,y)-plane.

7. Compute the integral $\iiint_D x^2 y^2 z\, dx\, dy\, dz$ where D is defined by $x^2 + y^2 < 1$, $0 < z < 1$.

8. Let D denote the set defined by: $x^2 + y^2 + z^2 < 1$, $y > 0$, $z > 0$. Compute the integrals:

   (a) $\iiint_D x^2\, dV;$ (b) $\iiint_D x^2 y\, dV;$ (c) $\iiint_D x^2 yz\, dV.$

9. The *centroid* of a region D in R^3 is a point $(\bar{x}_1, \bar{x}_2, \bar{x}_3)$ given by

$$\bar{x}_i = \frac{\iiint\limits_D x_i \, dV}{\iiint\limits_D dV} \qquad (i = 1,2,3).$$

Find the centroid of the domain bounded by the plane $z = 0$ and the paraboloid $x^2/a^2 + y^2/b^2 + z/c = 1$.

10. Find the centroid of the region bounded by the ellipsoid $x^2/a^2 + y^2/b^2 + z^2/c^2 = 1$ and restricted by $x \geq 0$, $y \geq 0$, $z \geq 0$.

11. The *moment of inertia* of a region D (of density 1) about the x-axis is the integral

$$I_x = \iiint\limits_D (y^2 + z^2) \, dx \, dy \, dz.$$

Find I_x for the D given in Problem 10.

12. Find the moment of inertia about the x-axis of the domain bounded by the ellipsoid $x^2/a^2 + y^2/b^2 + z^2/c^2 = 1$.

13. If $f_1(x)$ and $f_2(x)$ are integrable on a set D in R^3, then $f_1(x) + f_2(x)$ is also integrable on D, and

$$\iiint\limits_D f_1(x) \, dV + \iiint\limits_D f_2(x) \, dV = \iiint\limits_D (f_1(x) + f_2(x)) \, dV.$$

14. If $f \geq g$ and f, g are integrable over a bounded set D in R^n, then

$$\iint \cdots \int\limits_D f \, dV \geq \iint \cdots \int\limits_D g \, dV.$$

15. Prove the following *mean value theorem for integrals*: If $f(x)$ is a continuous function in a closed normal domain D in R^n, then there exists a point x^0 in the interior of D such that

$$\iint \cdots \int\limits_D f(x) \, dV = f(x^0) V(D) \qquad (V(D) = \text{volume of } D).$$

5. CHANGE OF VARIABLES IN MULTIPLE INTEGRALS

Let

$$x = f(u) \tag{1}$$

be a one-to-one mapping from an open set U in R^n onto an open set X in R^n.

Suppose that the mapping is continuously differentiable. That means that if we write (1) in terms of components

$$x_i = f_i(u_1,\dots,u_n), \qquad 1 \le i \le n,$$

then the f_i have first continuous derivatives in U. If the Jacobian

$$\frac{\partial(f_1,\dots,f_n)}{\partial(u_1,\dots,u_n)}$$

does not vanish at any point of U, then we say that the mapping f is a *diffeomorphism* from U onto X. By the inverse mapping theorem (Theorem 1 of Section 7.6), the inverse mapping

$$u = g(x)$$

is then also continuously differentiable.

THEOREM 1.   *Let $x = f(u)$ be a diffeomorphism from an open set U in R^n onto an open set X in R^n. Let D be a bounded domain contained in U and let $h(x)$ be an integrable function on the image $f(D)$ of D. Then the function*

$$h(f(u)) \left| \frac{\partial(f_1,\dots,f_n)}{\partial(u_1,\dots,u_n)} \right|$$

is integrable on D, and

$$\iint \dots \int_{f(D)} h(x)\, dx_1 \cdots dx_n = \iint \dots \int_D h(f(u)) \left| \frac{\partial(f_1,\dots,f_n)}{\partial(u_1,\dots,u_n)} \right| du_1 \cdots du_n. \quad (2)$$

Equation (2) is called the *change of variables' formula* for multiple integrals. The case $n = 1$ was proved in Section 4.6. (In that case, it was not necessary to assume that f is a diffeomorphism, only that f is continuously differentiable.) The proof for $n \ge 2$ is much harder and will not be given here in full generality. Instead, we shall give a proof that applies to a large class of domains D, including essentially all those that occur in applications. The proof depends on the notions of line and surface integrals. Since these notions will be developed only in the next chapter, we postpone the proof until then.

EXAMPLE 1.   Let (r,θ) be the polar coordinates of (x,y). Thus

$$x = r \cos \theta,$$
$$y = r \sin \theta. \quad (3)$$

Then (2) becomes

$$\iint_{D_{xy}} h(x,y) \, dx \, dy = \iint_{D_{r\theta}} h(r \cos \theta, r \sin \theta) r \, dr \, d\theta \tag{4}$$

where D_{xy} is the image of $D_{r\theta}$ under the mapping (3). Since the Jacobian of the transformation (3) is equal to r, Theorem 1 can be used only if the origin $(0,0)$ does not belong to D_{xy}. If $(0,0)$ lies in D_{xy}, let

$$D_{xy}{}^{\varepsilon} = D_{xy} \cap B_{\varepsilon} \qquad \text{where } B_{\varepsilon} = \{(x,y); \sqrt{x^2 + y^2} > \varepsilon\}.$$

Then

$$\iint_{D_{xy}{}^{\varepsilon}} h(x,y) \, dx \, dy = \iint_{D_{r\theta}{}^{\varepsilon}} h(r \cos \theta, r \sin \theta) r \, dr \, d\theta \tag{5}$$

where $D_{r\theta}{}^{\varepsilon}$ is the set whose image under (3) is $D_{xy}{}^{\varepsilon}$. It can now be shown that if $\varepsilon \to 0$, then both sides of (5) converge to the corresponding sides of (4). Thus, (4) is valid for any domain D_{xy}.

EXAMPLE 2.   Let (r,ϕ,θ) be the spherical coordinates of a point (x,y,z). Thus

$$x = r \sin \theta \cos \phi,$$
$$y = r \sin \theta \sin \phi,$$
$$z = r \cos \theta.$$

The Jacobian of this mapping is $r^2 \sin \theta$. Hence, (2) gives

$$\iiint_{D_{xyz}} h(x,y,z) \, dx \, dy \, dz = \iiint_{D_{r\phi\theta}} h(r \sin \theta \cos \phi, r \sin \theta \sin \phi, r \cos \theta) \tag{6}$$
$$\times r^2 \sin \theta \, dr \, d\theta \, d\phi,$$

provided the z-axis does not intersect D_{xyz}. If it intersects D_{xyz}, then we first apply (2) in $D_{xyz}^{\varepsilon} = D_{xyz} \cap C_{\varepsilon}$ where

$$C_{\varepsilon} = \{(x,y,z); x^2 + y^2 > \varepsilon\}$$

and then take $\varepsilon \to 0$.

PROBLEMS

1. Evaluate the integral $\iiint_{R}(x^2/a^2 + y^2/b^2 + z^2/c^2) \, dx \, dy \, dz$ over the domain bounded by the ellipsoid $x^2/a^2 + y^2/b^2 + z^2/c^2 = 1$; use the transformation $x = au$, $y = bv$, $z = cw$.

2. Compute $\iiint\limits_{R} (x^2 + y^2)\, dx\, dy\, dz$ for R as in the preceding problem; use the same transformation.

3. Calculate $\iint\limits_{R} dxdy/(x + y)$, where R is the region bounded by the lines $x + y = 1$, $x + y = 4$, $y = 0$, $x = 0$, using the transformation $x = u - uv$, $y = uv$.

4. Let R be the region bounded by the curves $x^2 - y^2 = 1$, $x^2 - y^2 = 4$, $x^2 + y^2 = 9$, $x^2 + y^2 = 16$ and lying in the quadrant $x \geq 0$, $y \geq 0$. Compute $\iint\limits_{R} xy\, dx\, dy$ by using the change of variables $u = x^2 - y^2$, $v = x^2 + y^2$.

5. Compute the integral $\iint\limits_{R} e^{-(x^2+y^2)}\, dx\, dy$, where R is the disc $x^2 + y^2 \leq a^2$, by using polar coordinates.

6. Compute the integral $\iiint\limits_{R} e^{(-x^2+y^2+z^2)}\, dx\, dy\, dz$, where R is the ball $x^2 + y^2 + z^2 \leq a^2$, by using spherical coordinates.

7. Let R be the region lying inside the circle $x^2 + y^2 = 2x$ and to the right of $x = 1$. Use polar coordinates to compute $\iint\limits_{R} \dfrac{x\, dx\, dy}{x^2 + y^2}$.

8. Let R be the region in the octant $x \geq 0$, $y \geq 0$, $z \geq 0$ that is bounded by $x + y + z = 1$. Use the transformation $x = u(1 - v)$, $y = uv(1 - w)$, $z = uvw$ in order to compute the integral $\iiint\limits_{R} \dfrac{dx\, dy\, dz}{y + z}$.

9. Compute the integral $\iiint\limits_{R} x^2 y\, dx\, dy\, dz$ in the region R given by $x^2 + y^2 < 1$, $0 \leq z \leq 1$, by using cylindrical coordinates (r, ϕ, z), that is,

$$x = r \cos \phi, \qquad y = r \sin \phi, \qquad z = z.$$

10. Let $x = f(u)$ be a continuously differentiable mapping from an open set U in R^n into R^n, and let S be a closed, bounded subset of U having content zero with respect to R^n. Prove that $f(S)$ has content zero in R^n. [*Hint:* For any $\varepsilon > 0$, there are n-dimensional cubes $R_1, \ldots, R_k$ that cover S, and such that $\sum V(R_j) < \varepsilon$. Let $\xi_i \in R_j$. If $x \in R_j$, then

$$\| f(x) - f(\xi_j) \| \leq M \| x - \xi_j \|.$$

Hence $f(R_j)$ is contained in a cube $\hat{R}_j$ with sides equal to M times the length of the diagonal of R_j. Therefore $\sum V(\hat{R}_j) < n^{n/2} M^n \varepsilon$.]

11. Denote by $V_n(r)$ the volume of the ball $\|x\| < r$ in R^n. Prove that $V_n(r) = c_n r^n$, c_n constant. Next, use the relation

$$V_{n+1}(r) = \int_{-r}^{r} d\rho \underset{x_1^2 + \cdots + x_n^2 < r^2 - \rho^2}{\iint \cdots \int} dx_1\, dx_2 \cdots dx_n$$

to show that

$$V_{n+1}(r) = 2V_n(1) \int_0^r (r^2 - s^2)^{n/2}\, ds.$$

Use this to show, by induction, that

$$V_{2n}(r) = \frac{\pi^n}{n!}\, r^{2n}, \qquad V_{2n-1}(r) = \frac{4^n \pi^{n-1} n!}{(2n)!}\, r^{2n-1}.$$

6. UNIFORM CONVERGENCE

Let $\{f_m(x)\}$ be a sequence of functions defined on a set S of R^n. Suppose there exists a function $f(x)$ defined on S such that, for any $\varepsilon > 0$ there is a positive integer n_0 for which

$$|f_m(x) - f(x)| < \varepsilon, \qquad \text{whenever } x \in S,\ m \geq n_0.$$

Then we say that $\{f_m(x)\}$ is *uniformly convergent* to $f(x)$ on S.

This definition extends the one given in Section 5.4 in case S is an interval in R^1. All the results derived in Section 5.4 extend almost word for word to functions defined on a set S in R^n. Thus, the uniform limit of continuous functions is continuous; a sequence $\{f_n(x)\}$ is uniformly convergent if, and only if, for any $\varepsilon > 0$ there is a positive integer n_0 such that

$$|f_m(x) - f_k(x)| < \varepsilon \qquad \text{whenever } x \in S,\ m \geq n_0,\ k \geq n_0.$$

An infinite series $\sum u_m(x)$ of functions $u_m(x)$ defined on S is said to be (uniformly) convergent if the sequence of its partial sums is (uniformly) convergent. The Weierstrass M-test is valid for the present case, the proof being the same as in one dimension.

The results of Section 5.5 also extend to functions of n variables. In particular:

THEOREM 1. *Let* $\sum\limits_{m=0}^{\infty} u_m(x)$ *be an infinite series of continuous functions* $u_m(x)$ *defined on a closed normal domain D in R^n. If the series is uniformly convergent in D to a function $S(x)$, then*

$$\iint \cdots \int_D S(x)\, dx_1 \cdots dx_n = \sum_{m=0}^{\infty} \iint \cdots \int_D u_m(x)\, dx_1 \cdots dx_n.$$

PROBLEMS

1. Prove Theorem 1.

7. IMPROPER INTEGRALS

Let D and G be two normal regions in R^n, and let $D \supset G$. We denote by $D-G$ the set of all points that belong to D but not to G. We shall have to deal, in this section, with integrals of the form

$$\iint \cdots \int_{D-G} f(x) \, dV$$

where $f(x)$ is piecewise continuous in D. Notice that the set $D-G$ need not be a region, but its boundary has content 0 since it is contained in $\partial G \cup \partial D$. The following theorem will be needed.

THEOREM 1.   *If D, G, and f are as above, then*

$$\iint \cdots \int_{D} f(x) \, dV - \iint \cdots \int_{G} f(x) \, dV = \iint \cdots \int_{D-G} f(x) \, dV. \tag{1}$$

Proof.   Define

$$f_1(x) = \begin{cases} f(x) & \text{if } x \in D-G, \\ 0 & \text{if } x \notin D-G, \end{cases}$$

$$f_2(x) = \begin{cases} f(x) & \text{if } x \in G, \\ 0 & \text{if } x \notin G, \end{cases}$$

$$f_3(x) = \begin{cases} f(x) & \text{if } x \in D, \\ 0 & \text{if } x \notin D. \end{cases}$$

Let R be an n-dimensional rectangle containing D. Since $f_3 = f_1 + f_2$ in R,

$$\iint \cdots \int_{R} f_1(x) \, dV + \iint \cdots \int_{R} f_2(x) \, dV = \iint \cdots \int_{R} f_3(x) \, dV.$$

Recalling the definition of the f_i, the last relation gives (1).

We shall denote by B_N the ball $\{x; \|x\| < N\}$ in R^n.

Let $f(x)$ be a function defined in the whole space R^n, and piecewise continuous in any bounded domain. Then, for any (bounded) normal domain D, the integral

$$\iint \cdots \int_D f(x)\, dx_1 \cdots dx_n$$

exists. Suppose there is a number I having the following property: For any $\varepsilon > 0$ there is a number N such that for any normal domain D,

$$\left| I - \iint \cdots \int_D f(x)\, dx_1 \cdots dx_n \right| < \varepsilon \qquad \text{if } D \supset B_N. \tag{2}$$

Then we say that f has *improper integral* on R^n, and the number I is called the improper integral of f on R^n. We denote this number by

$$\iint \cdots \int_{R^n} f(x)\, dx_1 \cdots dx_n, \quad \text{or} \quad \iint \cdots \int_{R^n} f(x)\, dV. \tag{3}$$

We also say that the improper integral in (3) is *convergent*.

Let $\{D_m\}$ be a monotone increasing sequence of sets, that is, $D_{m+1} \supset D_m$ for all m. If for any $N > 0$ there is a positive integer m such that

$$D_m \supset B_N,$$

then we say that the sequence $\{D_m\}$ is an *exhaustive*.

THEOREM 2. *The improper integral of f on R^n is convergent if, and only if, for any exhaustive sequence of normal domains D_m,*

$$\lim_{m \to \infty} \iint \cdots \int_{D_m} f(x)\, dx_1 \cdots dx_n \tag{4}$$

exists.

Proof. Suppose that f has improper integral I. Then, for any $\varepsilon > 0$ there is an $N > 0$ such that (2) holds. If $\{D_m\}$ is an exhaustive sequence of normal domains, then there is an m_0 such that $D_m \supset B_N$ if $m \geq m_0$. Applying (2) we conclude that

$$\left| I - \iint \cdots \int_{D_m} f(x)\, dx_1 \cdots dx_n \right| < \varepsilon \qquad \text{if } m \geq m_0.$$

This proves (4).

Suppose conversely that (4) holds for any exhaustive sequence of domains D_m. Let

$$I = \lim_{m \to \infty} \underset{D_m^0}{\iint \cdots \int} f(x) \, dx_1 \cdots dx_n,$$

where $\{D_m^0\}$ is a particular exhaustive sequence of normal domains. We claim that (2) holds. To prove it we shall suppose that (2) is false for some $\varepsilon > 0$, and derive a contradiction. Since (2) is false, there is certainly a normal domain E_1 such that

$$\left| I - \underset{E_1}{\iint \cdots \int} f(x) \, dx_1 \cdots dx_n \right| \geq \varepsilon.$$

Let $D_{N_1}^0 \supset E_1$. $D_{N_1}^0$ is contained in some ball $B_{N'}$. Since (2) is false for any N, there is a normal domain E_2 containing $B_{N'}$ (hence, containing $D_{N_1}^0$) such that

$$\left| I - \underset{E_2}{\iint \cdots \int} f(x) \, dx_1 \cdots dx_n \right| \geq \varepsilon.$$

Let $D_{N_2}^0 \supset E_2$. Again there is a normal domain E_3 such that $E_3 \supset D_{N_2}^0$ and

$$\left| I - \underset{E_3}{\iint \cdots \int} f(x) \, dx_1 \cdots dx_n \right| \geq \varepsilon.$$

Let $D_{N_3}^0 \supset E_3$. Continuing in this way step by step, we construct a sequence

$$\{E_1, D_{N_1}^0, E_2, D_{N_2}^0, E_3, D_{N_3}^0, \ldots\}.$$

Denote this sequence by $\{D_m\}$. It is an exhaustive sequence of normal domains. Therefore (4) must hold. On the other hand,

$$\left| I - \underset{D_{2m}}{\iint \cdots \int} f(x) \, dx_1 \cdots dx_n \right| \to 0 \qquad \text{if } m \to \infty$$

and

$$\left| I - \underset{D_{2m-1}}{\iint \cdots \int} f(x) \, dx_1 \cdots dx_n \right| \geq \varepsilon \qquad (m = 1, 2, \ldots).$$

But the last two relations contradict (4).

If the improper integral

$$\underset{R^n}{\iint \cdots \int} |f(x)| \, dx_1 \cdots dx_n \qquad (5)$$

is convergent then we say that the improper integral in (3) is *absolutely convergent*. If the improper integral in (3) is convergent but the improper integral in (5) is not convergent, then we say that the improper integral of (3) is *conditionally convergent*.

THEOREM 3.   *If the improper integral is absolutely convergent, then it is also convergent.*

Proof.   By Theorem 2, for any exhaustive sequence of normal domains D_m, the sequence of numbers

$$\iint\cdots\int_{D_m} |f(x)| \, dx_1\cdots dx_n$$

is convergent. Therefore it is also a Cauchy sequence. Using Theorem 1 we get, if $m > k$,

$$\left| \iint\cdots\int_{D_m} f(x) \, dx_1\cdots dx_n - \iint\cdots\int_{D_k} f(x) \, dx_1\cdots dx_n \right|$$

$$= \left| \iint\cdots\int_{D_m-D_k} f(x) \, dx_1\cdots dx_n \right| \le \iint\cdots\int_{D_m-D_k} |f(x)| \, dx_1\cdots dx_n$$

$$= \iint\cdots\int_{D_m} |f(x)| \, dx_1\cdots dx_n - \iint\cdots\int_{D_k} |f(x)| \, dx_1\cdots dx_n \to 0 \qquad \text{if } k \to \infty.$$

Consequently, the sequence of numbers

$$\iint\cdots\int_{D_m} f(x) \, dx_1\cdots dx_n$$

is a Cauchy sequence, and therefore it is convergent. Employing Theorem 2, we conclude that f has improper integral over R^n.

The usefulness of Theorem 3 is in the fact that it is easy to give simple sufficient conditions for absolute convergence. The next theorem serves as an important example.

THEOREM 4.   *If there exists a positive constant M and an exhaustive sequence $\{E_m\}$ of normal domains such that*

$$\iint\cdots\int_{E_m} |f(x)| \, dx_1\cdots dx_n \le M \qquad \textit{for all } m, \tag{6}$$

then the improper integral of f over R^n is absolutely convergent.

Proof.   Let $\{D_m\}$ be any exhaustive sequence of normal domains. Each domain D_m is contained in some domain E_k. Hence

$$\iint\cdots\int_{D_m} |f(x)| \, dx_1\cdots dx_n \le \iint\cdots\int_{E_k} |f(x)| \, dx_1\cdots dx_n \le M.$$

It follows that the sequence of numbers

$$\iint\cdots\int_{D_m} |f(x)| \, dx_1\cdots dx_n$$

is bounded. Since it is also monotone increasing, it is convergent. Now use Theorem 2 to conclude that the improper integral of $|f|$ is convergent.

The concept of improper integral of a piecewise continuous function over any unbounded domain G can be given analogously to the case in which $G = R^n$. We simply replace, in the condition (2), D by $D \cap G$ and B_N by $B_N \cap G$.

We shall now define the concept of an improper integral over a (bounded) normal domain D. Let y be a point contained in the closure $\bar{D}$ of D. Denote by $C_\delta(y)$ $(\delta \ge 0)$ the set $\{x ; \|x - y\| > \delta\}$ and let

$$D_\delta = D \cap C_\delta(y).$$

We denote by $\mathscr{D}_\delta$ the class of all normal domains G that contain D_δ and that are contained in some D_η, $0 < \eta < \delta$.

Now let $f(x)$ be a continuous function in $\bar{D}_\eta$ for any $\eta > 0$. Notice that $f(x)$ need not be defined at point y. For any $\delta > 0$ and for any normal domain G in $\mathscr{D}_\delta$, we can form the integral

$$\iint\cdots\int_{G} f(x) \, dx_1\cdots dx_n.$$

Suppose there exists a number I such that for any $\varepsilon > 0$ there is a $\delta > 0$ with the following property:

$$\left| I - \iint\cdots\int_{G} f(x) \, dx_1\cdots dx_n \right| < \varepsilon \qquad \text{whenever } G \in \mathscr{D}_\delta.$$

We then say that f has *improper integral* in (or on) D, and the number I is called the improper integral of f over D. We denote it by

$$\iint\cdots\int_{D} f(x) \, dx_1\cdots dx_n.$$

The analogs of Theorems 2, 3, and 4 are true, with similar proofs. In particular we have:

THEOREM 5. *If there is a positive constant M such that*

$$\iint \cdots \int_{D_\delta} |f(x)|\, dx_1 \cdots dx_n \leq M \tag{7}$$

for all $\delta > 0$ sufficiently small, then the improper integral of f over D is absolutely convergent (and therefore also convergent).

PROBLEMS

1. Let $f(x)$ and $g(y)$ be nonnegative functions on R^1, and suppose that the improper integrals $\int_{-\infty}^{\infty} f(x)\, dx$ and $\int_{-\infty}^{\infty} g(y)\, dy$ are convergent. Then also the improper integral $\iint_{R^2} f(x)g(y)\, dA$ is convergent, and it is equal to $\left(\int_{-\infty}^{\infty} f(x)\, dx\right) \cdot \left(\int_{-\infty}^{\infty} g(y)\, dy\right)$.

2. Let f, g be continuous functions in R^n, and let $f(x) \geq g(x) \geq 0$. Prove: (a) If the improper integral of f over R^n is convergent, then the improper integral of g over R^n is convergent. (b) If the improper integral of g over R^n is divergent (that is, is not convergent), then also the improper integral of f over R^n is divergent.

3. State the analogs of Theorems 2, 3, and 4 and the analog of Problem 2 for improper integrals over a bounded normal domain.

4. Let D be a bounded domain in R^n, and let $y \in D$. Find for each of the following improper integrals the value of α for which the integral converges:

    (a) $\displaystyle \iint_D \frac{dx_1\, dx_2}{[(x_1 - y_1)^2 + (x_2 - y_2)^2]^\alpha}$;

    (b) $\displaystyle \iiint_D \frac{dx_1\, dx_2\, dx_3}{[(x_1 - y_1)^2 + (x_2 - y_2)^2 + (x_3 - y_3)^2]^\alpha}$.

5. Let D be the domain $\{x; \|x\| > 1\}$ in R^n $(n = 2, 3)$ and let y be a point in R^n satisfying $\|y\| < 1$. Find the values of α for which the improper integrals in (a) and (b) of the preceding problem are convergent.

6. Let D be the unit disc in the plane and let (r, θ) be the polar coordinates of a point (x, y). Determine which of the following improper integrals is convergent:

    (a) $\displaystyle \iint_D \frac{x^2}{(x^2 + y^2)^{5/2}}\, dx\, dy$;   (b) $\displaystyle \iint_D \frac{\sin \theta}{r^{3/2}}\, dr\, d\theta$;

(c) $\iint\limits_D \frac{1}{r} \log \frac{1}{r} \, dx \, dy;$ (d) $\iint\limits_D \frac{x^2 + y^4}{(x^2 + y^2)^{3/2}} \, dx \, dy.$

7. Let $f(x,y,z)$ be a continuous positive function in the closed unit ball $D = \{(x,y,z); x^2 + y^2 + z^2 \le 1\}$. Let $r = \sqrt{x^2 + y^2 + z^2}$. Determine which of the following improper integrals is convergent:

(a) $\iiint\limits_D \frac{x^2 + y^2}{r^{9/2}} f(x,y,z) \, dV;$ (b) $\iiint\limits_D \frac{z^2}{r^6} f(x,y,z) \, dV;$

(c) $\iiint\limits_D \frac{f(x,y,z)}{r^3} \, dV;$ (d) $\iiint\limits_D \frac{x^2 y^2 z^2}{r^{17/2}} f(x,y,z) \, dV.$

8. Prove that $\int_0^\infty e^{-x^2} \, dx = \sqrt{\pi}/2$. [*Hint:* Let $I = \int_{-\infty}^\infty e^{-x^2} \, dx$. By Problem 1, $I^2 = \iint\limits_{R^2} e^{-x^2 - y^2} \, dx \, dy = \lim\limits_{N \to \infty} \iint\limits_{x^2 + y^2 < N^2} e^{-(x^2 + y^2)} \, dx \, dy$. Use polar coordinates to evaluate the last integral, and then take $N \to \infty$.]

9. Prove that the improper integral (3) is convergent if, and only if, for any $\varepsilon > 0$ there is a positive number N such that

$$\left| \iint\limits_D \cdots \int f(x) \, dx_1 \cdots dx_n - \iint\limits_G \cdots \int f(x) \, dx_1 \cdots dx_n \right| < \varepsilon$$

for any two normal domains containing B_N.

8. INTEGRALS DEPENDING ON A PARAMETER

Let D be a set in R^n and let S be a set in R^m. The set of all points (x,y) where $x \in D$, $y \in S$ is called the *product* of D and S, and it is denoted by $D \times S$. We may consider $D \times S$ as a subset of R^{m+n}. It is easily verified that the closure of $D \times S$ is $\bar{D} \times \bar{S}$, where $\bar{D}$ and $\bar{S}$ are the closures of D and S, respectively.

In this section we consider integrals of the form

$$\phi(y) = \iint\limits_D \cdots \int f(x,y) \, dx_1 \cdots dx_n. \tag{1}$$

THEOREM 1.   *Let D be a normal domain and let S be any set. If $f(x,y)$ is uniformly continuous in $D \times S$, then the integral $\phi(y)$ is a continuous function in S.*

Proof. Since $f(x,y)$ is uniformly continuous on $D \times S$, for any $\varepsilon > 0$ there is a $\delta > 0$ such that

$$|f(x,y) - f(x',y')| < \varepsilon \quad \text{if } \sqrt{\|x - x'\|^2 + \|y - y'\|^2} < \delta,$$

where (x,y) and (x',y') are any points of $D \times S$. Using this inequality we get

$$|\phi(y) - \phi(y')| = \left| \iint \cdots \int_D [f(x,y) - f(x,y')] \, dx_1 \cdots dx_n \right|$$

$$\leq \iint \cdots \int_D |f(x,y) - f(x,y')| \, dx_1 \cdots dx_n \leq \varepsilon V(D)$$

if $\|y - y'\| < \delta$, where $V(D)$ is the volume of D. This yields the continuity of ϕ.

Note that if $f(x,y)$ is defined and continuous on $\bar{D} \times \bar{S}$, then it is uniformly continuous on $\bar{D} \times S_0$ for any closed bounded subset S_0 of $\bar{S}$. Hence ϕ is continuous on S_0, and consequently also on $\bar{S}$.

The next theorem is concerned with the differentiability of $\phi(y)$.

THEOREM 2. *Let D be a normal domain in R^n and let S be an open set in R^m. Assume that $f(x,y)$ and its first partial derivatives $\partial f(x,y)/\partial y_i$ $(1 \leq i \leq m)$ are uniformly continuous functions in $D \times S$. Then the integral (1) has continuous first derivatives in S, and*

$$\frac{\partial \phi(y)}{\partial y_i} = \iint \cdots \int_D \frac{\partial f(x,y)}{\partial y_i} \, dx_1 \cdots dx_n \quad (i = 1,\ldots,m). \tag{2}$$

Formula (2) can be written in the form:

$$\frac{\partial}{\partial y_i} \iint \cdots \int_D f(x,y) \, dx_1 \cdots dx_n = \iint \cdots \int_D \frac{\partial}{\partial y_i} f(x,y) \, dx_1 \cdots dx_n. \tag{3}$$

It asserts that in order to differentiate an integral of a function depending on a parameter y, one may first differentiate the function with respect to the parameter and then integrate.

Proof. Let $y^0 \in S$, $h = (h_1, 0, 0, \ldots, 0)$. Then, by the mean value theorem,

$$\frac{\phi(y^0 + h) - \phi(y^0)}{h_1} = \iint \cdots \int_D \frac{f(x, y^0 + h) - f(x, y^0)}{h_1} \, dx_1 \cdots dx_n$$

$$= \iint \cdots \int_D \frac{\partial}{\partial y_1} f(x, y^0 + \theta h) \, dx_1 \cdots dx_n \quad (0 < \theta < 1);$$

θ depends on x. Since $\partial f(x,y)/\partial y_1$ is uniformly continuous in $D \times S$, for any $\varepsilon > 0$, there is a $\delta > 0$ such that

$$\left| \frac{\partial}{\partial y_1} f(x,y) - \frac{\partial}{\partial y_1} f(x,y^0) \right| < \varepsilon \qquad \text{if } \|y - y^0\| < \delta.$$

If $\|h\| = |h_1| < \delta$, then

$$\left| \frac{\phi(y^0 + h) - \phi(y^0)}{h_1} - \iint \cdots \int_D \frac{\partial}{\partial y_1} f(x,y^0) \, dx_1 \cdots dx_n \right|$$

$$= \left| \iint \cdots \int_D \left[\frac{\partial}{\partial y_1} f(x,y^0 + \theta h) - \frac{\partial}{\partial y_1} f(x,y^0) \right] dx_1 \cdots dx_n \right|$$

$$\leq \iint \cdots \int_D \left| \frac{\partial}{\partial y_1} f(x,y^0 + \theta h) - \frac{\partial}{\partial y_1} f(x,y^0) \right| dx_1 \cdots dx_n < \varepsilon V(D).$$

This proves that

$$\lim_{h_1 \to 0} \frac{\phi(y^0 + h) - \phi(y^0)}{h_1}$$

exists and is equal to $\iint \cdots \int_D (\partial/\partial y_1) f(x,y^0) \, dx_1 \cdots dx_n$.

We have thus proved (2) for $i = 1$. The proof for any other i is similar. We finally have to prove that $\partial\phi/\partial y_i$ are continuous functions. But this follows from (2) and from Theorem 1.

PROBLEMS

1. Let $\phi(y) = \int_0^1 \dfrac{y \, dx}{\sqrt{1 - y^2 x^2}}$, $0 < y < 1$. Use Theorem 2 to compute $\phi'(y)$.

2. Let $f(x,y)$, $\partial f(x,y)/\partial y$ be continuous functions for $a \leq x \leq b$, $a \leq y \leq b$. Prove that

$$\frac{d}{dy} \int_a^y f(x,y) dx = f(y,y) + \int_a^y \frac{\partial f(x,y)}{\partial y} \, dx$$

for all $a < y < b$.

3. Let $\phi(y) = \int_0^y x^n (y - x)^m dx$, where m and n are positive integers. Use Problem 2 to prove that

$$\phi^{(m)}(y) = \frac{m!}{n+1} y^{n+1}, \qquad \phi^{(j)}(0) = 0 \qquad \text{if } 0 \leq j \leq m - 1.$$

Deduce, by integrating $\phi^{(m)}(y)$ successively, that

$$\phi(y) = \frac{m!\, n!}{(m+n+1)!}\, y^{m+n+1}.$$

4. Let $\sigma(x_1,x_2)$ be a continuous function in a closed normal domain D of R^2. Consider the function

$$\phi(y_1,y_2) = \iint_D \sigma(x_1,x_2)\log \sqrt{(x_1 - y_1)^2 + (x_2 - y_2)^2}\ dx_1\, dx_2$$

for y in $R^2 - D$. Prove that $\phi(y_1,y_2)$ has two continuous derivatives, and

$$\frac{\partial^2 \phi}{\partial y_1{}^2} + \frac{\partial^2 \phi}{\partial y_2{}^2} = 0.$$

5. Let $\sigma(x)$ be a continuous function in a closed normal domain D of R^n, $n \geq 3$. Consider the function

$$\phi(y) = \iint \cdots \int_D \sigma(x)\, \|x - y\|^{2-n}\ dx_1 \cdots dx_n$$

for y in $R^n - D$. Prove that $\phi(y)$ has two continuous derivatives, and

$$\sum_{i=1}^{n} \frac{\partial^2 \phi}{\partial y_i{}^2} = 0.$$

6. Compute the derivative of $\phi(y) = \int_0^y e^{-x^2 y^2}\, dx$.

9. IMPROPER INTEGRALS DEPENDING ON A PARAMETER

Let $f(t,x)$ be a continuous function in (t,x) when $a \leq x \leq b$, $c \leq t < \infty$. Consider the improper integral

$$\phi(x) = \int_c^\infty f(t,x)\, dt. \tag{1}$$

We say that the improper integral is *uniformly convergent* on $[a,b]$ if for any $\varepsilon > 0$ there is a number N such that

$$\left| \phi(x) - \int_c^s f(t,x)\, dt \right| < \varepsilon \qquad \text{if } s \geq N,\ a \leq x \leq b. \tag{2}$$

The uniformity here is in the sense that N is independent of x.

THEOREM 1.   *If the integral (1) is uniformly convergent on* $[a,b]$, *then* $\phi(x)$ *is continuous in* $a \leq x \leq b$.

Proof.   By Theorem 1 of Section 8, the functions

$$\phi_m(x) = \int_c^m f(t,x)\,dt \qquad (m = 1,2,\ldots)$$

are continuous in $a \le x \le b$. The uniform convergence of the integral (1) implies the uniform convergence of $\{\phi_m(x)\}$ to $\phi(x)$. It follows that $\phi(x)$ is also continuous in $a \le x \le b$.

The following *Cauchy criterion* for uniform convergence holds:

THEOREM 2.   *The integral* (1) *is uniformly convergent if, and only if, for any $\varepsilon > 0$ there is a positive number N_0 such that*

$$\left| \int_N^M f(t,x)\,dt \right| < \varepsilon \qquad if\ M > N \ge N_0.$$

The following *Weierstrass M-test* for integrals holds:

THEOREM 3.   *If $|f(t,x)| \le g(t)$ and the improper integral $\int_c^\infty g(t)\,dt$ is convergent, then the integral* (1) *is uniformly convergent.*

The proofs of Theorems 2 and 3 are left to the student.

THEOREM 4.   *If the integral* (1) *is uniformly convergent on $[a,b]$, then*

$$\int_a^b \phi(x)\,dx = \int_c^\infty dt \int_a^b f(t,x)\,dx. \tag{3}$$

If we substitute $\phi(x)$ from (1) into (3), we get

$$\int_a^b dx \int_c^\infty f(t,x)\,dt = \int_c^\infty dt \int_a^b f(t,x)\,dx. \tag{4}$$

This means that the order of the iterated integral can be changed.

Proof.   Since the integral (1) is uniformly convergent, for any $\varepsilon > 0$ there is an $N > 0$ such that (2) holds. Hence

$$\left| \int_a^b \phi(x)\,dx - \int_a^b dx \int_c^s f(t,x)\,dt \right| < (b-a)\varepsilon.$$

On the other hand, by the corollary to Theorem 1 of Section 3,

$$\int_a^b dx \int_c^s f(t,x)\,dt = \int_c^s dt \int_a^b f(t,x)\,dx.$$

It follows that

$$\left| \int_c^s dt \int_a^b f(t,x)\, dx - \int_a^b \phi(x)\, dx \right| < (b-a)\varepsilon \quad \text{if } s > N. \tag{5}$$

Since for any $\varepsilon > 0$ there is an N for which (5) is satisfied, the improper integral

$$\int_c^\infty dt \int_a^b f(t,x)\, dx$$

exists, and it is equal to $\int_a^b \phi(x)\, dx$.

THEOREM 5.   *Suppose both $f(t,x)$ and $f_x(t,x)$ are continuous in (t,x) when $c \le t < \infty$, $a \le x \le b$. Suppose also that the integral (1) is convergent if $a \le x \le b$ and that the integral $\int_c^\infty f_x(t,x)\, dt$ is uniformly convergent on $[a,b]$. Then $\phi'(x)$ exists and*

$$\phi'(x) = \int_c^\infty f_x(t,x)\, dt. \tag{6}$$

Theorem 5 generalizes Theorem 2 of Section 8 (in case $m = n = 1$) to improper integrals.

Proof.   Define

$$g(x) = \int_c^\infty f_x(t,x)\, dt.$$

By Theorem 4, if $a < \xi \le b$,

$$\int_a^\xi g(x)\, dx = \int_c^\infty dt \int_a^\xi f_x(t,x)\, dx = \int_c^\infty [f(t,\xi) - f(t,a)]\, dt$$

$$= \phi(\xi) - \phi(a).$$

Since, by Theorem 1, $g(x)$ is continuous, $\phi'(x)$ exists and is equal to $g(x)$. This completes the proof.

Theorems 1–5 can be extended to improper multiple integrals depending on several variables, and also to improper integrals taken over bounded sets.

EXAMPLE 1.   The *gamma function* $\Gamma(x)$ is defined by the improper integral

$$\Gamma(x) = \int_0^\infty t^{x-1} e^{-t}\, dt \quad (x > 0). \tag{7}$$

The integral is improper both at $t = 0$ and $t = \infty$. If $x > 0$, then the integral

$$\int_0^1 t^{x-1} e^{-t} \, dt \tag{8}$$

is convergent. Moreover, for any $\eta > 0$, the integral is uniformly convergent if $\eta \le x < \infty$. Indeed, since

$$t^{x-1} e^{-t} < t^{\eta-1}$$

and

$$\int_0^1 t^{\eta-1} \, dt$$

is convergent, the analog of Theorem 3 (for improper integrals over bounded intervals) gives the uniform convergence of the integral (8).

Next, the integral

$$\int_1^\infty t^{x-1} e^{-t} \, dt$$

is uniformly convergent in any interval $0 \le x < A$, since

$$t^{x-1} e^{-t} \le t^{A-1} e^{-t}$$

and

$$\int_1^\infty t^{A-1} e^{-t} \, dt$$

is convergent.

It follows that the improper integral (7) is uniformly convergent in any interval $\eta \le x \le A$ ($\eta > 0$, $A > 0$). By Theorem 1, $\Gamma(x)$ is then a continuous function for $0 < x < \infty$.

Similarly, one shows that the integral

$$\int_0^\infty t^{x-1} (\log t) e^{-t} \, dt$$

is also uniformly convergent in every interval $\eta \le x \le A$. Applying Theorem 5 we conclude that $\Gamma'(x)$ exists and

$$\Gamma'(x) = \int_0^\infty t^{x-1} (\log t) e^{-t} \, dt \qquad (0 < x < \infty).$$

One can proceed similarly step by step to prove that

$$\Gamma^{(n)}(x) = \int_0^\infty t^{x-1} (\log t)^n e^{-t} \, dt \qquad (0 < x < \infty). \tag{9}$$

The gamma function satisfies the interesting relation

$$\Gamma(x + 1) = x\Gamma(x) \qquad (0 < x < \infty). \tag{10}$$

To prove it, we use integration by parts with $u = t^x$, $v' = e^{-t}$ to obtain

$$\int_\varepsilon^r t^x e^{-t}\, dt = \left[-t^x e^{-t}\right]_\varepsilon^r + \int_\varepsilon^r x t^{x-1} e^{-t}\, dt,$$

and then take $\varepsilon \to 0$, $r \to \infty$.

Since $\Gamma(1) = 1$, the relation (10) gives

$$\Gamma(n + 1) = n! \qquad (n = 1,2,\ldots). \tag{11}$$

EXAMPLE 2.   The *beta function* $B(p,q)$ is defined by

$$B(p,q) = \int_0^1 t^{p-1}(1 - t)^{q-1}\, dt \qquad (p > 0,\, q > 0). \tag{12}$$

This function is related to the gamma function by

$$B(p,q) = \frac{\Gamma(p)\Gamma(q)}{\Gamma(p + q)}. \tag{13}$$

To prove (13), we first substitute in the integral

$$\Gamma(p) = \int_0^\infty t^{p-1} e^{-t}\, dt,$$

$t = x^2$. We get

$$\Gamma(p) = \lim_{\substack{\varepsilon \to 0 \\ r \to \infty}} \int_\varepsilon^r t^{p-1} e^{-t}\, dt = \lim_{\substack{\varepsilon \to 0 \\ r \to \infty}} \int_{\sqrt{\varepsilon}}^{\sqrt{r}} 2x^{2p-1} e^{-x^2}\, dx$$

$$= 2\int_0^\infty x^{2p-1} e^{-x^2}\, dx.$$

Similarly,

$$\Gamma(q) = 2\int_0^\infty y^{2q-1} e^{-y^2}\, dy.$$

Denote the first quadrant of R^2 by R_0. Then, by Problem 1 of Section 7,

$$\Gamma(p)\Gamma(q) = 4\iint_{R_0} x^{2p-1} y^{2q-1} e^{-(x^2+y^2)}\, dx\, dy$$

$$= 4\lim_{\substack{\varepsilon \to 0 \\ N \to \infty}} \iint_{\substack{\varepsilon < x^2+y^2 < N \\ x > 0,\, y > 0}} x^{2p-1} y^{2q-1} e^{-(x^2+y^2)}\, dx\, dy.$$

Introducing polar coordinates, then writing the double integral as an iterated integral and letting $\varepsilon \to 0$, $N \to \infty$, we obtain

$$\left(2\int_0^\infty r^{2p+2q-1} e^{-r^2}\, dr\right)\left(2\int_0^{\pi/2} \sin^{2p-1}\theta \cos^{2q-1}\theta\, d\theta\right).$$

The first factor is equal to $\Gamma(p + q)$. The second factor reduces to $B(p,q)$ if one sets $x = \sin^2 \theta$.

EXAMPLE 3.   We shall prove that

$$\int_0^\infty \frac{\sin t}{t}\, dt = \frac{\pi}{2}. \tag{14}$$

Introduce the integral

$$f(x) = \int_0^\infty e^{-xt} \frac{\sin t}{t}\, dt \qquad \text{for } x \geq 0. \tag{15}$$

The integral is absolutely convergent if $x > 0$ and is convergent if $x = 0$. If $x \geq 0$, then, for any $M > N > 0$, integration by parts gives

$$\int_N^M \left(\frac{e^{-xt}}{t}\right) \sin t\, dt = \frac{e^{-xN}}{N} \cos N - \frac{e^{-xM}}{M} \cos M + \int_N^M \left(\frac{e^{-xt}}{t}\right)' \cos t\, dt.$$

Since

$$\left| \left(\frac{e^{-xt}}{t}\right)' \right| = \left| -\frac{xte^{-xt} + e^{-xt}}{t^2} \right| = \frac{1 + xt}{t^2} e^{-xt} \leq \frac{e^{xt}}{t^2} e^{-xt} = \frac{1}{t^2},$$

$$\left| \int_N^M e^{-xt} \frac{\sin t}{t}\, dt \right| \leq \frac{1}{N} + \frac{1}{M} + \int_N^M \frac{dt}{t^2} \leq \frac{3}{N}.$$

We can now use the Cauchy criterion for uniform convergence to conclude that the integral in (15) is uniformly convergent with respect to x, $0 \leq x < \infty$. By Theorem 1, $f(x)$ is then a continuous function and, in particular,

$$\int_0^\infty \frac{\sin t}{t}\, dt = f(0) = \lim_{x \to 0+} f(x) = \lim_{x \to 0+} \int_0^\infty e^{-xt} \frac{\sin t}{t}\, dt. \tag{16}$$

For any $\varepsilon > 0$,

$$|e^{-xt} \sin t| \leq e^{-\varepsilon t} \qquad \text{if } x \geq \varepsilon.$$

Since $e^{-\varepsilon t}$ is integrable over the interval $0 \leq t < \infty$, the Weierstrass M-test (Theorem 3) yields the uniform convergence of the integral

$$\int_0^\infty e^{-xt} \sin t\, dt \qquad \text{for } x \geq \varepsilon.$$

Using Theorem 5 we deduce that

$$f'(x) = -\int_0^\infty e^{-xt} \sin t\, dt.$$

But, for any $M > 0$,

$$-\int_0^M e^{-xt} \sin t \, dt = -\left[\frac{e^{-xt}(-x \sin t - \cos t)}{1 + x^2}\right]_0^M.$$

Taking $M \to \infty$ we get

$$f'(x) = -\frac{1}{1 + x^2} = -\frac{d}{dx} \tan^{-1} x.$$

Consequently,

$$f(x) = C - \tan^{-1} x, \qquad C \text{ constant.} \tag{17}$$

To find C, note that, for any $\varepsilon > 0$,

$$\left|\int_0^\varepsilon e^{-xt} \frac{\sin t}{t} \, dt\right| \le \int_0^\varepsilon e^{-xt} \, dt \le \int_0^\varepsilon dt = \varepsilon.$$

Also,

$$\left|\int_\varepsilon^\infty e^{-xt} \frac{\sin t}{t} \, dt\right| \le \int_\varepsilon^\infty e^{-xt} \, dt = \frac{e^{-x\varepsilon}}{x} < \varepsilon$$

if x is sufficiently large, say $x \ge K(\varepsilon)$. Hence

$$|f(x)| < 2\varepsilon \qquad \text{if } x \ge K(\varepsilon).$$

This shows that $f(x) \to 0$ if $x \to \infty$. Taking $x \to \infty$ in (17) and using the fact that $\tan^{-1} x \to \pi/2$ if $x \to \infty$, we find that $C = \pi/2$. If we now take $x \to 0$ in (17) and make use of (16), we obtain the assertion (14).

PROBLEMS

1. Prove that $\Gamma(1/2) = \sqrt{\pi}$. [*Hint:* Use the result of Problem 8 in Section 7 and substitute $x = \sqrt{t}$.]

2. Show that for any positive integer n,

$$\Gamma\left(n + \frac{1}{2}\right) = \frac{1 \cdot 3 \cdot 5 \cdots (2n - 1)}{2^n} \sqrt{n}.$$

3. Prove that for any $\alpha > 0$,

$$\int_0^\infty e^{-x^\alpha} \, dx = \frac{1}{\alpha} \Gamma\left(\frac{1}{\alpha}\right).$$

4. Prove that

$$\int_0^\infty \frac{e^{-at} \sin xt}{t} \, dt = \tan^{-1} \frac{x}{a} \qquad (a > 0).$$

[*Hint:* Denote the integral by $\phi(x)$. Prove $\phi'(x) = a/(a^2 + x^2)$.]

5. Prove that

$$\int_0^\infty e^{-t^2} \cos xt \, dt = \frac{\sqrt{\pi}}{2} e^{-x^2/4}.$$

[*Hint:* Denote the integral by $\phi(x)$. Prove: $\phi'/\phi = -1/(2x)$ and integrate.]

6. Prove by induction on n that

$$\int_0^\infty t^n e^{-xt} \, dt = \frac{n!}{x^{n+1}} \qquad (x > 0).$$

7. Prove that

$$\int_0^\infty \frac{1 - e^{-xt^2}}{t^2} \, dt = \sqrt{\pi x} \qquad (x > 0).$$

8. The function $F(s) = \int_0^\infty e^{-st} f(t) \, dt$ is called the *Laplace transform* of f. It is denoted by $(L\{f\})(s)$. Prove that if $f(t)$ and $f'(t)$ are continuous and bounded in $0 \le t < \infty$, then

$$(L\{f'\})(s) = s(L\{f\})(s) - f(0) \qquad (0 < s < \infty).$$

9. Show that

(a) $L\{1\} = \dfrac{1}{s} \quad (s > 0);$ (b) $L\{\sin at\} = \dfrac{a}{s^2 + a^2} \quad (s > 0);$

(c) $L\{\cosh at\} = \dfrac{a}{s^2 - a^2} \quad (s > a).$

10. If $\alpha > -1/2$, then

$$\int_0^1 \frac{x^{2\alpha}}{\sqrt{1 - x^2}} \, dx = \frac{\sqrt{\pi} \, \Gamma(\alpha + \frac{1}{2})}{2\Gamma(\alpha + 1)}.$$

11. Evaluate

$$\int_0^\infty \frac{1 - e^{-x}}{x^{1+\alpha}} \, dx \qquad (0 < \alpha < 1).$$

12. Prove that

$$\int_0^\infty e^{-x} \frac{1 - \cos xy}{x} \, dx = \frac{1}{2} \log(1 + y^2).$$

13. Prove that

$$\frac{2}{\pi} \int_0^\infty \frac{\sin xt}{t} \, dt = \operatorname{sgn} x.$$

14. Prove Theorem 2.

15. Prove Theorem 5.

9

LINE AND SURFACE INTEGRALS

1. LENGTH OF CURVES

Let γ be a curve in R^n given by $x = x(t) = (x_1(t),\ldots,x_n(t))$ where the parameter t varies in an interval $a \leq t \leq b$. Denote by π a partition

$$t_0 = a < t_1 < t_2 < \cdots < t_{k-1} < t_k = b \qquad (1)$$

of $[a,b]$, and by l_π the length of the polygonal line with vertices $x(t_0), x(t_1), \ldots, x(t_k)$. Thus

$$l_\pi = \sum_{i=1}^{k} \| x(t_i) - x(t_{i-1}) \|. \qquad (2)$$

Denote by $\Lambda(\gamma)$ the set of all the numbers l_π obtained when π varies over the set of all partitions of $[a,b]$. If $\Lambda(\gamma)$ is a bounded set, then we say that the curve γ is *rectifiable*. We then call the supremum of $\Lambda(\gamma)$ the *length* of γ, and denote it by $l(\gamma)$.

A curve γ, given by $x = x(t)$ for $a \leq t \leq b$, is said to be of *bounded variation* if each component $x_i(t)$ of $x(t)$ is a function of bounded variation in $[a,b]$.

298

THEOREM 1. *A curve γ is rectifiable if, and only if, it is of bounded variation.*

Proof. Suppose γ is rectifiable. Then there is a positive number M such that for any partition (1) of $[a,b]$, $l_\pi \leq M$. But then

$$\sum_{i=1}^{k} |x_j(t_i) - x_j(t_{i-1})| \leq \sum_{i=1}^{k} \|x(t_i) - x(t_{i-1})\| = l_\pi \leq M \qquad (j = 1,\ldots,n).$$

Hence $x_j(t)$ is of bounded variation. Suppose, conversely, that each component $x_j(t)$ is of bounded variation. Then, for any partition π of $[a,b]$, given by (1),

$$\sum_{i=1}^{k} |x_j(t_i) - x_j(t_{i-1})| \leq N_j \qquad (j = 1,\ldots,n)$$

where N_j are constants independent of π. Summing the inequalities

$$\|x(t_i) - x(t_{i-1})\| = \left\{ \sum_{j=1}^{n} (x_j(t_i) - x_j(t_{i-1}))^2 \right\}^{1/2} \leq \sum_{j=1}^{n} |x_j(t_i) - x_j(t_{i-1})|$$

over i, we get

$$l_\pi \leq \sum_{j=1}^{n} N_j.$$

This shows that the set $\Lambda(\gamma)$ is bounded, so that γ is rectifiable.

THEOREM 2. *Let γ by a piecewise continuously differentiable curve in R^n given by $x = x(t)$, $a \leq t \leq b$. Then γ is rectifiable and*

$$l(\gamma) = \int_a^b \|x'(t)\| \, dt. \qquad (3)$$

Proof.   Denote by $\tau_1,\ldots,\tau_{r-1}$ the points of discontinuity of the derivative $x'(t)$ of $x(t)$. Set $\tau_0 = a$, $\tau_r = b$. In each interval (τ_{i-1},τ_i) the components $x_j(t)$ of $x(t)$ are differentiable and their first derivatives are uniformly continuous, hence bounded. Using the mean value theorem we get, as in the proof of Theorem 1 of Section 8.1,

$$|x_j(t'') - x_j(t')| \leq M |t'' - t'| \qquad (a \leq t' \leq b, a \leq t'' \leq b).$$

Thus $x(t)$ is of bounded variation. By Theorem 1 it follows that γ is rectifiable. It remains to prove (3).

For any $\varepsilon > 0$, there is a partition π' of $[a,b]$, given by

$$t'_0 = a < t'_1 < t'_2 < \cdots < t'_{h-1} < t'_h = b$$

such that

$$l(\gamma) - \varepsilon < l_{\pi'} \le l(\gamma). \tag{4}$$

Since $x'(t)$ is uniformly continuous in each interval (τ_{i-1}, τ_i), there exists a $\delta > 0$ such that

$$\|x'(t) - x'(s)\| < \varepsilon \quad \text{if } \tau_{i-1} < t < \tau_i, \tau_{i-1} < s < \tau_i, |t - s| < \delta, 1 \le i \le r. \tag{5}$$

The function $\|x'(t)\|$ is integrable over $[a,b]$. Consequently, there is a $\delta' > 0$ such that for any partition (1) of $[a,b]$ whose mesh is less than δ',

$$\left| \sum_{m=1}^{k} \|x'(\bar{t}_m)\| (t_m - t_{m-1}) - \int_a^b \|x'(t)\| \, dt \right| < \varepsilon, \tag{6}$$

for any choice of $\bar{t}_m$ in (t_{m-1}, t_m) for which $x'(\bar{t}_m)$ exists.

Let π be a partition given by (1) such that (i) π is a refinement of π'; (ii) the points $\tau_1, \ldots, \tau_{r-1}$ occur among the points $t_1, t_2, \ldots, t_{k-1}$; (iii) the mesh of π is less than δ; (iv) the mesh of the partition π is less than δ'.

Since π is a refinement of π', it is eaily seen that $l_\pi \ge l_{\pi'}$. Hence, by (4),

$$l(\gamma) - \varepsilon \le l_\pi \le l(\gamma). \tag{7}$$

By the mean value theorem,

$$\|x(t_m) - x(t_{m-1})\| = \left\{ \sum_{j=1}^n (x_j(t_m) - x_j(t_{m-1}))^2 \right\}^{1/2}$$

$$= \left\{ \sum_{j=1}^n \left(\frac{dx_j(t_{m,j})}{dt} \right)^2 \right\}^{1/2} (t_m - t_{m-1}) \quad (t_{m-1} < t_{m,j} < t_m). \tag{8}$$

Recalling that the mesh of π is less than δ, and using (5), we can write

$$\left\{ \sum_{j=1}^n \left[\frac{dx_j(t_{m,j})}{dt} - \frac{dx_j(\bar{t}_m)}{dt} \right]^2 \right\}^{1/2} \le \sum_{j=1}^n \left| \frac{dx_j(t_{m,j})}{dt} - \frac{dx_j(\bar{t}_m)}{dt} \right| < n\varepsilon.$$

Hence, by (8) and the triangle inequality in R^n,

$$\|x(t_m) - x(t_{m-1})\| \le \left\{ \sum_{j=1}^n \left(\frac{dx_j(\bar{t}_m)}{dt} \right)^2 \right\}^{1/2} (t_m - t_{m-1}) + n\varepsilon(t_m - t_{m-1})$$

and

$$\|x(t_m) - x(t_{m-1})\| \ge \left\{ \sum_{j=1}^n \left(\frac{dx_j(\bar{t}_m)}{dt} \right)^2 \right\}^{1/2} (t_m - t_{m-1}) - n\varepsilon(t_m - t_{m-1}).$$

Summing over m, we obtain the inequalities:

$$l_\pi \leq \sum_{m=1}^{k} \|x'(\bar{t}_m)\| (t_m - t_{m-1}) + n(b-a)\varepsilon,$$

$$l_\pi \geq \sum_{m=1}^{k} \|x'(\bar{t}_m)\| (t_m - t_{m-1}) - n(b-a)\varepsilon.$$

Recalling that the mesh of π is less than δ', we can apply (6) and conclude that

$$\left| l_\pi - \int_a^b \|x'(t)\| \, dt \right| < n(b-a)\varepsilon + \varepsilon.$$

If we combine this with (7), we get

$$\left| l(\gamma) - \int_a^b \|x'(t)\| \, dt \right| \leq n(b-a)\varepsilon + \varepsilon + \varepsilon = C\varepsilon$$

where $C = 2 + n(b-a)$. Since ε is an arbitrary positive number, (3) follows.

Let γ be a curve in R^n given by $x(t)$ for $a \leq t \leq b$, and let δ be a curve in R^n given by $y(t)$ for $\alpha \leq t \leq \beta$. Suppose there exists a continuous and strictly monotone function $p(t)$ with domain $[a,b]$ and range $[\alpha,\beta]$ such that $y(p(t)) = x(t)$ for $a \leq t \leq b$. Then we say that the curve δ is *equivalent* to the curve γ, and that δ is obtained from γ by *reparametrization*. The function $p(t)$ is called a *reparametrization* of γ. Obviously, γ is then also equivalent to δ, with the reparametrization p^{-1} (the inverse of p).

If $p(t)$ is strictly monotone increasing, then we say that γ and δ have the same *orientation*. If $p(t)$ is strictly monotone decreasing, then we say that γ has *opposite orientation* to δ.

A reparametrization $p(t)$ is called *regular* if $p'(t)$ exists and is continuous and if $p'(t) \neq 0$ for all $a \leq t \leq b$.

If δ is obtained by regular reparametrization of γ, then, at each point t where $y'(p(t))$ exists,

$$x'(t) = y'(p(t))p'(t).$$

Hence, γ has the same orientation to δ (opposite orientation to δ) if the tangent vector of γ at t is a multiple of the tangent vector of δ at $p(t)$ by a positive (negative) scalar.

Let γ be a piecewise continuously differentiable curve given by $x = x(t)$, $a \leq t \leq b$. Denote by $s(r)$ the length of the curve $x(t)$, $a \leq t \leq r$. By Theorem 2,

$$s(t) = \int^t \|x'(\tau)\| \, d\tau. \tag{9}$$

If γ is continuously differentiable and if $x'(t) \neq 0$ for all $a \leq t \leq b$, then $s'(t)$ is continuous and

$$\frac{ds(t)}{dt} = \|x'(t)\| \neq 0 \quad (a \leq t \leq b). \tag{10}$$

Thus $s(t)$ is a regular reparametrization of γ. A curve δ obtained by reparametrizing a curve γ by $s(t)$ is called a *curve parametrized by its length*.

PROBLEMS

1. Let δ be a curve equivalent to a curve γ. Prove that δ is rectifiable if, and only if, γ is rectifiable, and in that case $l(\gamma) = l(\delta)$. [*Hint:* Use the definitions of a rectifiable curve and of a length of a curve.]

2. If γ is a continuous curve, then its range is a closed, bounded set. Is the same true when γ is piecewise continuous?

3. Let $x(t)$ ($a \leq t \leq b$) be a rectifiable curve, and let $s(c,r)$ denote the length of the curve $x(t)$, $c \leq t \leq r$. Prove that $s(a,r) = s(a,c) + s(c,r)$. [*Hint:* Recall the proof of Theorem 1 in Section 2.8.]

4. Let $x(t)$ ($a \leq t \leq b$) be a rectifiable curve, and denote by $s(r)$ the length of the curve $x(t)$, $a \leq t \leq r$. Prove that (i) $s(t)$ is monotone increasing, and (ii) if $x(t)$ is continuous in $[a,b]$, then $s(t)$ is continuous in $[a,b]$. [*Hint:* To prove (ii), recall the proof of Theorem 3 in Section 2.8.]

5. Let $x(s)$ be a continuously differentiable curve parametrized by its length. Prove that its tangent vector $x'(s)$ has unit length.

6. Find the length of the cycloid $x = a(t - \sin t)$, $y = a(1 - \cos t)$ from $t = 0$ to $t = 2\pi$.

7. Find the length of the circular helix $x = a \cos t$, $y = a \sin t$, $z = bt$, from $t = 0$ to $t = r$.

8. Find the length of the curve $x = \cosh t$, $y = \sinh t$, $z = t$ from $t = 0$ to $t = c$.

9. Find the length of the curve $x = a(\cos t + t \sin t)$, $y = a(\sin t - t \cos t)$ from $t = 0$ to $t = 2\pi$.

10. If a continuously differentiable curve is given in polar coordinates by $r = f(\theta)$ ($\alpha \leq \theta \leq \beta$), then its length is given by

$$\int_\alpha^\beta \sqrt{r'^2 + r^2}\, d\theta.$$

11. Find the length of the curve $r = a(1 - \cos \theta)$, $0 \leq \theta \leq 2\pi$.

12. Find the length of the curve $x_1 = t + 1$, $x_2 = 2t$, $x_3 = 3t^2 - t$, $x_4 = t^2 + t + 2$ from $t = 0$ to $t = 1$.

2. LINE INTEGRALS

Let γ be a curve in R^n, given by $x = x(t) = (x_1(t),...,x_n(t))$ where $a \leq t \leq b$. We shall denote the range of γ by $|\gamma|$. Let $f(x)$ be a bounded function defined on $|\gamma|$. We shall define the concept of the line integral of f with respect to the x_h-coordinate of γ.

We begin with a partition

$$t_0 = a < t_1 < t_2 < \cdots < t_{k-1} < t_k = b \tag{1}$$

of $[a,b]$. In each interval $t_{i-1} \leq t \leq t_i$ we choose a point τ_i and then form the sum

$$T = \sum_{i=1}^{k} f(x(\tau_i))[x_h(t_i) - x_h(t_{i-1})].$$

Suppose there exists a number N such that the following is true: For any $\varepsilon > 0$ there is a $\delta > 0$ such that for any partition (1) and for any choice of τ_i in $[t_{i-1},t_i]$,

$$|T - N| < \varepsilon \qquad \text{if } \max(t_i - t_{i-1}) < \delta. \tag{2}$$

Then we say that the *line integral* of f with respect to the x_h-coordinate of γ exists and is equal to N. We then write N in the form

$$\int_{\gamma} f(x) \, dx_h. \tag{3}$$

The above definition of the line integral resembles the definition of the Riemann integral of a function. If $x_h(t)$ is a monotone increasing function, then one can also define a concept of the line integral that is analogous to the Darboux integral. Thus, for any partition (1), one forms two sums: an *upper sum*

$$S = \sum_{i=1}^{k} M_i[x_h(t_i) - x_h(t_{i-1})]$$

and a *lower sum*

$$s = \sum_{i=1}^{k} m_i[x_h(t_i) - x_h(t_{i-1})],$$

where M_i and m_i are the supremum and the infimum of $f(x(t))$ in the interval

$t_{i-1} \leq t \leq t_i$. Denote by J the infimum of all the numbers S obtained by taking all possible partitions. Denote by I the supremum of all the numbers s obtained by taking all possible partitions. By arguments similar to those given in Sections 4.1 and 4.2, one shows that $J \geq I$; furthermore, the line integral of f with respect to the x_h-coordinate of γ exists if, and only if, $J = I$, and in that case the line integral (3) is equal to J.

Let γ be a curve $x = x(t)$, $a \leq t \leq b$. The point $x(a)$ is called the *initial point* of γ and the point $x(b)$ is called the *terminal point* of γ. If $x(a) = x(b)$, then we say that γ is a *closed curve*. If $x(t') \neq x(t'')$ whenever $t' \neq t''$ and the pair (t', t'') is not (a,b) or (b,a), then we say that γ is a *simple curve*.

Let γ_1 and γ_2 be two curves, given by $x(t)$ for $a \leq t \leq b$ and $y(t)$ for $c \leq t \leq d$. We define $\gamma_1 + \gamma_2$ to be the curve given by

$$x(t) \quad \text{if } a \leq t \leq b, \qquad y(t + c - b) \quad \text{if } b \leq t \leq b + d - c.$$

If γ_1 and γ_2 are continuous and if $x(b) = y(c)$, then $\gamma_1 + \gamma_2$ is also continuous. If γ_1 and γ_2 are piecewise continuously differentiable, and if $x(b) = y(c)$, then $\gamma_1 + \gamma_2$ is also piecewise continuously differentiable.

The proofs of the following theorems are left to the student.

THEOREM 1. *If γ and δ are equivalent curves in R^n and if $f(x)$ is a bounded function defined on $|\gamma|$, then the line integral*

$$\int_\gamma f(x)\, dx_h$$

exists if, and only if, the line integral

$$\int_\delta f(x)\, dx_h$$

exists, and in that case

$\int_\gamma f(x)\, dx_h = \int_\delta f(x)\, dx_h$ *if γ and δ have the same orientation,*

$\int_\gamma f(x)\, dx_h = -\int_\delta f(x)\, dx_h$ *if γ and δ have opposite orientations.*

THEOREM 2. *Let γ_1 and γ_2 be curves in R^n and let $f(x)$ be a bounded function defined on $|\gamma_1| \cup |\gamma_2|$. Then, the line integrals*

$$\int_{\gamma_1} f(x)\, dx_h, \qquad \int_{\gamma_2} f(x)\, dx_h$$

exist if, and only if, the line integral $\int_{\gamma_1 + \gamma_2} f(x)\, dx_h$ exists, and in that case,

$$\int_{\gamma_1} f(x)\, dx_h + \int_{\gamma_2} f(x)\, dx_h = \int_{\gamma_1 + \gamma_2} f(x)\, dx_h. \tag{4}$$

If γ is a curve given by $x(t)$, $a \leq t \leq b$, then we denote by $-\gamma$ the curve given by $y(t)$, $a \leq t \leq b$, where

$$y(t) = x(b + a - t), \qquad a \leq t \leq b.$$

$-\gamma$ is equivalent to γ but it has opposite orientation to γ. In view of Theorem 1,

$$\int_{-\gamma} f(x) \, dx_h = -\int_{\gamma} f(x) \, dx_h. \tag{5}$$

Let γ_1 and γ_2 be two curves. The curve $\gamma_1 + (-\gamma_2)$ is denoted by $\gamma_1 - \gamma_2$. In view of (4) and (5), we have

$$\int_{\gamma_1 - \gamma_2} f(x) \, dx_h = \int_{\gamma_1} f(x) \, dx_h - \int_{\gamma_2} f(x) \, dx_h. \tag{6}$$

If γ_1 and γ_2 are continuous curves having the same terminal points, then $\gamma_1 - \gamma_2$ is a continuous curve.

One indicates the orientation of a curve $x(t)$ 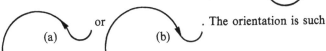 by

an arrow: or . The orientation is such

that if $t_2 > t_1$, then $x(t_2)$ is "ahead" of $x(t_1)$ along the curve, where the concept "ahead" is indicated by the arrow. Thus, if the orientation is as in (a), then the positions of $x(t_1)$, $x(t_2)$ are as follows:

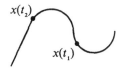

THEOREM 3. *Let γ be a piecewise continuously differentiable curve in R^n given by $x = x(t)$, $a \leq t \leq b$, and let $f(x)$ be a continuous function on $|\gamma|$. Then, for any $h = 1,\dots,n$, the line integral of f with respect to the x_h-coordinate of γ exists and*

$$\int_{\gamma} f(x) \, dx_h = \int_a^b f(x(t)) x_h'(t) \, dt. \tag{7}$$

Proof. We first assume that γ is continuously differentiable. Since $f(x(t)) x_h'(t)$ is a continuous function in $[a,b]$, the integral on the right of (7) exists. Hence, for any $\varepsilon > 0$ there is a $\delta > 0$ such that for any partition (1) with mesh $< \delta$,

$$\left| \sum_{i=1}^{k} f(x(\tau_i))x_h'(\tau_i)(t_i - t_{i-1}) - \int_a^b f(x(t))x_h'(t)\, dt \right| < \varepsilon, \qquad (8)$$

for any choice of τ_i in $t_{i-1} \le t \le t_i$.

Since $x_h'(t)$ is continuous in $[a,b]$, it is also uniformly continuous. Hence there is a $\delta' > 0$ such that

$$|x_h'(t) - x_h'(\tau)| < \frac{\varepsilon}{M(b-a)} \qquad \text{if } |t - \tau| < \delta',\, t \in [a,b],\, \tau \in [a,b], \qquad (9)$$

where M is a bound on the function $|f(x(t))|$ in $a \le t \le b$.

Using the mean value theorem and (9), we get

$$\left| \sum_{i=1}^{k} f(x(\tau_i))[x_h(t_i) - x_h(t_{i-1})] - \sum_{i=1}^{k} f(x(\tau_i))x_h'(\tau_i)(t_i - t_{i-1}) \right|$$

$$= \left| \sum_{i=1}^{k} f(x(\tau_i))[x_h'(\tilde{\tau}_i) - x_h'(\tau_i)](t_i - t_{i-1}) \right|$$

$$< \sum_{i=1}^{k} M \frac{\varepsilon}{M(b-a)}(t_i - t_{i-1}) = \varepsilon$$

provided $\max(t_i - t_{i-1}) < \delta'$; here $t_{i-1} < \tilde{\tau}_i < t_i$. Combining this with (8), we conclude that if the mesh of the partition (1) is less than $\min(\delta,\delta')$, then

$$\left| \sum_{i=1}^{k} f(x(\tau_i))[x_h(t_i) - x_h(t_{i-1})] - \int_a^b f(x(t))x_h'(t)\, dt \right| < 2\varepsilon.$$

This proves that the line integral

$$\int_\gamma f(x)\, dx_h \qquad (10)$$

exists and is equal to the right-hand side of (7).

So far we have assumed that γ is continuously differentiable. Now let γ be piecewise continuously differentiable and denote by $\tau_1, \tau_2, \ldots, \tau_{m-1}$ the points where $x'(t)$ is discontinuous. Set $\tau_0 = a$, $\tau_m = b$ and denote by γ_j the curve given by $x(t)$ for $\tau_{j-1} \le t \le \tau_j$. By what we already have proved,

$$\int_{\gamma_j} f(x)\, dx_h$$

exists and equals

$$\int_{\tau_{j-1}}^{\tau_j} f(x(t))x_h'(t)\, dt.$$

Using Theorem 2 we find that the line integral (10) exists and

$$\int_\gamma f(x)dx_h = \sum_{j=1}^{m} \int_{\tau_{j-1}}^{\tau_j} f(x(t))x_h'(t)\, dt = \int_a^b f(x(t))x_h'(t)\, dt.$$

From the proof of Theorem 3 we get:

COROLLARY. *If γ is only assumed to be continuous, and if a particular coordinate $x_h(t)$ of $x(t)$ is piecewise continuously differentiable, then the line integral $\int_\gamma f(x) \, dx_h$ exists and (7) holds.*

Let $f = (f_1,...,f_n)$ be a vector valued function defined on the range $|\gamma|$ of a curve γ. The sum

$$\sum_{h=1}^{n} \int_\gamma f_h(x) \, dx_h$$

is denoted by

$$\int_\gamma \sum_{h=1}^{n} f_h(x) \, dx_h, \quad \text{or} \int_\gamma f \cdot dx.$$

It is called the *line integral of f with respect to γ.*

PROBLEMS

1. Prove Theorem 1.

2. Prove Theorem 2. [*Hint:* Recall the proofs of Theorems 1 and 2 in Section 4.3.]

3. Evaluate the following line integrals for curves in the (x,y)-plane:

   (a) $\int_\gamma y^2 \, dx$; γ is the straight line initiating at $(0,0)$ and terminating at $(2,2)$.

   (b) $\int_\gamma y^2 \, dx + x^2 \, dy$; $\gamma: x = \sqrt{1 - y^2}, 0 \le y \le 1$.

   (c) $\int_\gamma y \, dx + x \, dy$; $\gamma: y = x^2, 0 \le x \le 2$.

   (d) $\int_\gamma \dfrac{y \, dx - x \, dy}{x^2 + y^2}$; $\gamma: x = \cos^3 t, y = \sin^3 t, 0 \le t \le \pi/2$. [*Hint:* substitute $u = \tan^3 t$.]

   (e) $\int_\gamma y^2 \, dx - x \, dy$; $\gamma: y^2 = 4x, 0 \le x \le 1$.

   (f) $\int_\gamma x^2 \, dy$; $\gamma: y = x^3 - 3x^2 + 2x, 0 \le x \le 2$.

   (g) $\int_\gamma y \sin x \, dx - x \cos y \, dy$; $\gamma: x = y, 0 \le y \le \pi/2$.

4. Evaluate the following line integrals for curves in the (x,y,z)-space:

(a) $\int_\gamma 3y\,dx + x\,dy - 2z\,dz$; γ: $y = \sqrt{1-x^2}$, $z = 1$, $-1 \le x \le 1$.

(b) $\int_\gamma (xy + y^2 - xyz)\,dx$; γ: $y = x^2$, $z = 0$, $-1 \le x \le 2$.

(c) $\int_\gamma (z + y)\,dx + (x + z)\,dy + (y + x)\,dz$; γ the polygonal line with vertices $(0,0,0)$, $(1,0,0)$, $(1,1,0)$, $(1,1,1)$.

(d) $\int_\gamma y\,dx - x\,dy + dz$; γ: $x = a\sin t$, $y = a\cos t$, $z = t$, $0 \le t \le \pi/2$.

(e) $\int_\gamma y\,dx + \sqrt{z}\,dy + 2x^2\,dz$; γ: $x = t$, $y = t^2$, $z = t^3 + 1$, $0 \le t \le 2$.

5. Let $f(x) = (x_1 + x_2, x_2 + x_3, x_3 - x_4, x_1)$, γ: $x_i = t^i$ $(1 \le i \le 4)$, $0 \le t \le 1$. Find $\int_\gamma f \cdot dx$.

6. Let γ be a piecewise continuously differentiable curve with length $l(\gamma)$, and let $f(x)$ be a continuous function defined on $|\gamma|$. Prove that

$$\left| \int_\gamma f \cdot dx \right| \le \left(\max_{|\gamma|} \|f\| \right) l(\gamma).$$

3. INTEGRAL ALONG A CURVE

Let γ be a rectifiable curve in R^n given by $x(t)$, $a \le t \le b$, and let $f(x)$ be a bounded function on $|\gamma|$. For any partition

$$t_0 = a < t_1 < t_2 < \cdots < t_{k-1} < t_k = b \tag{1}$$

of $[a,b]$, we form sums

$$T = \sum_{i=1}^{k} f(x(\tau_i)) l(\gamma_i)$$

where $l(\gamma_i)$ is the length of the curve γ_i given by $x(t)$, $t_{i-1} \le t \le t_i$, and τ_i is any point in the interval $[t_{i-1}, t_i]$.

Suppose there exists a number N having the following property: For any $\varepsilon > 0$ there is a $\delta > 0$ such that for any partition (1) of $[a,b]$ with mesh $< \delta$, and for any τ_i in $[t_{i-1}, t_i]$, the sum T satisfies: $|T - N| < \varepsilon$. Then we say that the *integral of f along γ* exists and is equal to N. We then write N in the form

$$\int_\gamma f(x)\,ds. \tag{2}$$

THEOREM 1. *If γ and δ are equivalent curves, and if the integral of f along γ exists, then the integral of f along δ also exists, and the two integrals are equal.*

The proof is left to the student.
Theorem 1 is analogous to Theorem 1 of Section 2. The analog of Theorem 2 of Section 2 is also valid. We shall now state an analog of Theorem 3 of Section 2.

THEOREM 2. *Let γ be a piecewise continuously differentiable curve in R^n, given by $x = x(t)$, $a \le t \le b$, and let $f(x)$ be a continuous function on $|\gamma|$. Then the integral of f along γ exists, and*

$$\int_\gamma f(x) \, ds = \int_a^b f(x(t)) \, \|x'(t)\| \, dt. \tag{3}$$

Suppose γ is parametrized by its length, that is, γ is given by $x = x(s)$, $0 \le s \le l(\gamma)$, and s is the length of the curve $x(\sigma)$, $0 \le \sigma \le s$. Then [by (9) of Section 1], at all the points s where $x'(s)$ exists, $\|x'(s)\| = 1$. Therefore, (3) reduces to

$$\int_\gamma f(x) \, ds = \int_0^{l(\gamma)} f(x(s)) \, ds. \tag{4}$$

Notice that the right- and left-hand sides coincide also directly by their definitions.

Proof. Suppose first that γ is continuously differentiable. Then $\|x'(t)\|$ is a continuous function in $a \le t \le b$. It is then also uniformly continuous. Thus, for any $\varepsilon > 0$ there is a $\delta > 0$ such that

$$\|x'(t) - x'(\tau)\| < \frac{\varepsilon}{M(b-a)} \quad \text{if } |t - \tau| < \delta, \, a \le t \le b, \, a \le \tau \le b, \tag{5}$$

where M is a bound on the function $f(x(t))$, $a \le t \le b$.

Since $f(x(t)) \, \|x'(t)\|$ is continuous, it is also integrable. Consequently, there is a $\delta' > 0$ such that for any partition (1) of $[a,b]$, and for any $t_{i-1} \le \tau_i \le t_i$,

$$\left| \sum_{i=1}^k f(x(\tau_i)) \, \|x'(\tau_i)\| \, (t_i - t_{i-1}) - \int_a^b f(x(t)) \, \|x'(t)\| \, dt \right| < \varepsilon \tag{6}$$

provided $\max(t_i - t_{i-1}) < \delta'$.

Consider now any sum

$$T = \sum_{i=1}^{k} f(x(\tau_i)) l(\gamma_i)$$

corresponding to a partition (1) of $[a,b]$ and τ_i in $[t_{i-1}, t_i]$. By Theorem 2 of Section 1,

$$l(\gamma_i) = \int_{t_{i-1}}^{t_i} \|x'(t)\| \, dt = \int_{t_{i-1}}^{t_i} \|x'(\tau_i) + [x'(t) - x'(\tau_i)]\| \, dt.$$

Using the triangle inequality, we get

$$-\int_{t_{i-1}}^{t_i} \|x'(t) - x'(\tau_i)\| \, dt \le l(\gamma_i) - \int_{t_{i-1}}^{t_i} \|x'(\tau_i)\| \, dt$$

$$\le \int_{t_{i-1}}^{t_i} \|x'(t) - x'(\tau_i)\| \, dt.$$

If the mesh of the partition (1) is less than δ, then we can use (5) to obtain

$$|l(\gamma_i) - \|x'(\tau_i)\| (t_i - t_{i-1})| < \frac{\varepsilon}{M(b-a)} (t_i - t_{i-1}).$$

It follows that

$$\left| T - \sum_{i=1}^{k} f(x(\tau_i)) \|x'(\tau_i)\| (t_i - t_{i-1}) \right| = \left| \sum_{i=1}^{k} f(x(\tau_i))[l(\gamma_i) - \|x'(\tau_i)\| (t_i - t_{i-1})] \right|$$

$$\le \sum_{i=1}^{k} M \frac{\varepsilon}{M(b-a)} (t_i - t_{i-1}) = \varepsilon.$$

Combining this with (6), we get

$$\left| T - \int_a^b f(x(t)) \|x'(t)\| \, dt \right| < 2\varepsilon$$

provided the mesh of the partition (1) is less than $\min(\delta, \delta')$. This proves the assertion of the theorem.

So far we have assumed that γ is continuously differentiable. If γ is only piecewise continuously differentiable, we apply the above proof to each portion of γ that is continuously differentiable, and then sum up over the finite number of these portions.

Let $f = (f_1, \dots, f_n)$ be a continuous vector valued function on the range $|\gamma|$ of a piecewise continuously differentiable curve γ, parametrized by its length: $x = x(s)$, $0 \le s \le l(\gamma)$. From Theorem 3 of Section 2,

$$\int_\gamma f \cdot dx = \sum_{h=1}^{n} \int f_h \, dx_h = \int_0^{l(\gamma)} \left[\sum_{h=1}^{n} f_h(x(s)) x_h'(s) \right] ds = \int_0^{l(\gamma)} f(x(s)) \cdot x'(s) \, ds.$$

Motivated by formula (4), we denote the last integral by

$$\int_\gamma f(x) \cdot x' \, ds.$$

We thus obtain the formula

$$\int_\gamma f \cdot dx = \int_\gamma f(x) \cdot x' \, ds = \int_0^{l(\gamma)} f(x(s)) \cdot x'(s) \, ds. \tag{7}$$

If $n = 3$ and f is the force acting on a particle moving along γ, the *work* done by the force is, by definition, the integral on the right-hand side of (7).

PROBLEMS

1. Prove Theorem 1.

2. If a particle moves along a curve $r = r(t)$ in R^3 under the action of force f, which is always orthogonal to the curve [that is, to the tangent vector $r'(t)$], then the work done is zero.

3. A particle of mass m moves along a curve $r = r(t)$ $(a \le t \le b)$ in R^3. The force f acting on it is given by Newton's second law: $f = m d^2 r / dt^2$. Show that the work done is equal to the gain in kinetic energy, that is, $(1/2)mv^2(b) - (1/2)mv^2(a)$, where $v(t) = \|dr(t)/dt\|$.

4. A particle of weight w descends from $(0,2,0)$ to $(4,0,0)$ along the parabola $8y = (x - 4)^2$, $z = 0$. It is acted on by both gravity and a horizontal force $(y,0,0)$. Find the total work done by these forces.

4. INDEPENDENCE OF LINE INTEGRALS ON THE CURVES

Let $\phi(x)$ be a continuously differentiable function defined in an open set D of R^n. The vector valued function

$$\left(\frac{\partial \phi}{\partial x_1}, \dots, \frac{\partial \phi}{\partial x_n} \right)$$

is called the *gradient* of ϕ, and is denoted by grad ϕ, or $\nabla \phi$.

Let $f = (f_1, \dots, f_n)$ be a vector valued function defined in an open set D of R^n. If there exists a continuously differentiable function ϕ in D such that $f = $ grad ϕ, then we say that f is a *gradient field*, or a *potential field*, and we call ϕ the *potential* of f. If $f = $ grad ϕ and ϕ is twice continuously differentiable, then $\partial f_i / \partial x_j = \partial f_j / \partial x_i$, since both sides are equal to $\partial^2 \phi / \partial x_i \partial x_j$.

THEOREM 1. *Let f be a gradient field in D, with potential ϕ. Then for any piecewise continuously differentiable curve γ in D with initial point x^0 and terminal point x^1,*

$$\int_{\gamma} f \cdot dx = \phi(x^1) - \phi(x^0). \tag{1}$$

Proof.   Let γ be given by $x(t)$, $a \le t \le b$. Then $x(a) = x^0$, $x(b) = x^1$. By Theorem 3 of Section 2,

$$\int_{\gamma} f \cdot dx = \int_{\gamma} \sum_{h=1}^{n} \frac{\partial \phi}{\partial x_h} \, dx_h = \int_{a}^{b} \left[\sum_{h=1}^{n} \frac{\partial \phi(x(t))}{\partial x_h} \frac{dx_h(t)}{dt} \right] dt = \int_{a}^{b} \frac{d}{dt} \phi(x(t)) \, dt$$

$$= \phi(x(b)) - \phi(x(a)) = \phi(x^1) - \phi(x^0).$$

Let γ_1 and γ_2 be two piecewise continuously differentiable curves lying in D and having the same initial point and the same terminal point. The assertion (1) implies that

$$\int_{\gamma_1} f \cdot dx = \int_{\gamma_2} f \cdot dx. \tag{2}$$

We express this property by saying that the line integral of f is *independent of the curve* (or *independent of the path*).

Let γ be any closed, piecewise continuously differentiable curve in D. Choose two points x^0 and x^1 on γ. We then can write $\gamma = \gamma_1 - \gamma_2$ where γ_1 consists of the one portion of γ between x^0 and x^1, γ_2 consists of the complementary portion of γ_1, and the orientation of γ is the same as that of γ_1 but opposite to that of γ_2. By (6) of Section 2,

$$\int_{\gamma} f \cdot dx = \int_{\gamma_1} f \cdot dx - \int_{\gamma_2} f \cdot dx. \tag{3}$$

Note that γ_1 and γ_2 have the same initial point and the same terminal point.

Figure 9.1 describes the relation $\gamma = \gamma_1 - \gamma_2$.

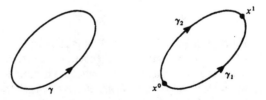

Figure 9.1

Conversely, for any two piecewise continuously differentiable curves γ_1 and γ_2 in D having the same initial point and the same terminal point, the curve $\gamma = \gamma_1 - \gamma_2$ is closed and (3) holds. We conclude that the line integral of f is independent of the curve if, and only if,

$$\int_\gamma f \cdot dx = 0 \qquad (4)$$

for any piecewise continuously differentiable, closed curve γ in D.

We shall now prove the converse of Theorem 1.

THEOREM 2. *Let $f = (f_1,...,f_n)$ be a continuous vector valued function defined in a domain D of R^n. If the line integral of f is independent of the curve, then f is a gradient field.*

Proof. Fix a point x^0 in D. For any x in D, let γ be a polygonal curve connecting x^0 to x and contained in D. Define

$$\phi(x) = \int_\gamma f \cdot dx. \qquad (5)$$

By assumption, $\phi(x)$ is independent of γ and is thus a function of x alone. We shall prove that ϕ is differentiable and

$$\frac{\partial \phi(x)}{\partial x_i} = f_i(x) \qquad (1 \le i \le n). \qquad (6)$$

Let $y = (y_1,...,y_n)$ be any point in D, and denote by γ a polygonal curve connecting x^0 to y. Let $h = (h_1,0,0,...,0)$, $h_1 > 0$, and denote by γ_1 the curve connecting y to $y + h$. Thus, γ_1 is given by

$$x_1(t) = y_1 + t,$$
$$x_i(t) = y_i \qquad (2 \le i \le n),$$

for $0 \le t \le h_1$. See Figure 9.2. By Theorem 2 of Section 2,

$$\phi(y + h) - \phi(y) = \int_{\gamma + \gamma_1} f \cdot dx - \int_\gamma f \cdot dx = \int_{\gamma_1} f \cdot dx.$$

By Theorem 3 of Section 2,

$$\int_{\gamma_1} f \cdot dx = \int_0^{h_1} \left[\sum_{i=1}^n f_i(x(t)) \frac{dx_i(t)}{dt} \right] dt = \int_0^{h_1} f_1(x(t))\, dt.$$

Using the mean value theorem for integrals we then get

$$\frac{\phi(y + h) - \phi(y)}{h_1} = \frac{1}{h_1} \int_0^{h_1} f_1(x(t))\, dt = f_1(y + \theta h), \qquad 0 < \theta < 1.$$

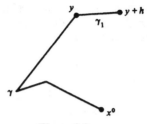

Figure 9.2

Similarly we obtain, in case $h_1 < 0$,

$$\frac{\phi(y + h) - \phi(y)}{h_1} = f_1(y + \theta_1 h), \qquad 0 < \theta_1 < 1.$$

It follows that $\partial\phi(y)/\partial x_1$ exists and is equal to $f_1(y)$. This proves (6) for $i = 1$. The proof for any other i is similar.

PROBLEMS

1. Determine for each of the following vector valued functions whether it is a potential field and, if it is, find its potential by computing line integrals:

   (a) $\left(\dfrac{1 + y^2}{x^3}, \; -\dfrac{1 + x^2}{x^2}\right)$;   (b) $(1 + xy, 1 - xz, 3 + z^2)$;

   (c) $\left(\dfrac{xy + 1}{y}, \dfrac{2y - x}{y^2}\right)$;   (d) $(2x + 2y, 2x - z^2, -2yz)$;

   (e) $(yz^2, xz^2 - 1, 2xyz - 2)$;   (f) $(x_2{}^2, x_3{}^2, x_4{}^2, x_1{}^2)$.

2. Let $\mathbf{f} = f_1\mathbf{i} + f_2\mathbf{j} + f_3\mathbf{k}$ be a continuously differentiable gradient field in some open set in R^3. Prove that curl $\mathbf{f} = 0$.

3. The Newtonian gravity field due to a mass M located at the origin of R^3 is $\mathbf{f} = -(M/r^3)\mathbf{r}$, where $\mathbf{r} = x\mathbf{i} + y\mathbf{j} + z\mathbf{k}$, $r = \|\mathbf{r}\|$. The field acts on a particle of mass m with force $m\mathbf{f}$. Prove:

   (a) $\mathbf{f}$ is a gradient field with potential M/r.

   (b) The work done by moving a particle of mass m from (x_0, y_0, z_0) into (x_1, y_1, z_1) is equal to $Mm/r_0 - Mm/r_1$, where

   $$r_0 = (x_0{}^2 + y_0{}^2 + z_0{}^2)^{1/2}, \, r_1 = (x_1{}^2 + y_1{}^2 + z_1{}^2)^{1/2}.$$

4. Prove that

$$f = \left(\frac{x_1}{r^n}, \frac{x_2}{r^n}, \ldots, \frac{x_n}{r^n}\right) \quad (n \geq 2)$$

is a gradient field.

5. Let $\sigma(\xi)$ be a continuous function in a closed normal domain D in R^n, $n \geq 2$. Prove that the vector valued function

$$\left(\iint \cdots \int_D \frac{(x_1 - \xi_1)\sigma(\xi)}{\|x - \xi\|^{n-2}} \, dV, \ldots, \iint \cdots \int_D \frac{(x_n - \xi_n)\sigma(\xi)}{\|x - \xi\|^{n-2}} \, dV\right)$$

is a gradient field in the exterior of D.

5. GREEN'S THEOREM

A continuous curve γ that is both simple and closed will be called a *simple closed curve*. A simple closed curve in R^2 is called a *Jordan curve*. *Jordan's theorem* asserts that such a curve divides the plane into two domains, one of which is bounded. More precisely, the complement of a Jordan's curve γ is the union $A \cup B$ of two domains A and B; A is bounded and B is unbounded. A is called the *interior* of γ and B is called the *exterior* of γ.

Even though Jordan's theorem is intuitively rather obvious, the known proofs are quite complicated.

Let γ be a Jordan's curve given by $P(t) = (x(t), y(t))$, $a \leq t \leq b$. As t increases, $P(t)$ moves along γ; see Figure 9.3. If it advances in such a way that A is to the left (as in Figure 9.3), then we say that it advances *counterclockwise* and that γ is *counterclockwise oriented*, or *positively oriented*, with respect to A. We then also say that γ is *negatively oriented* with respect to B. If $P(t)$ moves along γ in such a way that A is always to the right, then we say that γ is *clockwise oriented*, or *negatively oriented* with respect to A.

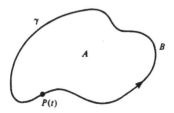

Figure 9.3

Recall that the vector $P'(t) = (x'(t), y'(t))$ is called the *tangent vector* to γ at $P(t)$. A unit vector that is orthogonal to $P'(t)$ is called a *normal* vector to γ at $P(t)$. There are two such vectors: the *outward normal*, which points into the exterior of γ, and the *inward normal*, which points into the interior of γ. The exterior normal $\mathbf{n}$ is such that if γ is positively oriented (with respect to A), then $P'(t)/\|P'(t)\|$ is obtained by rotating $\mathbf{n}$ counterclockwise by 90 degrees, as in Figure 9.4.

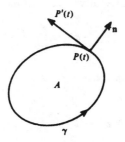

Figure 9.4

Let D be a domain in R^2 given by

$$\{(x,y); y_1(x) < y < y_2(x), \quad \text{for } a < x < b\}, \tag{1}$$

where $y_1(x)$ and $y_2(x)$ are continuous functions in $a \le x \le b$, and $y_2(x) > y_1(x)$ if $a < x < b$. Then we say that D is *x-projectable*. See, for example, Figure 9.5.

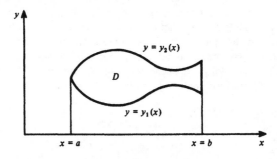

Figure 9.5

It is clear that the boundary of an x-projectable domain is a Jordan curve.
Similarly, one defines the concept of *y-projectable domain*. A domain D in R^2 is called a *simple domain* if it is both x-projectable and y-projectable. The following theorem is called *Green's theorem*.

THEOREM 1. *Let D be a simple domain in R^2 and denote by ∂D the boundary of D. Let γ be a Jordan curve with range ∂D, which is positively oriented with respect to D. Finally, let $M(x,y)$ and $N(x,y)$ be functions defined on D, which are uniformly continuous in D together with their partial derivatives $M_x(x,y)$ and $N_y(x,y)$. Then*

$$\iint_D \left(\frac{\partial M}{\partial x} + \frac{\partial N}{\partial y}\right) dx\, dy = \int_\gamma M\, dy - N\, dx. \tag{2}$$

Formula (2) is called *Green's formula*. The line integral on the right is often written in the more suggestive form:

$$\oint_{\partial D} M\, dy - N\, dx. \tag{3}$$

Proof. By Theorem 7 of Section 6.4, the uniformly continuous function N_y in D can be extended into a continuous function in $\bar{D} = D \cup \partial D$. By Theorem 2 of Section 8.3,

$$\iint_D \frac{\partial N}{\partial y}\, dx\, dy = \int_a^b dx \int_{y_1(x)}^{y_2(x)} \frac{\partial N}{\partial y}\, dy = \int_a^b N(x,y_2(x))\, dx - \int_a^b N(x,y_1(x))\, dx.$$

$$\tag{4}$$

We can write $\gamma = \gamma_1 + \gamma_2 + \gamma_3 + \gamma_4$, where γ_1 is the curve given by $y = y_1(x)$, $a \le x \le b$; γ_2 is the curve given by $y = y_2(b + a - x)$, $a \le x \le b$; γ_3 is the straight line lying on $x = a$ and connecting $(a,y_2(a))$ to $(a,y_1(a))$, and γ_4 is the straight line lying on $x = b$ and connecting $(b,y_1(b))$ to $(b,y_2(b))$. (If $y_1(a) = y_2(a)$ then γ_3 does not appear, and if $y_1(b) = y_2(b)$ then γ_4 does not appear.)

Note that $-\gamma_2$ is given by $y = y_2(x)$, $a \le x \le b$; see Figure 9.6.

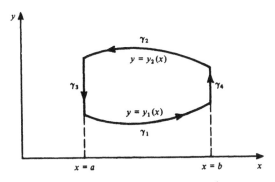

Figure 9.6

By the corollary to Theorem 3 of Section 2,

$$\int_a^b N(x,y_1(x))\,dx = \int_{\gamma_1} N(x,y)\,dx.$$

By the same corollary and by Theorem 1 of Section 2,

$$\int_a^b N(x,y_2(x))\,dx = \int_{-\gamma_2} N(x,y)\,dx = -\int_{\gamma_2} N(x,y)\,dx.$$

Using these relations in (4), we get

$$\iint_D \frac{\partial N}{\partial y}\,dx\,dy = -\int_{\gamma_1} N(x,y)\,dx - \int_{\gamma_2} N(x,y)\,dx. \tag{5}$$

Since γ_3 is given by $x = a$, $y = -t$ for $-y_2(a) \le t \le -y_1(a)$,

$$\int_{\gamma_3} N(x,y)\,dx = \int_{-y_2(a)}^{-y_1(a)} N(a,-t)\frac{dx}{dt}\,dt = 0.$$

Similarly,

$$\int_{\gamma_4} N(x,y)\,dx = 0.$$

We therefore can write (5) in the form

$$\iint_D \frac{\partial N}{\partial y}\,dx\,dy = -\int_{\gamma_1} N\,dx - \int_{\gamma_2} N\,dx - \int_{\gamma_3} N\,dx - \int_{\gamma_4} N\,dx = -\int_{\gamma} N\,dx.$$

Similarly, one can prove that

$$\iint_D \frac{\partial M}{\partial x}\,dx\,dy = \int_{\gamma} M\,dy,$$

and (2) follows.

DEFINITION.   A bounded domain D in R^2 is called a *Green's domain* if its boundary ∂D consists of a finite number of Jordan curves and if Green's formula (2) holds for all M, N as in Theorem 1.

Thus Theorem 1 asserts that any simple domain is a Green's domain.

Let D_1 and D_2 be Green's domains such that $D_1 \cap D_2$ is the empty set. Assume that $\partial D_1 \cap \partial D_2$ is a piecewise continuously differentiable curve δ. Denote by δ_0 the set of points of δ which are not the endpoints. (Refer to Figure 9.7.)

We then can assert:

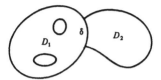

Figure 9.7

THEOREM 2.   *The domain $D_1 \cup D_2 \cup \delta_0$ is a Green's domain.*

Proof. Denote by δ_1 the curve δ, oriented counterclockwise with respect to D_1, and denote by δ_2 the curve δ, oriented clockwise with respect to D_1. ∂D_i ($i = 1,2$) consists of a finite number of Jordan curves $\gamma_{i,1},...,\gamma_{i,k_i}$ and of a Jordan curve $\gamma_i + \delta_i$, which contains D_i in its interior. The curves $\gamma_i, \gamma_{i,1},...,\gamma_{i,k_i}$ are taken to be positively oriented with respect to D_i. Set $\tilde{\gamma}_i = \gamma_i \cup \gamma_{i,1} \cup \cdots \cup \gamma_{i,k_i}$, as in Figure 9.8.

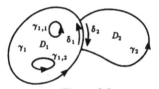

Figure 9.8

By Theorem 1,

$$\iint_{D_1} (M_x + N_y)\, dx\, dy = \int_{\tilde{\gamma}_1} M\, dy - N\, dx + \int_{\delta_1} M\, dy - N\, dx,$$

$$\iint_{D_2} (M_x + N_y)\, dx\, dy = \int_{\tilde{\gamma}_2} M\, dy - N\, dx + \int_{\delta_2} M\, dy - N\, dx.$$

If we add both sides and use the fact that

$$\int_{\delta_1} M\, dy = -\int_{\delta_2} M\, dy, \qquad \int_{\delta_1} N\, dx = -\int_{\delta_2} N\, dx,$$

then we obtain Green's formula for the domain $D_1 \cup D_2 \cup \delta_0$.

Theorems 1 and 2 can be used to construct a large class of Green's domains. Denote by $\mathscr{K}_1$ the class of all simple domains. By Theorem 1, any domain in $\mathscr{K}_1$ is a Green's domain. By applying Theorem 2 to pairs of domains D_1 and D_2 from $\mathscr{K}_1$, we obtain a larger class $\mathscr{K}_2$ of domains that are

Green's domains. Next we can apply Theorem 2 to pairs of domains D_1 and D_2 from $\mathcal{K}_2$, and thus get a still larger class $\mathcal{K}_3$ of Green's domains. Proceeding in this way step by step, we obtain a sequence of increasing classes $\mathcal{K}_m$ of domains. Denote by $\mathcal{K}$ the class consisting of all the domains belonging to at least one of the classes $\mathcal{K}_m$. We can state:

COROLLARY.   *Any domain in $\mathcal{K}$ is a Green's domain.*

It is known that any domain whose boundary consists of a finite number of simple closed rectifiable curves is a Green's domain. We shall not give the proof here, for two reasons: (a) the proof is very lengthy, and (b) in almost all applications of Green's formula, the domain in question already belongs to $\mathcal{K}$ (so that by the corollary to Theorem 2 it is a Green's domain).

A domain D in R^2 is called a *simply connected* domain if, for every Jordan curve in D, its interior lies in D.

A domain D in R^2 is called a *star domain* if there is a point (x^0, y^0) in D such that for any point (x, y) in D, the segment connecting it to (x^0, y^0) lies in D. The point (x^0, y^0) is called a *center* of the star domain.

We shall give an application of Green's theorem.

THEOREM 3.   *Let D be a star domain in R^2 and let $M(x,y)$ and $N(x,y)$ be continuously differentiable functions in D. Then the line integral $\int_\gamma M\,dx + N\,dy$ is independent of the curve if, and only if,*

$$\frac{\partial M}{\partial y} = \frac{\partial N}{\partial x}. \tag{6}$$

Proof. If the line integral $\int_\gamma M\,dx + N\,dy$ is independent of the curve then, by Theorem 2 of Section 4, there is a continuously differentiable function ϕ in D satisfying:

$$\phi_x = M, \qquad \phi_y = N.$$

Since M and N are continuously differentiable, $\phi_{xy} = \phi_{yx}$. Hence $M_y = \phi_{xy} = N_x$.

Suppose conversely that (6) holds. Let (x^0, y^0) be a center of the star domain D. For any point (x, y) in D, denote by γ the segment connecting (x^0, y^0) to (x, y), and define a function ϕ by

$$\phi(x,y) = \int_\gamma M\,dx + N\,dy. \tag{7}$$

For any h with small $|h|$, denote by γ_1 the segment connecting (x,y) to $(x + h,y)$, and denote by γ' the segment connecting (x^0,y^0) to $(x + h,y)$. Then

$$\phi(x + h,y) = \int_{\gamma'} M \, dx + N \, dy.$$

Since $\gamma + \gamma_1 - \gamma'$ is the clockwise-oriented boundary of the triangle Δ with vertices (x_0,y_0), (x,y), $(x + h,y)$, we can apply Theorem 1 to conclude that

$$\int_{\gamma+\gamma_1-\gamma'} M \, dx + N \, dy = \iint_{\Delta} \left(\frac{\partial N}{\partial x} - \frac{\partial M}{\partial y}\right) dx \, dy = 0.$$

Therefore

$$\phi(x + h,y) - \phi(x,y) = \int_{\gamma'} M \, dx + N \, dy - \int_{\gamma} M \, dx + N \, dy$$

$$= \int_{\gamma_1} M \, dx + N \, dy.$$

We now can proceed as in the proof of Theorem 2, Section 4, to deduce that $\partial\phi(x,y)/\partial x$ exists and is equal to M. Similarly, $\partial\phi(x,y)/\partial y$ exists and is equal to N. By Theorem 1 of Section 4, it follows that the line integral $\int M \, dx + N \, dy$ is independent of the path.

In the above proof we made use of Green's formula in a triangle. One can extend the proof to the case where D is any simple connected domain. In the definition of $\phi(x,y)$ in (7), one takes γ to be a polygonal line. In proving that $\phi(x,y)$ is defined independently of γ, one makes use of Green's formula in polygons. This is illustrated in Figure 9.9, where γ and γ' are two polygonal curves connecting (x^0,y^0) to (x,y). The arrows in the figure indicate the curve $\gamma - \gamma'$.

The proof that the domain bounded by a polygon is a Green's domain (and, in fact, belongs to $\mathcal{K}$) can be given by induction on the number of the sides of the polygon. We shall not give the details here.

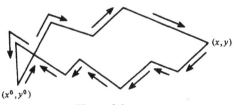

Figure 9.9

Theorem 3 can be given an interesting interpretation. Let M and N be continuously differentiable functions in a star domain D. Consider the problem of finding a solution ϕ in D of the following pair of equations:

$$\frac{\partial \phi}{\partial x} = M, \qquad \frac{\partial \phi}{\partial y} = N. \tag{8}$$

If such a solution exists, then, necessarily, $M_y = N_x$. Theorem 3 now asserts that the latter condition is not only necessary but also sufficient for (8) to have a solution.

In the next theorem we give a useful variant of Green's formula.

THEOREM 4.   *Let D be a simple domain in R^2 with a piecewise continuously differentiable boundary γ. Denote by n the outward normal to γ and by $\cos(x,n)$, $\cos(y,n)$ the cosines of the angles that n forms with the x-axis and the y-axis, respectively. Let M and N be as in Theorem 1. Then*

$$\iint\limits_{D} \left(\frac{\partial M}{\partial x} + \frac{\partial N}{\partial y} \right) dx\, dy = \int_0^{l(\gamma)} [M \cos(x,n) + N \cos(y,n)]\, ds \tag{9}$$

where $l(\gamma)$ is the length of γ and s is the length parameter.

The integral $\int_0^{l(\gamma)}$ is often written in the form $\int_{\partial D}$.

Proof.   Let $(x(s),y(s))$ be a variable point on γ, $0 \leq s \leq l(\gamma)$. Denote by ϕ the angle formed by the tangent $(x'(s),y'(s))$ and the x-axis, and denote by α the angle formed by the outward normal n and the x-axis; see Figure 9.10.

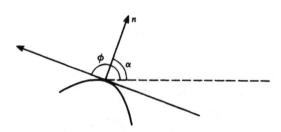

Figure 9.10

Note that $\alpha = \phi - \pi/2$. Hence

$$\cos(x,n) = \cos \alpha = \cos\left(\phi - \frac{\pi}{2} \right) = \cos\left(\frac{\pi}{2} - \phi \right) = \sin \phi = \frac{dy}{ds},$$

$$\cos(y,n) = \sin(x,n) = \sin \alpha = \sin\left(\phi - \frac{\pi}{2}\right) = -\sin\left(\frac{\pi}{2} - \phi\right)$$

$$= -\cos \phi = -\frac{dx}{ds}.$$

It follows that

$$\int_{\gamma} M \, dy - N \, dx = \int_{0}^{\gamma(s)} \left[M \frac{dy}{ds} - N \frac{dx}{ds} \right] ds$$

$$= \int_{0}^{\gamma(s)} [M \cos(x,n) + N \cos(y,n)] \, ds.$$

Substituting this into Green's formula, the assertion (9) follows.

PROBLEMS

1. Prove that a domain bounded by a polygon with five sides is a Green's domain.

2. Let $0 < \alpha < \beta$. Prove that the domain defined by $\alpha < x^2 + y^2 < \beta$ is a Green's domain.

3. Prove that a star domain is simply connected.

4. Use Green's formula to evaluate the following line integrals:
   (a) $\oint_{\partial D} 2xy \, dx - 3xy \, dy$; D is the square bounded by $x = 3, x = 5, y = 1, y = 3$.
   (b) $\oint_{\gamma} xy^2 \, dx + 2x^2y \, dy$; γ the clockwise-oriented ellipse $4x^2 + 9y^2 = 36$.
   (c) $\oint_{\partial D} e^x \sin y \, dx + e^x \cos y \, dy$; D any Green's domain.
   (d) $\int_{\gamma} (x^3 + y) \, dx - y^3 \, dy$; γ the circle $x^2 + y^2 = 1$ with clockwise orientation.
   (e) $\int_{\partial D} f(x) \, dx + g(y) \, dy$; D any simple domain.

5. We write $\Delta w = w_{xx} + w_{yy}$, $\nabla w = (w_x, w_y)$. Let u and v be functions whose first two derivatives are uniformly continuous in a Green's domain D with piecewise continuously differentiable boundary ∂D. Show that

$$\iint_D u \, \Delta v \, dx \, dy + \iint_D \nabla u \cdot \nabla v \, dx \, dy = \int_{\partial D} u \frac{\partial v}{\partial n} \, ds,$$

$$\iint_D (u \, \Delta v - v \, \Delta u) \, dx \, dy = \int_{\partial D} \left(u \frac{\partial v}{\partial n} - v \frac{\partial u}{\partial n} \right) ds$$

where n is the outward normal to ∂D.

6. A function u defined in an open set D of R^2 is called a *harmonic function* if its first two derivatives are continuous and $\Delta u = 0$ in D. Prove that if u is harmonic in D then, for any Green's domain G with piecewise continuously differentiable boundary ∂G such that $G \cup \partial G$ lies in D,

$$\int_{\partial G} \frac{\partial u}{\partial n}\, ds = 0,$$

$$\iint_{G} (u_x^2 + u_y^2)\, dx\, dy = \int_{\partial G} u\, \frac{\partial u}{\partial n}\, ds.$$

7. Let C be a simple domain. Show that

$$\text{area of } D = \oint_{\partial D} x\, dy = -\oint_{\partial D} y\, dx.$$

8. Let $M = -\dfrac{y}{x^2 + y^2}$, $N = \dfrac{x}{x^2 + y^2}$. Then $\int_C M\, dx + N\, dy$ is independent of the path in any simply connected domain not containing the origin. Show that if C is a circle $x^2 + y^2 = R^2$, then $\oint_C M\, dx + N\, dy = 2\pi$.

9. Show that the vector $\left(\dfrac{x}{x^2 + y^2}, \dfrac{y}{x^2 + y^2} \right)$ is a gradient field in the domain $G: x^2 + y^2 > 0$. (This domain is not simply connected.)

6. CHANGE OF VARIABLES IN DOUBLE INTEGRALS

Let D be a bounded domain in R^2 with piecewise continuously differentiable boundary ∂D. As mentioned in Section 5, D is a Green's domain. Since we have not proved it, we shall assume that D is a Green's domain.
Let

$$x = g(u,v), \qquad y = h(u,v) \tag{1}$$

be a diffeomorphism of an open set Ω_0 in R^2 onto an open set D_0 in R^2 containing the closure $\bar{D}$ of D. We shall assume that the image Ω of D, under the inverse of the mapping (1), is a Green's domain. (This assumption actually can be proved since, as will be seen later, the boundary of Ω is piecewise continuously differentiable.) We shall also assume throughout this section that the second mixed derivative of h exists and is continuous.

THEOREM 1.   *Under the foregoing assumptions on D, g, h, for any continuous function $f(x,y)$ in R^2,*

$$\iint_D f(x,y)\, dx\, dy = \iint_\Omega f(g(u,v),h(u,v)) \left| \frac{\partial(x,y)}{\partial(u,v)} \right|\, du\, dv, \tag{2}$$

where Ω is the set that is mapped onto D by the mapping (1).

Proof. Let

$$u = \phi(x,y), \qquad v = \psi(x,y) \tag{3}$$

be the inverse mapping to (1). Denote by γ a continuously differentiable curve

$$x = x(t), \qquad y = y(t) \qquad (a \le t \le b)$$

with range ∂D, positively oriented with respect to D. The image δ of γ is given by

$$u = \phi(x(t),y(t)),$$
$$v = \psi(x(t),y(t)).$$

Hence δ is also piecewise continuously differentiable. δ is the boundary of Ω, and it is either positively oriented with respect to Ω, or negatively oriented. Let

$$F(x,y) = \int_0^x f(\xi,y) \, d\xi. \tag{4}$$

Then $\partial F/\partial x = f$ in R^2.

Using Green's formula, we can write

$$\iint_D f(x,y) \, dx \, dy = \iint_D \frac{\partial F}{\partial x} \, dx \, dy = \int_\gamma F(x,y) \, dy$$

$$= \int_a^b F(x(t),y(t)) \frac{dy}{dt} \, dt = \int_a^b F(x(t),y(t)) \left[h_u \frac{du}{dt} + h_v \frac{dv}{dt} \right] dt$$

$$= \int_\delta F(g(u,v),h(u,v))h_u \, du + F(g(u,v),h(u,v))h_v \, dv.$$

Applying Green's formula, we get

$$\iint_D f(x,y) \, dx \, dy = \pm \iint_\Omega \left[\frac{\partial}{\partial u}(Fh_v) - \frac{\partial}{\partial v}(Fh_u) \right] du \, dv,$$

where the sign is $+$ $(-)$ if δ is positively (negatively) oriented with respect to Ω.

The last integrand is equal to

$$(F_u h_v + F h_{uv}) - (F_v h_u + F h_{uv}) = F_u h_v - F_v h_u$$

$$= (F_x g_u + F_y h_u)h_v - (F_x g_v + F_y h_v)h_u$$

$$= f \frac{\partial(g,h)}{\partial(u,v)}.$$

Hence

$$\iint_D f(x,y)\, dx = \pm \iint_\Omega f(g(u,v),h(u,v)) \frac{\partial(g,h)}{\partial(u,v)}\, du\, dv. \tag{5}$$

The sign $\pm$ is independent of f. Taking $f = 1$, we see that

$$\pm \iint_\Omega \frac{\partial(g,h)}{\partial(u,v)}\, du\, dv$$

must be a positive number. Since $\partial(g,h)/\partial(u,v)$ does not change sign in Ω, we conclude that the sign in $\pm\partial(g,h)/\partial(u,v)$ is such that this expression becomes $|\partial(g,h)/\partial(u,v)|$. Consequently, (5) reduces to (2).

Remarks.   As a byproduct of the proof of Theorem 1 we get the following result: The orientation of the image of γ is positive if, and only if, the Jacobian of the map is positive. Here γ is a positively oriented curve whose range is the boundary of D.

PROBLEMS

1.   Use the transformation $x = r \cos \theta$, $y = r \sin \theta$ to evaluate the integral

$$\iint_D \frac{x}{\sqrt{x^2 + y^2}}\, dx\, dy$$

where D is the domain bounded by $x^2 + y^2 = 1$, $x^2 + y^2 = 4$ and lying inside $x^2 + y^2 = 2x$.

2.   Let D be the triangle bounded by $x = y$, $x + y = 0$ and $x - 2y = 2$. Use the transformation $u = 2x - y$, $v = x - 2y$ to evaluate the integral

$$\iint_D \cos\left[\frac{\pi(2x - y)}{2(x - 2y)}\right]\, dx\, dy.$$

3.   Prove (2) in case f is only assumed to be continuous in $\bar{D}$, but D is a simple domain. [*Hint:* Extend f by 0 and prove that F, defined by (4), is continuous in D and $\partial F/\partial x = f$ in D.]

7. SURFACES AND AREA

A vector valued function from a region in R^2 into R^3 is called a *surface* in R^3, or, briefly, a surface. Thus, a surface S is given by

$$x = x(u,v),$$

$$y = y(u,v), \tag{1}$$

$$z = z(u,v)$$

where (u,v) vary in a region B of R^2. We usually shall take B to have a piece-wise continuously differentiable boundary. The variables u,v are called *parameters* and the functions in (1) are said to give a *parametric representation* of S. If these functions are uniformly continuous, then we say that S *is continuous*; if, in addition, the first derivatives of these functions are piecewise continuous then we say that S *is piecewise continuously differentiable*; finally, if these functions and their first derivatives are uniformly continuous in the interior of B, then we say that S is *continuously differentiable*.

A surface given by

$$z = f(x,y) \tag{2}$$

is called *z-projectable*. Similarly, one defines the concepts of y-projectable and x-projectable surfaces. A surface is called *projectable* if it is either z-, y-, or x-projectable. In Section 8.1 it was proved that the graph of a continuous projectable surface is of content 0 in R^3. (In Section 8.1 the region B was a closed domain.)

A plane π in R^3 is a set of points given by

$$ax + by + cz + d = 0$$

where $a^2 + b^2 + c^2 > 0$. If $c \neq 0$, then π is the graph of a z-projectable surface given by

$$z = -\frac{a}{c}x - \frac{b}{c}y - \frac{d}{c}.$$

Similarly, if $b \neq 0$, then the plane π is the graph of a y-projectable surface, and if $a \neq 0$, then it is the graph of an x-projectable surface.

The vector (a,b,c) is called the *normal* vector to the plane π. For any two points $P_0 = (x_0,y_0,z_0)$ and $P_1 = (x_1,y_1,z_1)$ in the plane, (a,b,c) is orthogonal to the vector $P_0 - P_1 = (x_0 - x_1, y_0 - y_1, z_0 - z_1)$. Note (see Figure 9.11) that the line through P_0, which is parallel to $P_0 - P_1$, lies in the plane.

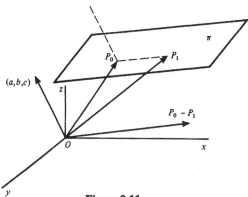

Figure 9.11

Consider the surface S given by (1). Assume that it is piecewise continuously differentiable. Let (u_0,v_0) be any point in the interior of B, and set $x(u_0,v_0) = x_0$, $y(u_0,v_0) = y_0$, $z(u_0,v_0) = z_0$, $P_0 = (x_0,y_0,z_0)$. Take any curve

$$u = u(t), \qquad v = v(t)$$

passing through (u_0,v_0); say, $u_0 = u(t_0)$, $v_0 = v(t_0)$. There corresponds to it a curve

$$
\begin{aligned}
x &= x(u(t),v(t)), \\
y &= y(u(t),v(t)), \\
z &= z(u(t),v(t))
\end{aligned}
\tag{3}
$$

in R^3, which passes through P_0, and which lies in S; see Figure 9.12.

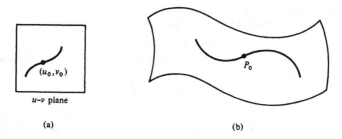

u–v plane

(a)   (b)

Figure 9.12

Suppose $x(u,v)$, $y(u,v)$, $z(u,v)$ are continuously differentiable at the point (u_0,v_0). The tangent vector to the curve (3) at the point P_0 is

$$(x_u u'(t_0) + x_v v'(t_0), y_u u'(t_0) + y_v v'(t_0), z_u u'(t_0) + z_v v'(t_0)) \tag{4}$$

where $x_u = x_u(P_0)$, $x_v = x_v(P_0)$, and so forth. We call it a *tangent vector* to S at P_0.

Consider the vector $n = (n_1,n_2,n_3)$ where

$$
\begin{aligned}
n_1 &= \frac{\partial(y,z)}{\partial(u,v)} = y_u z_v - y_v z_u, \\[4pt]
n_2 &= \frac{\partial(z,x)}{\partial(u,v)} = z_u x_v - z_v x_u, \\[4pt]
n_3 &= \frac{\partial(x,y)}{\partial(u,v)} = x_u y_v - x_v y_u.
\end{aligned}
\tag{5}
$$

It is easily verified that the scalar product of n with the vector (4) is zero.

Thus, n is orthogonal to any tangent vector to S at P_0. We call n a *normal* to S at P_0. Any vector λn (λ scalar) also is called a normal to S at P_0. If $n \neq 0$, then the two vectors $\pm n / \|n\|$ are called the *unit normals* to S at P_0.

If $n \neq 0$, then the plane passing through P_0 and having n for its normal is called the *tangent plane* to S at P_0. Its equation is

$$n_1 x + n_2 y + n_3 z - (n_1 x_0 + n_2 y_0 + n_3 z_0) = 0. \tag{6}$$

If S is given by (2), then

$$\left(-\frac{z_x}{\sqrt{1 + z_x^{\,2} + z_y^{\,2}}}, -\frac{z_y}{\sqrt{1 + z_x^{\,2} + z_y^{\,2}}}, \frac{1}{\sqrt{1 + z_x^{\,2} + z_y^{\,2}}} \right) \tag{7}$$

is a unit normal to S at P_0. The tangent plane to S at P_0 then is given by

$$z - z_0 = f_x(P_0)(x - x_0) + f_y(P_0)(y - y_0) \tag{8}$$

We want to define the concept of the area of S. We shall motivate it by looking first at the special case where S is given by $z = f(x,y)$, and f, f_x, and f_y are uniformly continuous in the interior of a rectangle $D: a \leq x \leq b$, $c \leq y \leq d$.

We form a partition of D, by

$$\begin{aligned} x_0 = a < x_1 < x_2 < \cdots < x_{k-1} < x_k = b, \\ y_0 = c < y_1 < y_2 < \cdots < y_{m-1} < y_m = d, \end{aligned} \tag{9}$$

and denote by R_{ij} the rectangle

$$x_{i-1} \leq x \leq x_i, \qquad y_{j-1} \leq y \leq y_j.$$

Denote by $f(R_{ij})$ the image of R_{ij} by f. Let (ξ_i, η_j) be any point in R_{ij}, and denote by $n = n(\xi_i, \eta_j)$ that unit normal to S at $P_{ij} = (\xi_i, \eta_j, f(\xi_i, \eta_j))$ that forms an acute angle with the z-axis.

Denote by $\hat{R}_{ij}$ the set of points on the tangent plane to S at P_{ij} whose projection on the (x,y)-plane is R_{ij}. Figure 9.13(b) shows cuts of $f(R_{ij})$ and $\hat{R}_{ij}$ by a plane orthogonal to the (x,y)-plane. The cut of $f(R_{ij})$ is a curve l, and the cut of $\hat{R}_{ij}$ is a straight segment l_0, tangent to l. Therefore, the length of l is "approximately" equal to the length of l_0, that is, the quotient of the lengths tends to zero if the mesh of the partition (9) converges to 0. This situation motivates us to state that the area of $f(R_{ij})$ should "approximately" be equal to the area of $\hat{R}_{ij}$.

Now $\hat{R}_{ij}$ (whose boundary is a quadrangle) lies on a plane that forms an angle γ_{ij} with the (x,y)-plane (see Figure 9.13(a)); that is, the angle formed by normals to these two planes is γ_{ij}. Denote by Λ the intersection of the two planes. If Λ is parallel to a side of $\hat{R}_{ij}$ (in which case the boundary of $\hat{R}_{ij}$ is a

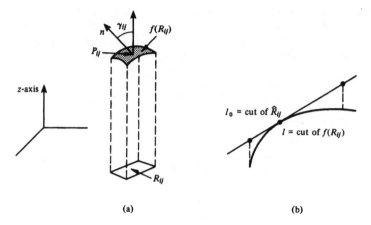

(a)   (b)

Figure 9.13

rectangle), then it is also parallel to a side of R_{ij}. But then $\hat{R}_{ij}$ is congruent to a rectangle $\tilde{R}_{ij}$ as in the following Figure 9.14, where Λ' is parallel to Λ.

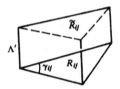

Figure 9.14

It follows that

$$\text{area of } \hat{R}_{ij} = \frac{A(R_{ij})}{\cos \gamma_{ij}} \quad (A(R_{ij}) = \text{area of } R_{ij}). \tag{10}$$

In general, given $\varepsilon > 0$ we can approximate $\hat{R}_{ij}$ by a union of mutually disjoint rectangles $\hat{R}_{ij,k}$ using a partition of a sufficiently small mesh, as described in Figure 9.15. The lines $\Lambda_1, \Lambda_2, \Lambda_3, \dots, \Lambda_n$ are taken to be parallel to line Λ. Each rectangle $\hat{R}_{ij,k}$ and its projection $R_{ij,k}$ on the (x,y)-plane can be treated as in the previous case. Therefore, by (10),

$$\text{area of } \hat{R}_{ij,k} = \frac{A(R_{ij,k})}{\cos \gamma_{ij}}.$$

The $R_{ij,k}$ are mutually disjoint rectangles, and their union forms an approximation to R_{ij}. We conclude that (10) is true, at least " approximately."

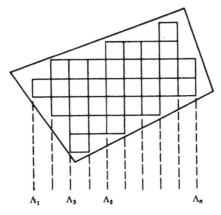

Figure 9.15

We have shown so far that the area of $f(R_{ij})$ should be given, approximately, by

$$T = \sum_{i,j} A(R_{ij})\sec \gamma_{ij}.$$

Recalling that (by (7))

$$\sec \gamma_{ij} = \sqrt{1 + (f_x(\xi_i,\eta_j))^2 + (f_y(\xi_i,\eta_j))^2},$$

we get

$$T = \sum_{i,j} \sqrt{1 + (f_x(\xi_i,\eta_j))^2 + (f_y(\xi_i,\eta_j))^2}(x_i - x_{i-1})(y_j - y_{j-1}). \tag{11}$$

From the results of Section 8.2, it follows that if the mesh of the partition (9) tends to zero then the sums T converge to the integral

$$\iint_D \sqrt{1 + f_x^2 + f_y^2}\, dx\, dy. \tag{12}$$

DEFINITION 1.   Let S be a surface given by $z = f(x,y)$ where f, f_x, and f_y are uniformly continuous in a bounded region D whose boundary ∂D is of content zero in R^2. Then the integral (12) is called the *surface area* of S, or the *area* of S.

We have motivated this definition in the case where D is an open rectangle. If D is an arbitrary bounded region, then we can approximate it by a union of mutually disjoint rectangles D_k (compare Figure 9.14). Since we should expect the area of $f(\bigcup_k D_k)$ to be the sums of the areas of $f(D_k)$, and since

$$\iint_{\bigcup_k D_k} = \sum_k \iint,$$

we are led to define the surface area of S by the integral (12).

So far we have defined the surface area of a z-projectable surface S given by (2). We want to extend the definition to any piecewise continuously differentiable surface S' given by the functions (1) defined on a region G. Suppose first that the graph of S coincides with the range of S', that S' is continuously differentiable, and that G is a domain. It is natural to define the area of S' to be the same as the area of S. The area of S is given by the integral (12). What we shall do, then, is transform this integral by change of variables and express it as an integral with respect to (u,v) in G. Let G_0 be an open set containing the closure of G and let D_0 be an open set containing the closure of D. We shall prove:

THEOREM 1.   *Suppose the map given by*

$$x = x(u,v), \qquad y = y(u,v) \tag{13}$$

is a diffeomorphism from G_0 onto D_0, mapping G onto D. Let

$$E = x_u^2 + y_u^2 + z_u^2,$$
$$F = x_u x_v + y_u y_v + z_u z_v,$$
$$G = x_v^2 + y_v^2 + z_v^2.$$

Then the integral (12) is equal to the integral

$$\iint_G \sqrt{EG - F^2} \, du \, dv. \tag{14}$$

Proof.   Denote by

$$u = u(x,y), \qquad v = v(x,y)$$

the inverse of (13). Then

$$x = x(u(x,y), v(x,y)),$$
$$y = y(u(x,y), v(x,y)).$$

Differentiating these two relations with respect to x, we get

$$1 = x_u u_x + x_v v_x,$$
$$0 = y_u u_x + y_v v_x.$$

Solving for u_x, v_x and using the notation (5), we get

$$u_x = \frac{1}{n_3}\, y_v, \qquad v_x = -\frac{1}{n_3}\, y_u. \tag{15}$$

Differentiating the relation $z = z(u(x,y),v(x,y))$ with respect to x and using (15), we obtain

$$z_x = z_u u_x + z_v v_x = -\frac{n_1}{n_3}. \tag{16}$$

Similarly,

$$u_y = -\frac{1}{n_3}\, x_v, \qquad v_y = \frac{1}{n_3}\, x_u,$$

and

$$z_y = -\frac{n_2}{n_3}. \tag{17}$$

Noting that

$$\frac{\partial(x,y)}{\partial(u,v)} = n_3 \tag{18}$$

and employing Theorem 1 of Section 6, we conclude, upon using (16) and (17), that

$$\iint\limits_{D} \sqrt{1 + z_x{}^2 + z_y{}^2}\; dx\, dy = \iint\limits_{D} \sqrt{n_1{}^2 + n_2{}^2 + n_3{}^2}\; \frac{dx\, dy}{n_3}$$

$$= \iint\limits_{G} \sqrt{n_1{}^2 + n_2{}^2 + n_3{}^2}\; du\, dv.$$

Since, as is easily verified,

$$n_1{}^2 + n_2{}^2 + n_3{}^2 = EG - F^2, \tag{19}$$

the assertion of Theorem 1 follows.

Suppose the map

$$(u,v) \to (x(u,v),y(u,v),z(u,v)) \qquad ((u,v) \in G)$$

is one to one. Then we say that the surface S' is *simple* (or *without self-intersections*). If S is a z-projectable surface given by $z = f(x,y)$ and having the range of S' for its graph, then $z(u,v)$ is determined by $(x(u,v),y(u,v))$:

$$z(u,v) = f(x(u,v),y(u,v)).$$

It follows that the map (13) is one to one. Since $n_3 \neq 0$ when S is z-projectable, we also have, by (18),

$$\frac{\partial(x,y)}{\partial(u,v)} \neq 0.$$

Hence the map (13) is a diffeomorphism from G onto D. Applying Theorem 1 we conclude:

COROLLARY.   *Let S be a continuously differentiable z-projectable surface. If S' is a simple surface and if its range coincides with the graph of S, then the map (13) is a diffeomorphism and, consequently, the integral (12) is equal to the integral (14).*

DEFINITION 2.   Let S be a surface given by (1), where $x(u,v), y(u,v)$, and $z(u,v)$ are continuous in the closure of a bounded region G, and their first derivatives are discontinuous only on a set of content zero. The *surface area* of S (or, simply, the *area* of S) is defined to be the integral (14). This integral may be taken also as an improper integral.

We have justified this definition in case G is a domain, S is simple and continuously differentiable, and its range coincides with the graph of a z-projectable surface. This definition can be justified also in case of a continuously differentiable S whose range coincides with the graph of either a y-projectable surface or an x-projectable surface. We now can motivate this definition for any surface by breaking it into portions whose range coincides with the range of a continuously differentiable surface. (The argument is similar to the one used following Definition 1.)

Definition 2 generalizes Definition 1.

PROBLEMS

1.   Write each of the following surfaces in the form $f(x,y,z) = 0$:
  (a) $x = au \cos v,\ y = bu \sin v,\ z = u$;
  (b) $x = a \sin u \cosh v,\ y = b \cos u \cosh v,\ z = c \sinh v$;
  (c) $x = r \cos \theta,\ y = r \sin \theta,\ z = r^2/2 \sin 2\theta$.

2.   Find a normal and the tangent plane to each of the following surfaces:
  (a) $x^2/a^2 + y^2/b^2 + z^2/c^2 = 1$ at (x_0, y_0, z_0);
  (b) $x = au \cos v,\ y = bu \sin v,\ z = u^2 \cos 2v$ at (u_0, v_0);
  (c) $x^4 + y^4 - z^4 = 1$ at $(1,1,1)$.

3. Let $u = u(t)$, $v = v(t)$ ($\alpha \le t \le \beta$) be a piecewise continuously differentiable curve. It defines a curve γ on the surface S given by (1). Prove that the length $l(\gamma)$ of γ is given by

$$l(\gamma) = \int_\alpha^\beta \sqrt{E(u')^2 + 2Fu'v' + G(v')^2}\, dt, \qquad (20)$$

where E, F, G are defined as in Theorem 1.

4. The sphere S with center 0 and radius 1 is given by $x = a \sin\theta \cos\phi$, $y = a \sin\theta \sin\phi$, $z = a \cos\theta$, where $0 \le \phi < 2\pi$, $0 \le \theta < \pi$. Show that a curve γ on S, given by $\theta = \theta(t)$, $\phi = \phi(t)$ for $\alpha \le t \le \beta$ has length

$$l(\gamma) = a \int_\alpha^\beta \sqrt{(\sin^2\theta)(\phi')^2 + (\theta')^2}\, dt.$$

5. Find the area of each of the following surfaces:
   (a) $z = x^2 + y^2$, $x^2 + y^2 \le 1$;
   (b) $z = \sqrt{x^2 + y^2}$, $x^2 + y^2 \le 1$;
   (c) $z = xy$, $x^2 + y^2 \le 1$.

6. Find the area of an ellipsoid:

$$\frac{x^2}{a^2} + \frac{y^2}{b^2} + \frac{z^2}{c^2} = 1.$$

7. Find the area of a torus: $x = (a + b \cos\phi)\cos\theta$, $y = (a + b \cos\phi)\sin\theta$, $z = b \sin\phi$ where $0 \le \phi \le \pi$, $0 \le \theta < \pi/2$.

8. Find the area of the part of the cone $x^2 + y^2 = z^2$ lying inside the cylinder $x^2 + y^2 = 2ax$.

9. Find the area of the surface given by $x = r \cos\theta$, $y = r \sin\theta$, $z = \theta$, where $0 \le r \le 1$, $0 \le \theta \le 2\pi$.

10. Find the area of the part of the cylinder $x^2 + z^2 = a^2$ lying inside the cylinder $x^2 + y^2 = ax$.

11. Find the area of the part of the sphere $x^2 + y^2 + z^2 = c^2$ inside the paraboloid $x^2/a^2 + y^2/b^2 = 2(z + c)$, provided $0 < b \le a \le c$.

12. Find the area of the part of the sphere $x^2 + y^2 + z^2 = a^2$ inside the cylinder $x^2 + y^2 = ay$.

8. SURFACE INTEGRALS

Let S be a surface given by uniformly continuously differentiable functions

$$x = x(u,v), \qquad y = y(u,v), \qquad z = z(u,v) \qquad ((u,v) \in G), \qquad (1)$$

and let f be a continuous function in the closure of the range of S. We want to introduce a concept of a surface integral of f that is analogous to the concept of an integral along a curve (introduced in Section 3).

Suppose G is a rectangle and S is simple, and form a partition

$$u_0 = a < u_1 < u_2 < \cdots < u_{k-1} < u_k = b$$
$$v_0 = c < v_1 < v_2 < \cdots < v_{m-1} < v_m = d \tag{2}$$

of G. There corresponds to it a partition of S as shown in Figure 9.16. To each rectangle

$$R_{ij}: u_{i-1} \leq u \leq u_i, \; v_{j-1} \leq v \leq v_j$$

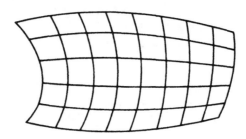

Figure 9.16

there corresponds a surface S_{ij}. Denote its area by $A(S_{ij})$. Then

$$A(S_{ij}) = \iint\limits_{R_{ij}} \sqrt{EG - F^2} \, du \, dv \tag{3}$$

where

$$E = x_u^2 + y_u^2 + z_u^2,$$
$$F = x_u x_v + y_u y_v + z_u z_v, \tag{4}$$
$$G = x_v^2 + y_v^2 + z_v^2.$$

Applying the mean value theorem for double integrals (see Problem 15 of Section 8.4), we get

$$A(S_{ij}) = \sqrt{E(\hat{P}_{ij})G(\hat{P}_{ij}) - F^2(\hat{P}_{ij})} \, (u_i - u_{i-1})(v_j - v_{j-1}) \tag{5}$$

where $\hat{P}_{ij}$ is some point of R_{ij}.

Let P_{ij} be any point in S_{ij}, and form the sum

$$T = \sum f(P_{ij})A(S_{ij}). \tag{6}$$

If we substitute $A(S_{ij})$ from (5) into the last sum, we obtain

$$T = \sum_{i,j} f(P_{ij})\sqrt{E(\hat{P}_{ij})G(\hat{P}_{ij}) - F^2(\hat{P}_{ij})}(u_i - u_{i-1})(v_j - v_{j-1}).$$

THEOREM 1. *Let a simple surface S be given by* (1), *where* x_u, x_v, y_u, y_v, z_u, z_v *are uniformly continuous in an open rectangle G. Then for any* $\varepsilon > 0$, *there is a* $\delta > 0$ *such that for any partition* (2) *of G with mesh* $< \delta$, *and for any sum T given by* (6),

$$\left| T - \iint\limits_{G} f(x(u,v),y(u,v),z(u,v))\sqrt{EG - F^2}\, du\, dv \right| < \varepsilon. \tag{7}$$

The proof can be given analogously to the proof of Theorem 2 in Section 3. We shall omit the details.

The concept of a surface integral that we have in mind is that which is given by lim T, as the mesh goes to 0. In view of Theorem 1, this limit coincides with the integral in (7), when G is a rectangle and S is simple. This motivates the following definition:

The *surface integral of f over S* is the integral

$$\iint\limits_{G} f(x,y,z)\sqrt{EG - F^2}\, du\, dv, \tag{8}$$

where x, y, z are the functions of u, v given by Equations (1) of the surface. This integral may be taken also as an improper integral.

One denotes the surface integral also by

$$\iint\limits_{S} f\, dS$$

where, by (8) and (19) of Section 7,

$$dS = \sqrt{EG - F^2}\, du\, dv = \sqrt{n_1{}^2 + n_2{}^2 + n_3{}^2}\, du\, dv. \tag{9}$$

dS is called the *element of surface area*. If S is z-projectable, then

$$dS = \sqrt{1 + z_x{}^2 + z_y{}^2}\, dx\, dy = \sec \gamma\, dx\, dy, \tag{10}$$

where γ is the acute angle formed by the normal to S and the z-axis.

PROBLEMS

1. Prove Theorem 1.

2. Let S be a continuously differentiable surface given by (1), and let $\tilde{S}$ be a continuously differentiable surface given by $x = \tilde{x}(\tilde{u},\tilde{v})$, $y = \tilde{y}(\tilde{u},\tilde{v})$, $z = \tilde{z}(\tilde{u},\tilde{v})$, where $(\tilde{u},\tilde{v})$ varies in $\tilde{G}$. Assume that there is a diffeomorphism $\tilde{u} = \tilde{u}(u,v)$, $\tilde{v} = \tilde{v}(u,v)$ from an open set G_0 onto an open set $\tilde{G}_0$, mapping G onto $\tilde{G}$, where G_0 contains the closure of G and $\tilde{G}_0$ contains the closure of $\tilde{G}$. Prove that if S and $\tilde{S}$ have the same range, then

$$\iint_S f \, dS = \iint_S f \, d\tilde{S}.$$

3. Compute the surface integrals:

   (a) $\iint_S x^2 z \, dS$, $S: x^2 + y^2 = 1, 0 \leq z \leq 1$;

   (b) $\iint_S x^2 \, dS$, $S: x^2 + y^2 = 1, 0 \leq z \leq 1$;

   (c) $\iint_S (y^2 + z^2) \, dS$, $S: x^2 + y^2 + z^2 = 1, z \geq 0$;

   (d) $\iint_S (x^2 + y^2)z \, dS$, $S: x^2 + y^2 + z^2 = 4, z \geq 1$;

   (e) $\iint_S \dfrac{dS}{\sqrt{z - y + 1}}$, $S: 2z = x^2 + 2y, 0 \leq x \leq 1, 0 \leq y \leq 1$.

4. Let S be the sphere $x^2 + y^2 + z^2 = a^2$. The Newtonian potential at a point $(0,0,c)$ formed by a mass with constant density σ on S is

$$U(c) = \iint_S \frac{\sigma}{[x^2 + y^2 + (z - c)^2]^{1/2}} \, dS.$$

   Prove that $U(c) = \text{const.}$ if $c < a$, $U(c) = \text{const.}/c$ if $c > a$. In evaluating the integral, use polar coordinates and the substitution $t^2 = a^2 + c^2 - 2ac \cos \phi$.

5. The Newtonian potential at $(0,0,-a)$ due to a mass with constant density σ on the hemisphere $S: x^2 + y^2 + z^2 = a^2, z \geq 0$, is

$$U = \iint_S \frac{\sigma}{[x^2 + y^2 + (z + a)^2]^{1/2}} \, dS.$$

Compute U.

6. Let $\sigma(x,y,z)$ be a continuous function, and let S be a continuously differentiable surface. The Newtonian potential at a point (x,y,z) outside S due to a mass distribution σ on S is the integral

$$U(x,y,z) = \iint\limits_{S} \frac{\sigma(\xi,\eta,\zeta)}{[(x-\xi)^2 + (y-\eta)^2 + (z-\zeta)^2]^{1/2}}\, dS.$$

Prove that U is harmonic, that is, $U_{xx} + U_{yy} + U_{zz} = 0$.

9. THE DIVERGENCE THEOREM

Recall that a z-projectable surface given by

$$z = f(x,y), \qquad (x,y) \in D, \qquad D \text{ a closed domain,}$$

is said to be continuously differentiable if f is continuous in D and f_x, f_y are uniformly continuous in the interior of D.

Similarly, one defines the concept of continuously differentiable y-projectable and x-projectable surfaces.

Recall that a normal region G in R^3 is a region whose boundary S can be written in the form $C_1 \cup C_2 \cup \cdots \cup C_k$, where each C_j is the graph of a projectable surface S_j, and $C_i \cap C_j$ $(i \neq j)$ contains no interior points of either S_i or S_j. If all these projectable surfaces are continuously differentiable, then we say that G is a *normal region with piecewise continuously differentiable boundary*. Let G be such a region and let $P_0 = (x_0, y_0, z_0)$ be an interior point of one of the surfaces S_j. Then the two unit normals to S at P_0 are well defined. The one pointing into the interior of G is called the *inward* (or *inner*) *normal*, and the one pointing into the exterior of G is called the *outward* (or *outer*) *normal*.

Let G be a domain in R^3 given by

$$\{(x,y,z) \colon \phi_1(x,y) < z < \phi_2(x,y), (x,y) \in D\} \tag{1}$$

where D is a domain in R^2, $\phi_1(x,y) < \phi_2(x,y)$ for all (x,y) in D. Denote by $\overline{D}$ the closure of D. If the surfaces

$$S_1 \colon z = \phi_1(x,y), \qquad (x,y) \in \overline{D},$$

$$S_2 \colon z = \phi_2(x,y), \qquad (x,y) \in \overline{D}$$

are piecewise continuously differentiable, then we call G a *z-projectable* domain. Similarly, one defines the concept of x-projectable and y-projectable domains. A domain G is called *simple* if it is x-projectable, y-projectable, and z-projectable.

The following theorem is called the *divergence theorem*. It is analogous to Green's theorem in the version given in Theorem 4 of Section 5.

THEOREM 1. *Let P, Q, R be continuous functions in the closure of a simple domain G in R^3, and assume that the derivatives P_x, Q_y, R_z exist and are uniformly continuous in G. Denote by S the boundary of G. Then*

$$\iiint\limits_{G} \left(\frac{\partial P}{\partial x} + \frac{\partial Q}{\partial y} + \frac{\partial R}{\partial z} \right) dV = \iint\limits_{S} [P \cos(x,n) + Q \cos(y,n) + R \cos(z,n)] \, dS, \quad (2)$$

where n is the outward normal to S and $\cos(x,n)$, $\cos(y,n)$, $\cos(z,n)$ are the cosines of the angles formed by n and the x-axis, y-axis, and z-axis, respectively.

Proof. We shall first prove that

$$\iiint\limits_{G} \frac{\partial R}{\partial z} \, dV = \iint\limits_{S} R \cos(z,n) \, dS. \quad (3)$$

G is given by (1). By Theorem 6 of Section 8.4,

$$\iiint\limits_{G} \frac{\partial R}{\partial z} \, dV = \iint\limits_{D} dA \int_{\phi_1(x,y)}^{\phi_2(x,y)} \frac{\partial R(x,y,z)}{\partial z} \, dz$$

$$= \iint\limits_{D} R(x,y,\phi_2(x,y)) \, dA - \iint\limits_{D} R(x,y,\phi_1(x,y)) \, dA. \quad (4)$$

Refer to Figure 9.17. Denote by S_3 the set of points of S that do not belong to $S_1 \cup S_2$. The outward normal at each point of S_3 is orthogonal to the z-axis. Hence

$$\iint\limits_{S} R \cos(z,n) \, dS = \iint\limits_{S_2} R \cos(z,n) \, dS + \iint\limits_{S_1} R \cos(z,n) \, dS$$

$$= \iint\limits_{S_2} R \cos \gamma \, dS - \iint\limits_{S_1} R \cos \gamma \, dS,$$

where γ is the acute angle formed by the normal to S and the z-axis. From the definition of a surface integral [in particular, (10) of Section 8], we get

$$\iint\limits_{S_i} R \cos \gamma \, dS = \iint\limits_{D} R (x, y, \phi_i (x, y)) \, dx \, dy \qquad (i = 1,2).$$

Thus

$$\iint\limits_{S} R \cos(z,n) \, dS = \iint\limits_{D} R(x,y,\phi_2(x,y)) \, dx \, dy - \iint\limits_{D} R(x,y,\phi_1(x,y)) \, dx \, dy.$$

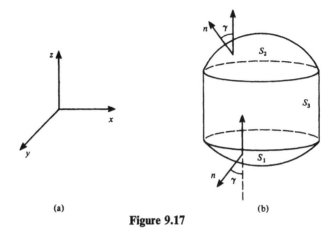

Figure 9.17

Comparing this with (4), the equality (3) follows. Similarly, one proves that

$$\iiint_G \frac{\partial P}{\partial x}\, dV = \iint_S P \cos(x,n)\, dS,$$

$$\iiint_G \frac{\partial Q}{\partial y}\, dV = \iint_S Q \cos(y,n)\, dS.$$

Combining these equalities with (3), (2) follows.

Remark. The divergence theorem is valid also for any normal domain with piecewise continuously differentiable boundary. However, the proof is very lengthy, and will not be given here.

Introduce the vector field

$$\mathbf{F} = P\mathbf{i} + Q\mathbf{j} + R\mathbf{k}.$$

Denote the outward normal to S by $\mathbf{n}$. Then

$$\operatorname{div} \mathbf{F} = P_x + Q_y + R_z,$$

$$\mathbf{F} \cdot \mathbf{n} = P \cos(x,n) + Q \cos(y,n) + R \cos(z,n).$$

The assertion (2) therefore can be written in the form

$$\iiint_G \operatorname{div} \mathbf{F}\, dV = \iint_S \mathbf{F} \cdot \mathbf{n}\, dS. \qquad (5)$$

The divergence theorem derives its name from this form.

EXAMPLE 1.   Let G be a domain in R^3 for which the divergence theorem is valid, and denote the boundary of G by S. Let O be the origin of R^3 and denote by ψ the angle formed by the outward normal $\mathbf{n}$ to S at $P = (x,y,z)$ and the vector $\mathbf{r} = \overrightarrow{OP} = x\mathbf{i} + y\mathbf{j} + z\mathbf{k}$. Thus, $\mathbf{r} \cdot \mathbf{n} = \cos\psi$. Applying (5) with $\mathbf{F} = \mathbf{r}$, we get

$$\iiint\limits_G 3\, dV = \iint\limits_S r \cos\psi\, dS, \qquad \text{where } r = \sqrt{x^2 + y^2 + z^2}.$$

Consequently, the volume V of G is given by the formula:

$$V = \frac{1}{3} \iint\limits_S r \cos\psi\, dA. \tag{6}$$

It will be shown, in Example 3 below, that the divergence theorem holds in a ball. If G is a ball with center 0 and radius R, then $r = R$ and $\cos\psi = 1$ in the integral of (6). We thus obtain the relation

$$V = \frac{R}{3} A,$$

where A is the area of the sphere.

EXAMPLE 2.   Let $\Delta u = u_{xx} + u_{yy} + u_{zz}$, $\nabla u = (u_x, u_y, u_z)$. A function u whose first two derivatives are continuous and $\Delta u = 0$ is called a *harmonic function*. Let u be a function having first uniformly continuous derivatives in a domain G, and let v be a function whose first two derivatives are uniformly continuous in G. Applying the divergence theorem with

$$P = uv_x, \qquad Q = uv_y, \qquad R = uv_z$$

and using the relation

$$v_x \cos(x,n) + v_y \cos(y,n) + v_z \cos(z,n) = \frac{\partial v}{\partial n},$$

we get *Green's first identity*:

$$\iiint\limits_G u\, \Delta v\, dV + \iiint\limits_G \nabla u \cdot \nabla v\, dV = \iint\limits_S u \frac{\partial v}{\partial n}\, dS. \tag{7}$$

Assume next that the first two derivatives of u are uniformly continuous in G. Then we also have

$$\iiint\limits_G v\, \Delta u\, dV + \iiint\limits_G \nabla v \cdot \nabla u\, dV = \iint\limits_S v \frac{\partial u}{\partial n}\, dS.$$

Comparing this with (7), we get *Green's second identity*:

$$\iiint\limits_{G} (u\,\Delta v - v\,\Delta u)\,dV = \iint\limits_{S} \left(u\,\frac{\partial v}{\partial n} - v\,\frac{\partial u}{\partial n}\right) dS. \tag{8}$$

EXAMPLE 3.   We shall prove that the divergence theorem is valid in a ball $x^2 + y^2 + z^2 < R^2$. Denote the ball by G and its boundary by S. For any $\varepsilon > 0$, let G_ε be the intersection of G with the half-space $z > \varepsilon$, and let S_ε be the intersection of S with the half-space $z > \varepsilon$. Similarly, let $G_{-\varepsilon}$ and $S_{-\varepsilon}$ denote the intersections of G and S, respectively, with the half-space $z < -\varepsilon$. Note that Theorem 1 does not apply to G since S is given by

$$z = \pm\sqrt{R^2 - x^2 - y^2}$$

and, therefore, z_x and z_y become unbounded as $x^2 + y^2 \to R^2$. However, the divergence theorem applies in G_ε and in $G_{-\varepsilon}$. Denoting by σ_ε the part of the boundary of G_ε lying on $z = \varepsilon$, we have

$$\iiint\limits_{G_\varepsilon} \frac{\partial R}{\partial z}\,dV = \iint\limits_{S_\varepsilon} R\,\cos(z,n)\,dS + \iint\limits_{\sigma_\varepsilon} R\,\cos(z,n)\,dS. \tag{9}$$

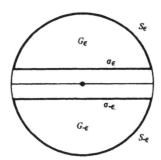

Figure 9.18

Similarly,

$$\iiint\limits_{G_{-\varepsilon}} \frac{\partial R}{\partial z}\,dV = \iint\limits_{S_{-\varepsilon}} R\,\cos(z,n)\,dS + \iint\limits_{\sigma_{-\varepsilon}} R\,\cos(z,n)\,dS, \tag{10}$$

where $\sigma_{-\varepsilon}$ is the part of the boundary of $G_{-\varepsilon}$ lying on $z = -\varepsilon$.

Observing that $\cos(z,n) = 1$ on σ_ε, $\cos(z,n) = -1$ on $\sigma_{-\varepsilon}$, one can show that

$$\lim_{\varepsilon \to 0}\left(\iint_{\sigma_\varepsilon} R\cos(z,n)\,dS + \iint_{\sigma_{-\varepsilon}} R\cos(z,n)\,dS \right) = 0. \tag{11}$$

Denoting by S_0 (S_{-0}) the intersection of S with $z > 0$ $(z < 0)$, one can show further that

$$\text{the improper integrals } \iint_{S_{\pm 0}} R\cos(z,n)\,dS \text{ exist,} \tag{12}$$

so that

$$\lim_{\varepsilon \to 0}\left(\iint_{S_\varepsilon} R\cos(z,n)\,dS + \iint_{S_{-\varepsilon}} R\cos(z,n)\,dS \right) = \iint_{S} R\cos(z,n)\,dS. \tag{13}$$

Finally, one can show that

$$\lim_{\varepsilon \to 0}\left(\iiint_{G_\varepsilon} \frac{\partial R}{\partial z}\,dV + \iiint_{G_{-\varepsilon}} \frac{\partial R}{\partial z}\,dV \right) = \iiint_{G} \frac{\partial R}{\partial z}\,dV. \tag{14}$$

Adding (9) and (10) and taking $\varepsilon \to 0$, we obtain, upon using (11), (13), and (14), the equality (3). The analogous equalities for P and Q follow in the same way.

PROBLEMS

1. Prove the assertion (11).

2. Prove the assertion (12).

3. Prove the assertion (14).

4. Use the divergence theorem to evaluate

$$\iint [x^2 \cos(x,n) + y^2 \cos(y,n) + z^2 \cos(z,n)]\,dS,$$

where S is given by $x^2 + y^2 + z^2 = 2az$.

5. Use the divergence theorem to evaluate

$$\iint_{S} \mathbf{v} \cdot \mathbf{n}\,dS, \quad \mathbf{v} = x^2\mathbf{i} + y^2\mathbf{j} + z^2\mathbf{k}, \qquad \mathbf{n} \text{ outward normal,}$$

where S is the boundary of the cube $0 \le x \le 1, 0 \le y \le 1, 0 \le z \le 1$.

6. Use the divergence theorem to compute $\iiint\limits_{G} \text{div } \mathbf{F} \, dV$, where $\mathbf{F} = x\mathbf{i} - y\mathbf{j} + 2z\mathbf{k}$, $G: x^2 + y^2 + (z-1)^2 < 1$.

7. Let B_2 and B_1 be open balls in R^3 such that the closure of B_1 is contained in B_2. Prove that the divergence theorem holds in the region $B_2 - B_1$.

8. Let G be a simple domain with boundary S, and let $\mathbf{F}$ be a twice continuously differentiable vector in a domain containing $G \cup S$. Prove that $\iint\limits_{s} (\text{curl } \mathbf{F}) \cdot \mathbf{n} \, dS = 0$.

9. Let G be a domain given by (1), where ϕ_1, ϕ_2 have uniformly continuous derivatives in D. Prove that the volume V of G is given by

$$V = \iint\limits_{D} z \, dx \, dy.$$

10. If u is harmonic in a simple domain G with boundary S, and if its first two derivatives are uniformly continuous in G, then

$$\iiint\limits_{G} (u_x{}^2 + u_y{}^2 + u_z{}^2) \, dS = \iint\limits_{S} u \frac{\partial u}{\partial n} \, dS.$$

If, in particular, $u = 0$ on S, then $u = 0$ throughout G.

11. Let G be a ball with boundary S. Let $Q = (\xi, \eta, \zeta)$ be a point in G and denote by C_ε a ball with center Q and radius ε. Set $G_\varepsilon = G - C_\varepsilon$ (G_ε consists of all the points that belong to G but do not belong to C_ε). By Problem 7, we can apply the divergence theorem in G_ε, for all sufficiently small $\varepsilon > 0$. Green's second identity is then also valid. Use it with $v = 1/r, r = ((x - \xi)^2 + (y - \eta)^2 + (z - \zeta)^2)^{1/2}$. Show, upon taking $\varepsilon \to 0$, that

$$u(Q) = -\frac{1}{4\pi} \iiint\limits_{G} \frac{\Delta u}{r} \, dV + \frac{1}{4\pi} \iint\limits_{S} \frac{\partial u}{\partial r} \frac{1}{r} \, dS - \frac{1}{4\pi} \iint\limits_{S} u \frac{\partial}{\partial n} \frac{1}{r} \, dS. \quad (15)$$

This is called *Green's third identity*. It actually holds for any domain G for which the divergence theorem is valid.

12. Use (15) to prove that a harmonic function has derivatives of all orders. [*Hint:* r has derivatives of all orders.]

13. Let u be a harmonic function in a domain G and let B be a ball with center Q and boundary S, such that $B \cup S \subset G$. Prove that

$$u(Q) = \frac{1}{A(S)} \iint\limits_{S} u \, dS \qquad (A(S) = \text{area of } S).$$

This assertion is called the *Gauss theorem*, or the *mean value theorem* for harmonic functions.

14. Find $\mathbf{F} = g(r)\mathbf{r}$ ($\mathbf{r} = x\mathbf{i} + y\mathbf{j} + z\mathbf{k}$) such that div $\mathbf{F} = r^m$ ($m \neq -3, m \geq 0$). Use it to prove that

$$\iiint\limits_{G} r^m \, dV = \frac{1}{m+3} \iint\limits_{S} r^m \mathbf{r} \cdot \mathbf{n} \, dS,$$

where G is a simple domain with boundary S.

15. Consider a sequence of spheres S_n in R^3, with center P_n and radius r_n, such that $P_n \to P$, $r_n \to 0$ as $n \to \infty$. Let $\mathbf{F}$ be a continuously differentiable vector field in a neighborhood of P. Prove that

$$(\text{div } \mathbf{F})(P) = \lim_{n \to \infty} \frac{1}{V_n} \iint\limits_{S_n} \mathbf{F} \cdot \mathbf{n} \, dS$$

where V_n is the volume of the ball with boundary S_n.

10. CHANGE OF VARIABLES IN TRIPLE INTEGRALS

Let G be a normal domain in R^3 with piecewise continuously differentiable boundary S, for which the divergence theorem holds. (By the remark following the proof of Theorem 1 of Section 9, the divergence theorem always holds in such a domain.)

Let

$$x = f(u,v,w),$$

$$y = g(u,v,w), \qquad\qquad (1)$$

$$z = h(u,v,w)$$

be a diffeomorphism from a domain G_0' in R^3 onto a domain G_0 containing $G \cup S$. Denote by G' the image of G and by S' the image of S under the inverse of (1). We shall assume that the functions f, g, h have continuous first two derivatives in G_0', and that the divergence theorem holds in G'.

THEOREM 1. *Let the foregoing assumptions hold, and let k be a continuous function in R^3. Then*

$$\iiint\limits_{G} k(x,y,z) \, dx \, dy \, dz$$

$$= \iiint\limits_{G'} k(f(u,v,w), g(u,v,w), h(u,v,w)) \left| \frac{\partial(x,y,z)}{\partial(u,v,w)} \right| du \, dv \, dw. \qquad (2)$$

The proof is analogous to the proof of Theoerem 1 in Section 6. In that proof we have used Green's theorem. Here we shall use the divergence theorem.

Proof. Let

$$u = u(x,y,z),$$
$$v = v(x,y,z), \tag{3}$$
$$w = w(x,y,z)$$

by the inverse of the mapping (1). The boundary S is given by piecewise continuously differentiable functions

$$x = x(s,t),$$
$$y = y(s,t), \tag{4}$$
$$z = z(s,t),$$

where (s,t) varies in some region A. If we substitute the functions of (4) into (3), we see that the image S' of S is given by piecewise continuously differentiable functions $u = u(s,t)$, $v = v(s,t)$, $w = w(s,t)$.

Introduce the function

$$K(x,y,z) = \int_0^z k(x,y,\zeta)\, d\zeta.$$

It satisfies

$$\frac{\partial K}{\partial z} = k \text{ in } G.$$

Set

$$n_1 = \frac{\partial(y,z)}{\partial(s,t)}, \qquad n_2 = \frac{\partial(z,x)}{\partial(s,t)}, \qquad n_3 = \frac{\partial(x,y)}{\partial(s,t)}, \qquad N = \sqrt{n_1{}^2 + n_2{}^2 + n_3{}^2}.$$

Similarly, set

$$n_1' = \frac{\partial(v,w)}{\partial(s,t)}, \qquad n_2' = \frac{\partial(w,u)}{\partial(s,t)}, \qquad n_3' = \frac{\partial(u,v)}{\partial(s,t)}, \qquad N' = \sqrt{(n_1')^2 + (n_2')^2 + (n_3')^2}.$$

We claim that

$$n_3 = \frac{\partial(f,g)}{\partial(v,w)}\, n_1' + \frac{\partial(f,g)}{\partial(w,u)}\, n_2' + \frac{\partial(f,g)}{\partial(u,v)}\, n_3'. \tag{5}$$

Indeed, on S', $u = u(s,t)$, $v = v(s,t)$, $w = w(s,t)$. Substituting this into (1), we get

$$x = f(u(s,t),v(s,t),w(s,t)),$$

$$y = g(u(s,t),v(s,t),w(s,t)).$$

Differentiating with respect to s and t, we obtain

$$x_s = f_u u_s + f_v v_s + f_w w_s,$$

$$x_t = f_u u_t + f_v v_t + f_w w_t,$$

$$y_s = g_u u_s + g_v v_s + g_w w_s,$$

$$y_t = g_u u_t + g_v v_t + g_w w_t.$$

Substituting these expressions into

$$n_3 = x_s y_t - x_t y_s,$$

(5) follows.

The vectors $v = (n_1,n_2,n_3)$ and $v' = (n_1',n_2',n_3')$ are normal to S and S', respectively. To be definite let us suppose that v is outward and v' is inward. Then

$$n = \frac{v}{N} \qquad \text{and} \qquad n' = -\frac{v'}{N'} \tag{6}$$

are the outward (unit) normals to S and S', respectively.

By the divergence theorem,

$$\iiint\limits_G k \, dx \, dy \, dz = \iiint\limits_G \frac{\partial K}{\partial z} \, dx \, dy \, dz = \iint\limits_S K \cos(x,n) \, dS = \iint\limits_S K \frac{n_3}{N} \, dS.$$

By (9) of Section 8, $dS = N \, ds \, dt$. Hence

$$\iiint\limits_G k \, dx \, dy \, dz = \iint\limits_A K \, n_3 \, ds \, dt.$$

Substituting n_3 from (5) into the last integral, we get

$$\iiint\limits_G k \, dx \, dy \, dz = \iint\limits_A K \left[\frac{\partial(f,g)}{\partial(v,w)} n_1' + \frac{\partial(f,g)}{\partial(w,u)} n_2' + \frac{\partial(f,g)}{\partial(u,v)} n_3' \right] ds \, dt.$$

Since the elements of surface area dS' on S' is given by

$$dS' = N' \, ds \, dt,$$

and since

$$n_1' = -N' \cos(x,n'), \qquad n_2' = -N' \cos(y,n'), \qquad n_3' = -N' \cos(z,n'),$$

we get

$$\iiint\limits_{G} k \, dx \, dy \, dz$$

$$= -\iint\limits_{S'} K \left[\frac{\partial(f,g)}{\partial(v,w)} \cos(x,n') + \frac{\partial(f,g)}{\partial(w,u)} \cos(y,n') + \frac{\partial(f,g)}{\partial(u,v)} \cos(z,n') \right] dS'.$$

Evaluating the last surface integral by the divergence theorem, and using the relation

$$\frac{\partial}{\partial u} \left[K \frac{\partial(f,g)}{\partial(v,w)} \right] + \frac{\partial}{\partial v} \left[K \frac{\partial(f,g)}{\partial(w,u)} \right] + \frac{\partial}{\partial w} \left[K \frac{\partial(f,g)}{\partial(u,v)} \right] = kJ \qquad (7)$$

where

$$J = \frac{\partial(x,y,z)}{\partial(u,v,w)},$$

we get

$$\iiint\limits_{G} k \, dx \, dy \, dz = - \iiint\limits_{G'} kJ \, du \, dv \, dw. \qquad (8)$$

Since $J \neq 0$ in G', either $J > 0$ in G' or $J < 0$ in G'. Taking $k = 1$ in (8), we see that $J < 0$. Therefore, $-J = |J|$. Using this relation in (8), the assertion (2) follows.

We have completed the proof of Theorem 1 in case (6) holds. The proof for any of the other cases (where $n = \pm v/N$, $n' = \pm v'/N'$) is similar.

PROBLEMS

1. Complete the details of the proof of (5).

2. Prove (7).

3. Let G be the region bounded by the paraboloids $z = x^2 + y^2$, $z = 4x^2 + 4y^2$ and lying between the planes $z = 1$ and $z = 4$. Use the transformation

$$x = \frac{v}{u} \cos w, \qquad y = \frac{v}{u} \sin w, \qquad z = v^2$$

to compute the integral

$$\iiint\limits_{G} (x^2 + y^2) \, dx \, dy \, dz.$$

4. Let G be a region given by

$$x \geq 0, y \geq 0, z \geq 0, 1 \leq x^2 + y^2 \leq 4, 9 \leq x^2 + y^2 + z^2 \leq 16.$$

Use the transformation $x = v \cos w, y = v \sin w, z = \sqrt{u - v^2}$ to compute the integral $\iiint\limits_{G} z \, dx \, dy \, dz$.

11. .STOKES' THEOREM

Let S be a surface in R^3 given by

$$x = f(u,v),$$

$$y = g(u,v), \qquad\qquad (1)$$

$$z = h(u,v),$$

where f, g, h are defined and continuous in a bounded, closed domain D, and their first two derivatives are uniformly continuous in the interior of D. We assume that S is simple, that is, the mapping (1) from D into R^3 is one to one. We also assume that S is *regular*, that is, it has nonvanishing normal vectors at each point of S. For simplicity of notation, we shall denote the range of S also by S. Finally, we assume that D is a Green's domain with piecewise continuously differentiable boundary ∂D, given by $u = u(t)$, $v = v(t), \alpha \leq t \leq \beta$.

Suppose it is possible to assign to each point (x,y,z) of S a unit normal $\mathbf{n} = \mathbf{n}(x,y,z)$ in such a way that $\mathbf{n}(x,y,z)$ is a continuous vector field on S. Then we say that S is *orientable* and that the vector field $\mathbf{n}(x,y,z)$ defines an *orientation* on S. The vector field $-\mathbf{n}(x,y,z)$ is said to define the *opposite orientation* to that of $\mathbf{n}(x,y,z)$.

Let C_* be any simple closed curve lying on S. The orientation $\mathbf{n}(x,y,z)$ determines an orientation on C_* as follows: If one walks along C_* with his feet on C_* and his head at the end of the vector $\mathbf{n}$, then the domain S_* on S bounded by C_* is always to his left; see Figure 9.19. We call this orientation of C_* the orientation *induced* by the orientation of S (given by $\mathbf{n}(x,y,z)$). Note that $-\mathbf{n}(x,y,z)$ induces the opposite orientation on C_*. The above definition of the induced orientation is not quite rigorous. One can give a rigorous definition, but this involves developing tools that go beyond the scope of this book.

Denote by C the curve

$$x = f(u(t),v(t)),$$

$$y = g(u(t),v(t)),$$

$$z = h(u(t),v(t)),$$

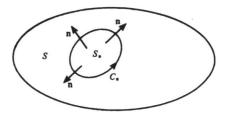

Figure 9.19

where $\alpha \le t \le \beta$. The range of C is the boundary of S. We fix on ∂D the positive orientation (that is, the counterclockwise orientation). This determines an orientation of C. We choose on S the orientation that induces on C the already given orientation. To the orientation of S thus determined there corresponds a continuous vector field of unit normals. We denote this vector field by $\mathbf{n} = \mathbf{n}(x,y,z)$, and write

$$\mathbf{n} = (\cos(x,n),\cos(y,n),\cos(z,n)).$$

We can now state *Stokes' theorem*:

THEOREM 1.   *Let D, S, C be as above and let P, Q, R be functions having first continuous derivatives in an open set containing S. Then*

$$\iint\limits_{S} \left[\left(\frac{\partial R}{\partial y} - \frac{\partial Q}{\partial z} \right)\cos(x,n) + \left(\frac{\partial P}{\partial z} - \frac{\partial R}{\partial x} \right)\cos(y,n) + \left(\frac{\partial Q}{\partial x} - \frac{\partial P}{\partial y} \right)\cos(z,n) \right] dS$$

$$= \int_{C} P\, dx + Q\, dy + R\, dz. \tag{2}$$

If S is contained in the plane $z = 0$, then Stokes' theorem reduces to Green's theorem.

Proof.   Let

$$n_1 = \frac{\partial(y,z)}{\partial(u,v)}, \qquad n_2 = \frac{\partial(z,x)}{\partial(u,v)}, \qquad n_3 = \frac{\partial(x,y)}{\partial(u,v)}, \qquad N = \sqrt{n_1{}^2 + n_2{}^2 + n_3{}^2}.$$

We claim that for all points of S,

$$(n_1,n_2,n_3) = N\mathbf{n} = (N\cos(x,n), N\cos(y,n), N\cos(z,n)). \tag{3}$$

To prove (3) write $\mathbf{v} = (n_1,n_2,n_3)/N$. Then $\mathbf{v} - \mathbf{n}$ is a continuous vector field on S and, at each point of S, either $\|\mathbf{v} - \mathbf{n}\| = 2$ or $\|\mathbf{v} - \mathbf{n}\| = 0$. It follows that either $\|\mathbf{v} - \mathbf{n}\| = 2$ at all the points of S or $\|\mathbf{v} - \mathbf{n}\| = 0$ at all the points of

S. Consequently, it is sufficient to prove that (3) holds at one point $E_0 = (x_0, y_0, z_0)$ of S. We take E_0 to be a point of C, at which C is continuously differentiable. Since $(n_1, n_2, n_3) \neq 0$ at E_0, one of the components n_i is $\neq 0$ at E_0. It will be enough to consider the case where $n_3(E_0) \neq 0$. By continuity, $n_3 \neq 0$ at all the points of S lying in some neighborhood of E_0. Denote this set of points by S'.

Let (u_0, v_0) be a point on ∂D whose image by (1) is E_0. Let Γ_0 be a counterclockwise oriented curve in D, having a common curve Γ_0' with ∂D. Denote by Δ_0 and Δ_0' the images of Γ_0 and Γ_0', respectively, under the mapping $x = f(u,v)$, $y = g(u,v)$. Denote by C_0 and C_0' the images of Γ_0 and Γ_0', respectively, by (1). Thus, the projection of C_0 on the (x,y)-plane is Δ_0. See Figure 9.20.

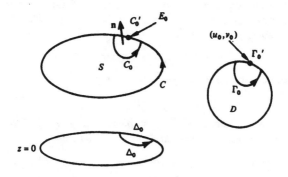

Figure 9.20

Suppose first that $n_3 > 0$ at E_0, so that $n_3 > 0$ at all the points of S'. Choose Γ_0 in a sufficiently small neighborhood of (u_0, v_0) such that all the points of S surrounded by C_0 lie in S'. Since

$$\frac{\partial(x,y)}{\partial(u,v)} = n_3 > 0 \text{ inside } \Gamma_0,$$

the remark made at the end of Section 6 implies that the orientation of Δ_0, determined by the mapping $x = f(u,v)$, $y = g(u,v)$, is the same as that of Γ_0, that is, counterclockwise. But then, the same is true of the orientation of C_0, since Δ_0 is obtained from C_0 by projection on the (x,y)-plane. Since C and C_0 have a common part C_0', we conclude that the orientation of C is counterclockwise. Since the orientation n of S was defined so as to induce this orientation of C, n must form an acute angle with the z-axis. Thus $\cos(z,n) > 0$. We have thus proved that if $n_3 > 0$ at E_0, then $\cos(z,n) > 0$ at E_3. Similarly, one shows that if $n_3 < 0$, then $\cos(z,n) < 0$. This completes the proof of (3).

We proceed to prove (2). We first consider the special case where $Q = R = 0$. Then (2) reduces to

$$\iint_S \left[\frac{\partial P}{\partial z} \cos(y,n) - \frac{\partial P}{\partial y} \cos(z,n) \right] dS = \int_C P \, dx. \tag{4}$$

By (3) and by (9) of Section 8,

$$\iint_S \left[\frac{\partial P}{\partial z} \cos(y,n) - \frac{\partial P}{\partial y} \cos(z,n) \right] dS = \iint_D \left(\frac{\partial P}{\partial z} n_2 - \frac{\partial P}{\partial y} n_3 \right) du \, dv. \tag{5}$$

Using the relations

$$P_u = P_x x_u + P_y y_u + P_z z_u,$$
$$P_v = P_x x_v + P_y y_v + P_z z_v,$$

we find that

$$P_u x_v - P_v x_u = P_y (y_u x_v - y_v x_u) + P_z (z_u x_v - z_v x_u) = P_z n_2 - P_y n_3.$$

Substituting this into the second integral of (5), we find, upon using Green's theorem,

$$\iint_S \left[\frac{\partial P}{\partial z} \cos(y,n) - \frac{\partial P}{\partial y} \cos(z,n) \right] dS = \iint_D \left[\frac{\partial}{\partial u} \left(P \frac{\partial x}{\partial v} \right) - \frac{\partial}{\partial v} \left(P \frac{\partial x}{\partial u} \right) \right] du \, dv$$

$$= \int_{\partial D} P \frac{\partial x}{\partial u} \, du + P \frac{\partial x}{\partial v} \, dv$$

$$= \int_\alpha^\beta \left[P \frac{\partial x}{\partial u} \frac{du}{dt} + P \frac{\partial x}{\partial v} \frac{dv}{dt} \right] dt$$

$$= \int_\alpha^\beta P \frac{dx}{dt} \, dt = \int_C P \, dx.$$

Thus (4) is proved.

Having proved (2) in the special case of $Q = R = 0$, we now can prove it, in the same way, in the special cases $P = R = 0$ and $P = Q = 0$. Combining the three special cases, the assertion (2) follows.

Stokes' theorem can be extended to some piecewise continuously differentiable surfaces. We give one example.

Let $S = S_1 \cup S_2$ where each surface S_i ($i = 1,2$) satisfies the conditions of Theorem 1. Assume that $S_1 \cap S_2$ is a curve C_0 and the orientations of S_1 and S_2 induce opposite orientations on C_0. If we apply Stokes' theorem to the surfaces S_1 and S_2 separately, and add the corresponding surface integrals and the corresponding line integrals, then the line integrals on C_0 cancel out, and (2) follows; see Figure 9.21. This situation arises when S is a sphere and S_1, S_2 are the northern and southern hemispheres.

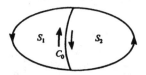

Figure 9.21

Stokes' formula (2) can be written in a more compact form, using vector notation. Introduce the vectors

$$\mathbf{F} = P\mathbf{i} + Q\mathbf{j} + R\mathbf{k}, \qquad \mathbf{T} = \frac{dx}{dt}\mathbf{i} + \frac{dy}{dt}\mathbf{j} + \frac{dz}{dt}\mathbf{k},$$

and the symbol

$$d\mathbf{T} = (dx)\mathbf{i} + (dy)\mathbf{j} + (dz)\mathbf{k}.$$

Then (2) reduces to

$$\iint_S (\operatorname{curl} \mathbf{F}) \cdot \mathbf{n} \, dS = \int_C \mathbf{F} \cdot d\mathbf{T} = \int_\alpha^\beta \mathbf{F} \cdot \mathbf{T} \, dt; \qquad (6)$$

here, $\mathbf{F} \cdot d\mathbf{T} = P \, dx + Q \, dy + R \, dz$.

We shall give an application of Stokes' theorem that is the analog of Theorem 3 of Section 5 in three dimensions. We first recall (see Theorems 1, 2 of Section 4) that a vector field $\mathbf{F} = (P, Q, R)$ is a gradient field in a domain G if, and only if, the line integrals of $\mathbf{F}$ are independent of the curve. Now, if $\mathbf{F}$ is a gradient field, then there is a function ϕ such that

$$\frac{\partial \phi}{\partial x} = P, \qquad \frac{\partial \phi}{\partial y} = Q, \qquad \frac{\partial \phi}{\partial z} = R. \qquad (7)$$

Assuming $\mathbf{F}$ to be continuously differentiable, we deduce that

$$\frac{\partial R}{\partial y} - \frac{\partial Q}{\partial z} = 0, \qquad \frac{\partial P}{\partial z} - \frac{\partial R}{\partial x} = 0, \qquad \frac{\partial Q}{\partial x} - \frac{\partial P}{\partial y} = 0, \qquad (8)$$

that is, $\operatorname{curl} \mathbf{F} = 0$. We shall prove the converse, provided the domain G is a *star domain*, that is, there is a point O in G (called a *center* of G) such that for any other point E in G, the line segment connecting E to O is contained in G.

THEOREM 2.   *Let G be a star domain and let P, Q, and R be continuously differentiable in G, satisfying* (8). *Then there is a function φ satisfying* (7) *in G.*

Proof.   Let O be a center of the star domain G. For any point E in G, denote by γ_E the line segment connecting O to E, and define

$$\phi(E) = \int_{\gamma_E} P\,dx + Q\,dy + R\,dz.$$

We shall prove that ϕ satisfies (7). Let $E = (\xi,\eta,\zeta)$, $F = (\xi + h,\eta,\zeta)$, where h is sufficiently small and positive. Denote by γ' the line segment connecting E to F. Let C be the curve $\gamma_E + \gamma' - \gamma_F$ and denote by S the planar surface bounded by the triangle with vertices O, E, F; see Figure 9.22. Applying Stokes' formula in S we find that

$$\int_C P\,dx + Q\,dy + R\,dz = 0.$$

We also have

$$\int_{\gamma'} P\,dx + Q\,dy + R\,dz = \int_{\xi}^{\xi+h} P(\xi',\eta,\zeta)\,d\xi'.$$

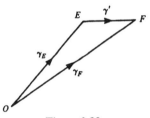

Figure 9.22

From the last two equations and from the definition of ϕ we deduce that

$$\frac{\phi(\xi + h,\eta,\zeta) - \phi(\xi,\eta,\zeta)}{h} = \frac{1}{h}\int_{\xi}^{\xi+h} P(\xi',\eta,\zeta)\,d\xi' = P(\xi + \theta h,\eta,\zeta) \quad (0 < \theta < 1).$$

It follows that

$$\lim_{h \to 0+} \frac{\phi(\xi + h,\eta,\zeta) - \phi(\xi,\eta,\zeta)}{h} = P(\xi,\eta,\zeta).$$

Similarly, one proves that

$$\lim_{h \to 0-} \frac{\phi(\xi + h,\eta,\zeta) - \phi(\xi,\eta,\zeta)}{h} = P(\xi,\eta,\zeta).$$

Consequently, ϕ_x exists and is equal to P. The proofs that $\phi_y = Q$ and $\phi_z = R$ are similar.

DEFINITION. A domain G in R^n is said to be *simply connected* if it has the following property: Given any simple, closed, continuous curve $\gamma: x = h(t)$, $a \le t \le b$, with range in G, there is a continuous function $x = F(s,t)$ defined for $0 \le s \le 1$, $a \le t \le b$, such that: (i) $F(0,t) = h(t)$, $a \le t \le b$; (ii) $F(1,t) = P$, $a \le t \le b$, where P is some point in G, and (iii) $F(s,t)$ lies in G for all $0 \le s \le 1$, $a \le t \le b$. If $n = 2$, then this definition can be shown to be equivalent to the definition given in Section 5.

Defining curves γ_s by $x = F(s,t)$, $a \le t \le b$, we say that the family $\{\gamma_s\}$ gives a *continuous deformation* of γ into a point P.

Theorem 2 is valid in any simply connected domain G in R^3. The essential step in the proof is showing that the line integral of ϕ over a polygonal curve is independent of the curve (compare the remark following the proof of Theorem 3 in Section 5). Here one needs to use Stokes' theorem. The details are very lengthy, and will be omitted.

The surfaces occurring in Stokes' theorem are orientable. Not all surfaces are orientable. We give here an example of a nonorientable surface, called the *Möbius strip*. To construct it, take a strip as in Figure 9.23. Denote by B' a point lying on the back of the strip, behind B. Give a half twist by turning the edge B such that B' appears in front. Now bend and bring together C_1 to D_1 and C_2 to D_2, as in Figure 9.24.

Figure 9.23

The Möbius strip is nonorientable. Indeed, if we fix $\mathbf{n}(R)$ at some point R and then move this normal continuously along a curve that crosses the line C_1C_2 once, when we arrive back at R the normal is $-\mathbf{n}(R)$.

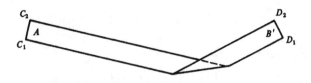

Figure 9.24

PROBLEMS

1. Use Stokes' theorem to evaluate the following line integrals:

    (a) $\int_C -3y\,dx + 3x\,dy + dz$, C is the circle $x^2 + y^2 = 1$, $z = 2$ oriented so that y increases for positive x;

    (b) $\int_C y\,dx + z\,dy + x\,dz$, C the intersection of $x + y = \alpha$, $x^2 + y^2 + z^2 = \alpha(x + y)$, $\alpha > 0$, oriented counterclockwise as viewed from the origin;

    (c) $\int_C 2xy^2\,dx + 2x^2yz\,dy + (x^2y^2 - 2z)\,dz$, $C: x = \cos t$, $y = \sin t$, $z = \sin t$, $0 \le t \le 2\pi$.

2. Let C be a curve in which the cylinder $x^2 + y^2 = a^2$ intersects a plane $\alpha x + \beta y = 1$. Use Stokes' theorem to show that

    (a) $\int_C z(x^2 - 1)\,dy + y(x + 1)\,dz = 0$.

    (b) $\int_C y(z - 1)\,dx + x(z + 1)\,dy + 2\pi a^2$, if C is suitably oriented.

3. Let S and C be as in Stokes' theorem. Let u be a continuously differentiable function in an open set containing S and let v be a function whose first two derivatives exist and are continuous in an open set containing S. Prove that

    $$\iint_S \mathbf{n} \cdot (\nabla u \times \nabla v)\,dS = \int_C u\nabla v \cdot d\mathbf{T}.$$

4. Consider a sequence of discs S_m in R^3 with center P and radius $r_m, r_m \to 0$ if $m \to \infty$; see Figure 9.25. Denote by $\mathbf{n}_m$ the unit normal to S_m that induces counterclockwise orientation on the boundary C_m of S_m. Assume that $\mathbf{n}_m \to \mathbf{n}$ if $m \to \infty$. Let $\mathbf{u}$ be a continuously differentiable vector field defined in a neighborhood of P. Use (6) to prove that

    $$\text{curl } \mathbf{u}(P) \cdot \mathbf{n} = \lim_{m \to \infty} \frac{1}{\pi r_m^2} \oint_{C_m} \mathbf{u} \cdot d\mathbf{T}.$$

Figure 9.25

5. Verify Stokes' formula for the vector $(P,Q,R) = (z,x,y)$ with S being the hemisphere $x^2 + y^2 + z^2 = a^2$, $z \geq 0$.

6. Use the proof of Theorem 2 to find a potential ϕ for each of the following gradient fields:

   (a) $(2x/z)\mathbf{i} + (2y/z)\mathbf{j} + \left(1 - \left(\dfrac{x^2 + y^2}{z^2}\right)\right)\mathbf{k}$; take $O = (1,1,1)$;

   (b) $e^{-xy}(y - xy^2 + yz)\mathbf{i} + e^{-xy}(x - x^2y + xz)\mathbf{j} - e^{-xy}\mathbf{k}$; take $O = (0,0,0)$;

   (c) $2xy\mathbf{i} + (x^2 + \log z)\mathbf{j} + (y/z)\mathbf{k}$; take $O = (0,0,1)$.

7. Evaluate the integral $\int_C \sin yz\, dx + xz \cos yz\, dy + xy \cos yz\, dz$ where C is the helix $x = \cos t, y = \sin t, z = t, 0 \leq t \leq 2\pi$. [*Hint:* Show that the integral is independent of the curve.]

8. Let $\mathbf{u} = -\dfrac{y}{x^2 + y^2}\mathbf{i} + \dfrac{x}{x^2 + y^2}\mathbf{j} + z\mathbf{k}$. $\mathbf{u}$ is continuously differentiable in the domain $G: 1 < x^2 + y^2 < 5, -1 < z < 1$. Show that curl $\mathbf{u} = 0$ in G, but $\int_C \mathbf{u} \cdot d\mathbf{T} \neq 0$ when C is the circle $x^2 + y^2 = 4$, $z = 0$. Explain why this does not contradict Theorem 2.

10

SELECTED TOPICS

This chapter contains several independent topics.

In Section 1 we derive a formula for a finite sum of numbers and apply it to derive asymptotic formulas. In particular, Stirling's formula for $n!$, as $n \to \infty$, is obtained.

In Section 2 we consider an important class of divergent series, called asymptotic series. The basic theory for such series is developed.

In Section 3 we study the asymptotic behavior of integrals $\int_0^x \phi(x,t)\, dt$ as $x \to \infty$. A general theorem is proved, and then is applied to extend Stirling's formula.

In Section 4 we treat Fourier series and establish a fundmental convergence theorem. This theorem then is used to derive many interesting formulas for sums of series.

In Section 5 we deal with the problem of computing the extremum of a function $f(x_1,...,x_n)$ when the x_i are subject to some restrictions.

In Section 6, the divergence theorem, proved in Chapter 9 for $n = 3$, is proved for any n.

Section 7 deals very briefly with harmonic functions.

In Section 8 we consider ordinary differential equations. Some methods of solving such equations are given. Next, a fundamental theorem is proved, asserting the existence and uniqueness of solutions.

Finally, in Section 9 we define the Darboux–Stieltjes integral and prove some basic theorems. This integral is an important generalization of the Darboux integral considered in Chapter 4.

1. EULER'S SUMMATION FORMULA

Let $f(x)$ be a monotone decreasing, positive function defined in the interval $0 \leq x < \infty$, and set $f_n = f(n)$. Then, by the integral test (Theorem 8 of Section 5.1), the series $\sum f_n$ is convergent if, and only if, the integral $\int_0^\infty f(x)\,dx$ is convergent. We shall now evaluate the nonnegative expression

$$f_0 + f_1 + \cdots + f_n - \int_0^n f(x)\,dx,$$

assuming $f(x)$ to be continuously differentiable.

We begin with the relations

$$\sum_{k=1}^{n} \int_{k-1}^{k} kf'(x)\,dx = \sum_{k=1}^{n} k(f_k - f_{k-1}) = -(f_0 + f_1 + \cdots + f_n) + (n+1)f_n.$$

If x belongs to the interval $(k-1, k)$, $k = [x] + 1$. Hence

$$f_0 + f_1 + \cdots + f_n = (n+1)f_n - \int_0^n ([x] + 1)f'(x)\,dx. \qquad (1)$$

By integration by parts we get

$$(n+1)f_n = nf_n + f_n = \int_0^n xf'(x)\,dx + \int_0^n f(x)\,dx + f_n.$$

Substituting this into (1), we conclude that

$$\sum_{k=0}^{n} f_k - \int_0^n f(x)\,dx = f_n + \int_0^n (x - [x] - 1)f'(x)\,dx,$$

or

$$\sum_{k=0}^{n} f_k - \int_0^n f(x)\,dx = \frac{1}{2}(f_0 + f_n) + \int_0^n \left(x - [x] - \frac{1}{2}\right)f'(x)\,dx. \qquad (2)$$

This is a special case of *Euler's summation formula*. The general formula is obtained by subjecting the last integral to any number of integrations by parts, assuming f to have derivatives of all orders. We now shall use Euler's formula (2) to derive some interesting results.

Euler's constant: Take $f(x) = 1/(1 + x)$ and apply (2) with n replaced by $n - 1$:

$$1 + \frac{1}{2} + \cdots + \frac{1}{n} - \log n = \frac{1}{2} + \frac{1}{2n} - \int_0^{n-1} \frac{P(x)}{(1+x)^2} \, dx$$

where

$$P(x) = x - [x] - \frac{1}{2}.$$

The last integral is convergent since $|P(x)| \le 1/2$. Noting that $P(x + 1) = P(x)$, we can write that integral also in the form

$$\int_1^n \frac{P(x)}{x^2} \, dx.$$

It follows that

$$\lim_{n \to \infty} \left(1 + \frac{1}{2} + \cdots + \frac{1}{n} - \log n\right) = \gamma \tag{3}$$

where

$$\gamma = \frac{1}{2} - \int_1^\infty \frac{P(x)}{x^2} \, dx.$$

γ is called *Euler's constant*. Its decimal expansion begins with 0.57721.

Stirling's formula: This is the formula:

$$\lim_{n \to \infty} \frac{n!}{(n/e)^n \sqrt{2\pi n}} = 1. \tag{4}$$

To derive it, we use Euler's summation formula (2) with $f(x) = \log(1 + x)$, and obtain

$$\log 1 + \log 2 + \cdots + \log n = \int_1^n \log x \, dx + \frac{1}{2} \log n + \int_1^n \frac{P(x)}{x} \, dx.$$

Since $\int \log x \, dx = x \log x - x$, we get

$$\log n! = \left(n + \frac{1}{2}\right) \log n - (n - 1) + \int_1^n \frac{P(x)}{x} \, dx. \tag{5}$$

We claim that the improper integral

$$\int_1^\infty \frac{P(x)}{x} \, dx \tag{6}$$

is convergent. To prove it, note that

$$\int_{k}^{k+1} \frac{P(x)}{x}\, dx = \int_{0}^{1} \frac{P(x)}{k+x}\, dx = \int_{0}^{1} \frac{x - 1/2}{k+x}\, dx = \int_{0}^{1} \left(1 - \frac{k+1/2}{k+x}\right) dx$$

$$= 1 - \left(k + \frac{1}{2}\right)\log\left(1 + \frac{1}{k}\right),$$

and

$$\left(k + \frac{1}{2}\right)\log\left(1 + \frac{1}{k}\right) = \left(k + \frac{1}{2}\right)\left(\frac{1}{k} - \frac{1}{2k^2} + \frac{1}{3k^3} - \cdots\right)$$

$$= 1 + \frac{1}{2k} - \frac{1}{2k} - \frac{1}{4k^2} + \frac{1}{3k^2} + \cdots.$$

It follows that

$$\left|\int_{k}^{k+1} \frac{P(x)}{x}\, dx\right| \le \frac{C}{k^2} \qquad (C \text{ constant}).$$

Therefore, for any $y > z > k_0$, k_0 positive integer,

$$\left|\int_{z}^{y} \frac{P(x)}{x}\, dx\right| \le \int_{z}^{[z]+1} \frac{|P(x)|}{x}\, dx + \int_{[y]}^{y} \frac{|P(x)|}{x}\, dx + \sum_{k=k_0}^{[y]-1} \frac{C}{k^2}$$

$$\le \frac{C'}{k_0} + \sum_{k=k_0}^{\infty} \frac{C}{k^2} \to 0 \qquad \text{if } k_0 \to \infty \qquad (C' \text{ constant}).$$

Applying the criterion of Cauchy (Theorem 1 of Section 5.8), (6) follows. From (5) and (6) we see that

$$\log n! = \left(n + \frac{1}{2}\right)\log n - n + \rho_n, \qquad \rho_n = 1 + \int_{1}^{n} \frac{P(x)}{x}\, dx \to \rho \text{ as } n \to \infty.$$

$$(7)$$

To find ρ, write

$$2\log(2 \cdot 4 \cdots 2n) = 2\log[2^n(1 \cdot 2 \cdot 3 \cdots n)] = 2n\log 2 + 2\log n!$$

$$= 2n\log 2 + (2n+1)\log n - 2n + 2\rho_n$$

$$= (2n+1)\log 2n - 2n - \log 2 + 2\rho_n,$$

and

$$\log(2n+1)! = \left(2n + \frac{3}{2}\right)\log(2n+1) - (2n+1) + \rho_{2n+1}.$$

By subtraction

$$\log \frac{2 \cdot 4 \cdot 6 \cdot \cdots \cdot 2n}{1 \cdot 3 \cdot 5 \cdot \cdots \cdot (2n-1)} \cdot \frac{1}{2n+1} = (2n+1)\log\left(1 - \frac{1}{2n+1}\right)$$
$$- \frac{1}{2}\log(2n+1) + 1 - \log 2 + 2\rho_n - \rho_{2n+1}.$$

Transforming $(1/2) \log 1/(2n+1)$ from the left to the right, we get

$$\log \frac{2 \cdot 4 \cdot 6 \cdot \cdots \cdot (2n-2) \cdot 2n}{1 \cdot 3 \cdot 5 \cdot \cdots \cdot (2n-1)\sqrt{2n+1}}$$
$$= (2n+1)\log\left(1 - \frac{1}{2n+1}\right) + 1 - \log 2 + 2\rho_n - \rho_{2n+1}. \quad (8)$$

We shall need *Wallis' formula*

$$\frac{\pi}{2} = \lim_{n \to \infty} \left[\frac{2 \cdot 4 \cdot \cdots \cdot 2n}{1 \cdot 3 \cdot \cdots \cdot (2n-1)}\right]^2 \frac{1}{2n+1}. \quad (9)$$

Assuming (9) for the moment, and taking $n \to \infty$ in (8), we obtain

$$\log \sqrt{\frac{\pi}{2}} = -1 + 1 - \log 2 + 2\rho - \rho.$$

Hence, $\rho = \log \sqrt{2\pi}$. We therefore can write (7) in the form

$$\log n! = \left(n + \frac{1}{2}\right)\log n - n + \log\sqrt{2\pi} - \varepsilon_n, \quad (10)$$

where

$$\varepsilon_n = \int_n^\infty \frac{P(x)}{x} \, dx \to 0 \qquad \text{if } n \to \infty. \quad (11)$$

Stirling's formula (4) follows immediately from (10) and (11).

It remains to prove Wallis' formula. We recall (see Problem 4 of Section 4.6) that

$$\int_0^{\pi/2} \sin^{2n-1} x \, dx = \frac{2 \cdot 4 \cdot 6 \cdot \cdots \cdot (2n-2)}{1 \cdot 3 \cdot 5 \cdot \cdots \cdot (2n-1)},$$
$$\int_0^{\pi/2} \sin^{2n} x \, dx = \frac{1 \cdot 3 \cdot 5 \cdot \cdots \cdot (2n-1)}{2 \cdot 4 \cdot 6 \cdot \cdots \cdot (2n)} \cdot \frac{\pi}{2}. \quad (12)$$

The inequalities

$$\sin^{2n+1} x < \sin^{2n} x < \sin^{2n-1} x$$

hold if $0 < x < \pi/2$. Integrating over the interval $(0, \pi/2)$ and using (12), we get

$$\frac{2 \cdot 4 \cdot 6 \cdots \cdot (2n)}{1 \cdot 3 \cdot 5 \cdots \cdot (2n+1)} < \frac{1 \cdot 3 \cdot 5 \cdots \cdot (2n-1)}{2 \cdot 4 \cdot 6 \cdots \cdot (2n)} \frac{\pi}{2} < \frac{2 \cdot 4 \cdot 6 \cdots \cdot (2n-2)}{1 \cdot 3 \cdot 5 \cdots \cdot (2n-1)}.$$

Hence

$$\frac{2n}{2n+1} \cdot \frac{\pi}{2} < \left[\frac{2 \cdot 4 \cdot 6 \cdots \cdot (2n)}{1 \cdot 3 \cdot 5 \cdots \cdot (2n-1)} \right]^2 \frac{1}{2n+1} < \frac{\pi}{2}.$$

Since the left-hand side converges to $\pi/2$ as $n \to \infty$, the same is necessarily true of the expression in the middle. This completes the proof of (9).

DEFINITION.   Let $\{f_n\}$ and $\{g_n\}$ be two sequences. If $\lim f_n/g_n = 1$, then we write $f_n \sim g_n$ and say that the sequence $\{f_n\}$ is *asymptotically equivalent* to the sequence $\{g_n\}$.

Stirling's formula can be written in the form

$$n! \sim \left(\frac{n}{e} \right)^n \sqrt{2\pi n}. \tag{13}$$

PROBLEMS

1. Prove that $\displaystyle\lim_{n \to \infty} \frac{(n!)^{1/n}}{n} = \frac{1}{e}$

2. Show that $\displaystyle\frac{1}{2^{2n}} \binom{2n}{n} \sim \frac{1}{\sqrt{\pi n}}$

3. Prove that for any positive integer p

    $$1^p + 2^p + \cdots + n^p = \frac{n^{p+1}}{p+1} + \frac{n^p}{2} + C(n) n^{p-1}, \qquad |C(n)| \le C,$$

    where C is a constant independent of n.

4. Prove that for any $0 < s < 1$

    $$\lim_{n \to \infty} \left(1 + \frac{1}{2^s} + \cdots + \frac{1}{n^s} - \frac{n^{1-s}}{1-s} \right)$$

    exists.

2. ASYMPTOTIC SERIES

Consider a power series in x^{-1}:

$$a_0 + \frac{a_1}{x} + \frac{a_2}{x^2} + \cdots + \frac{a_n}{x^n} + \cdots. \tag{1}$$

The series may be either convergent or divergent. Let $f(x)$ be a function defined in an interval $x_0 \leq x < \infty$. Suppose that, for any nonnegative integer n,

$$\lim_{x \to \infty} \left[f(x) - \left(a_0 + \frac{a_1}{x} + \frac{a_2}{x^2} + \cdots + \frac{a_n}{x^n} \right) \right] x^n = 0. \tag{2}$$

Then we say that the series (1) is an *asymptotic series* for $f(x)$ (as $x \to \infty$), and that $f(x)$ has the *asymptotic expansion* (or asymptotic series) (1) (as $x \to \infty$), and we write

$$f(x) \sim a_0 + \frac{a_1}{x} + \frac{a_2}{x^2} + \cdots + \frac{a_n}{x^n} + \cdots. \tag{3}$$

If (2) is only known for $n = 0, 1, \ldots, m$, then we write

$$f(x) \sim a_0 + \frac{a_1}{x} + \frac{a_2}{x^2} + \cdots + \frac{a_m}{x^m}$$

and say that $f(x)$ has the asymptotic expansion

$$a_0 + \frac{a_1}{x} + \frac{a_2}{x^2} + \cdots + \frac{a_m}{x^m}.$$

Even though asymptotic series do not converge in general, they are nevertheless important for the purpose of approximating functions.

EXAMPLE 1.   For any positive integer n, $x^n e^{-x} \to 0$ as $x \to \infty$. This implies that

$$e^{-x} \sim 0 + \frac{0}{x} + \frac{0}{x^2} + \cdots + \frac{0}{x^n} + \cdots.$$

EXAMPLE 2.   If $\lim\limits_{x \to \infty} f(x)$ does not exist, then $f(x)$ does not have asymptotic expansion since (2) already is false for $n = 0$.

EXAMPLE 3.   Let

$$f(x) = e^x \int_x^\infty \frac{e^{-t}}{t} \, dt = \int_x^\infty \frac{e^{x-t}}{t} \, dt.$$

By integration by parts

$$f(x) = \left[-\frac{e^{x-t}}{t} \right]_x^\infty - \int_x^\infty \frac{e^{x-t}}{t^2} \, dt = \frac{1}{x} - \int_x^\infty \frac{e^{x-t}}{t^2} \, dt.$$

Integrating by parts the last integral, we get

$$f(x) = \frac{1}{x} - \frac{1}{x^2} + 2\int_x^\infty \frac{e^{x-t}}{t^3}\,dt.$$

Proceeding by induction, one can prove that

$$f(x) = \frac{1}{x} - \frac{1}{x^2} + \frac{2!}{x^3} - \frac{3!}{x^4} + \cdots + (-1)^{n-1}\frac{(n-1)!}{x^n} + (-1)^n n! \int_x^\infty \frac{e^{x-t}}{t^{n+1}}\,dt.$$

Since

$$x^n \int_x^\infty \frac{e^{x-t}}{t^{n+1}}\,dt \le x^n \int_x^\infty \frac{e^{x-t}}{x^{n+1}}\,dt = \frac{1}{x}\int_x^\infty e^{x-t}\,dt = \frac{1}{x} \to 0 \text{ if } x \to \infty,$$

we conclude that

$$f(x) \sim \frac{1}{x} - \frac{1}{x^2} + \frac{2!}{x^3} - \frac{3!}{4!} + \cdots + (-1)^{n-1}\frac{(n-1)!}{x^n} + \cdots.$$

Notice that the asymptotic series on the right is not a convergent series.

The following theorem shows that if an asymptotic expansion exists, then it is unique.

THEOREM 1. *If*

$$f(x) \sim a_0 + \frac{a_1}{x} + \frac{a_2}{x^2} + \cdots + \frac{a_n}{x^n} + \cdots,$$

$$f(x) \sim b_0 + \frac{b_1}{x} + \frac{b_2}{x^2} + \cdots + \frac{b_n}{x^n} + \cdots,$$

then $a_0 = b_0, a_1 = b_1, a_2 = b_2,\ldots,a_n = b_n,\ldots.$

Proof. From (2) it follows that

$$a_n = \lim_{x\to\infty} \left[f(x) - a_0 - \frac{a_1}{x} + \cdots - \frac{a_{n-1}}{x^{n-1}} \right]. \tag{4}$$

Similarly,

$$b_n = \lim_{x\to\infty} \left[f(x) - b_0 - \frac{b_1}{x} - \cdots - \frac{b_{n-1}}{x^{n-1}} \right]. \tag{5}$$

Taking $n = 0$, we get $a_0 = \lim_{x\to\infty} f(x) = b_0$. Suppose we already have proved that $a_k = b_k$ for $k = 0,1,\ldots n-1$. From (4) and (5) we then deduce that $a_n = b_n$. This completes the proof of the theorem by induction.

THEOREM 2. *Let*

$$f(x) \sim a_0 + \frac{a_1}{x} + \frac{a_2}{x^2} + \cdots + \frac{a_n}{x^n} + \cdots,$$

$$g(x) \sim b_0 + \frac{b_1}{x} + \frac{b_2}{x^2} + \cdots + \frac{b_n}{x^n} + \cdots,$$

and let λ, μ *be any real numbers. Then*

$$\lambda f(x) + \mu g(x) \sim \lambda a_0 + \mu b_0$$

$$+ \frac{\lambda a_1 + \mu b_1}{x} + \frac{\lambda a_2 + \mu b_2}{x^2} + \cdots + \frac{\lambda a_n + \mu b_n}{x^n} + \cdots, \quad (6)$$

$$f(x)g(x) \sim c_0 + \frac{c_1}{x} + \frac{c_2}{x^2} + \cdots + \frac{c_n}{x^n} + \cdots, \quad (7)$$

where

$$c_n = a_0 b_n + a_1 b_{n-1} + a_2 b_{n-2} + \cdots + a_{n-1} b_1 + a_n b_0.$$

Proof.   The proof of (6) is left to the student. To prove (7) we write

$$f(x) = a_0 + \frac{a_1}{x} + \cdots + \frac{a_{n-1}}{x^{n-1}} + \frac{a_n + \delta(x)}{x^n},$$

$$g(x) = b_0 + \frac{b_1}{x} + \cdots + \frac{b_{n-1}}{x^{n-1}} + \frac{b_n + \eta(x)}{x^n}.$$

From (2) we see that $\delta(x) \to 0$ if $x \to \infty$. Similarly, $\eta(x) \to 0$ if $x \to \infty$. Hence

$$\left[f(x)g(x) - \left(c_0 + \frac{c_1}{x} + \cdots + \frac{c_n}{x^n} \right) \right] x^n$$

$$= a_0 \eta + b_0 \delta + \frac{a_1(b_n + \eta) + a_2 b_{n-1} + \cdots + a_{n-1} b_2 + (a_n + \delta) b_1}{x}$$

$$+ \cdots + \frac{(a_n + \delta)(b_n + \eta)}{x^n} \to 0 \qquad \text{if } x \to \infty.$$

THEOREM 3. *Let*

$$f(x) \sim a_0 + \frac{a_1}{x} + \frac{a_2}{x^2} + \cdots + \frac{a_n}{x^n} + \cdots. \quad (8)$$

If $f(x)$ is continuous for $x \geq x_0$, then

$$\int_x^\infty \left[f(t) - a_0 - \frac{a_1}{t} \right] dt \sim \frac{a_2}{x} + \frac{a_3}{2x^2} + \cdots + \frac{a_{n+1}}{nx^n} + \cdots. \tag{9}$$

If $f'(x)$ is continuous and has asymptotic expansion, then

$$f'(x) = -\frac{a_1}{x^2} - \frac{2a_2}{x^3} - \cdots - \frac{(n-1)a_{n-1}}{x^n} - \cdots. \tag{10}$$

The function $f(x) = e^{-x} \sin(e^x)$ has asymptotic expansion:

$$e^{-x} \sin(e^x) \sim 0 + \frac{0}{x} + \frac{0}{x^2} + \cdots.$$

Its derivative $f'(x) = \cos(e^x) - e^{-x} \sin(e^x)$ does not have asymptotic expansion since $\lim_{x \to \infty} f'(x)$ does not exist. This example shows that the assumption made in Theorem 3 that $f'(x)$ has asymptotic expansion cannot be omitted.

Proof of Theorem 3.   Since $t^2(f(t) - a_0 - a_1/t) \to a_2$ as $t \to \infty$,

$$\left| f(t) - a_0 - \frac{a_1}{t} \right| \leq \frac{C}{t^2} \quad (C \text{ constant}).$$

Therefore the integral in (9) is convergent. Denote it by $\phi(x)$. For any positive integer n

$$\left[f(t) - a_0 - \frac{a_1}{t} \right] - \frac{a_2}{t^2} - \cdots - \frac{a_{n+1}}{t^{n+1}} = \frac{\delta(t)}{t^{n+1}}$$

where $\delta(t) \to 0$ if $t \to \infty$. Integrating both sides in the interval $x < t < N$, and taking $N \to \infty$, we get

$$\phi(x) - \frac{a_2}{x} - \cdots - \frac{a_{n+1}}{nx^n} = \int_x^\infty \frac{\delta(t)}{t^{n+1}} \, dt.$$

For any $\varepsilon > 0$ there is an $\bar{x}$ such that $\bar{x} > x_0$ and $|\delta(t)| < \varepsilon$ if $t > \bar{x}$. It follows that

$$\left| \int_x^\infty \frac{\delta(t)}{t^{n+1}} \, dt \right| \leq \varepsilon \int_x^\infty \frac{dt}{t^{n+1}} = \frac{\varepsilon}{nx^n} \quad \text{if } x > \bar{x}.$$

Consequently,

$$\overline{\lim_{x \to \infty}} \left| \left[\phi(x) - \frac{a_2}{x} - \cdots - \frac{a_{n+1}}{nx^n} \right] x^n \right| \leq \frac{\varepsilon}{n}.$$

Since ε is arbitrary,

$$\lim_{x \to \infty} \left[\phi(x) - \frac{a_2}{x} - \cdots - \frac{a_{n+1}}{nx^n} \right] x^n = 0.$$

This completes the proof of (9).

To prove (10), let

$$f'(x) \sim b_0 + \frac{b_1}{x} + \frac{b_2}{x^2} + \cdots + \frac{b_n}{x^n} + \cdots. \tag{11}$$

Then

$$
\begin{aligned}
f(x) &= \int_{x_0}^{x} f'(t)\, dt + f(x_0) \\
&= \int_{x_0}^{x} \left(b_0 + \frac{b_1}{t} \right) dt + f(x_0) + \int_{x_0}^{x} \left[f'(t) - b_0 - \frac{b_1}{t} \right] dt \\
&= b_0 x + b_1 \log x + B - \int_{x}^{\infty} \left[f'(t) - b_0 - \frac{b_1}{t} \right] dt \tag{12}
\end{aligned}
$$

where

$$B = -b_0 x_0 - b_1 \log x_0 + f(x_0) + \int_{x_0}^{\infty} \left[f'(t) - b_0 - \frac{b_1}{t} \right] dt.$$

By the first part of Theorem 3, the last integral on the right-hand side of (12) has asymptotic expansion

$$\frac{b_2}{x} + \frac{b_3}{2x} + \cdots + \frac{b_{n+1}}{nx^n} + \cdots.$$

Therefore

$$\lim_{x \to \infty} [f(x) - b_0 x - b_1 \log x - B] = 0,$$

$$\lim_{x \to \infty} \left[f(x) - b_0 x - b_1 \log x - B - \frac{b_2}{x} - \frac{b_3}{2x} - \cdots - \frac{b_{n+1}}{nx^n} \right] x^n = 0$$

$$(n = 1, 2, \ldots).$$

Comparing this with (2) (for $n = 0, 1, 2, \ldots$), we find that $b_0 = b_1 = 0$, and $b_n = -(n-1)a_{n-1}$ for $n = 1, 2, \ldots$. Thus (11) reduces to (10).

Notation. If

$$\frac{F(x) - f(x)}{g(x)} \sim a_0 + \frac{a_1}{x} + \frac{a_2}{x^2} + \cdots + \frac{a_n}{x^n} + \cdots,$$

then we write

$$F(x) \sim f(x) + g(x)\left(a_0 + \frac{a_1}{x} + \frac{a_2}{x^2} + \cdots + \frac{a_n}{x^n} + \cdots \right). \tag{13}$$

PROBLEMS

1. Find asymptotic expansions (as $x \to \infty$) for

   (a) $\dfrac{1}{x^2 + 1}$;

   (b) $\dfrac{\cos x}{x^2 + 1}$;

   (c) $\dfrac{1}{x^3 + x + 1}$;

   (d) $\dfrac{\sin x}{x^2 + 1}$.

2. If $f(x) = \displaystyle\sum_{n=0}^{\infty} a_n x^n$ and the series is convergent for $|x| \leq 1$, then

$$f\left(\frac{1}{x}\right) \sim a_0 + \frac{a_1}{x} + \cdots + \frac{a_n}{x^n} + \cdots.$$

3. If $f(x) \sim a_0 + a_1/x + a_2/x^2$ and $a_0 \neq 0$, then

$$\frac{1}{f} \sim \frac{1}{a_0} - \frac{a_1}{a_0^2} \cdot \frac{1}{x} + \frac{a_1^2 - a_0 a_2}{a_0^3} \cdot \frac{1}{x^2}.$$

4. Prove that

$$\exp\left(a_0 + \frac{a_1}{x} + \frac{a_2}{x^2}\right) \sim (\exp a_0)\left(1 + \frac{a_1}{x} + \frac{\frac{1}{2}a_1^2 + a_2}{x^2}\right).$$

5. Show that

$$\int_0^{\infty} \frac{e^{-u}}{u + x}\, du \sim \frac{1}{x} - \frac{1}{x^2} + \frac{2!}{x^3} - \cdots + (-1)^n \frac{n!}{x^{n+1}} + \cdots.$$

$$\left[Hint: \quad \frac{1}{u + x} = \frac{1}{x} - \frac{u}{x^2} + \frac{u^2}{x^3} - \cdots + (-1)^n \frac{u^n}{x^{n+1}} + (-1)^{n+1} \frac{u^{n+1}}{x^{n+2}} \frac{1}{1 + (u/x)}. \right]$$

6. Show that

$$\int_x^{\infty} e^{-t^2}\, dt \sim e^{-x^2}\left(\frac{1}{2x} - \frac{1}{2^2 x^3} + \frac{1 \cdot 3}{2^3 x^5} - \frac{1 \cdot 3 \cdot 5}{2^4 x^7} + \cdots\right).$$

7. Show that

$$\int_x^{\infty} e^{-t} t^{a-1}\, dt \sim e^{-x} x^a\left[\frac{1}{x} + \frac{a - 1}{x^2} + \frac{(a - 1)(a - 2)}{x^3} + \cdots\right].$$

3. ASYMPTOTIC BEHAVIOR OF INTEGRALS DEPENDING ON A PARAMETER

Let $F(x)$ and $G(x)$ be functions defined for all x sufficiently large. If

$$\lim_{x \to \infty} \frac{F(x)}{G(x)} = 1, \tag{1}$$

then we write

$$F \sim G \qquad \text{as } x \to \infty, \tag{2}$$

and say that $F(x)$ is *asymptotically equal* to $G(x)$, or, briefly, $F(x)$ is *asymptotic* to $G(x)$, as $x \to \infty$. The notation (2) fits in with the notation (13) of Section 2 specialized to $g = G$, $f = 0$, $a_0 = 1$, and $a_n = 0$ if $n = 1,2,\ldots$. Note that we have not assumed that $\lim_{x \to \infty} F(x)$ and $\lim_{x \to \infty} G(x)$ exist.

EXAMPLE 1. $\sinh x \sim \dfrac{1}{2} e^x$ as $x \to \infty$.

EXAMPLE 2.   For any $x > 0$, denote by $\pi(x)$ the number of prime numbers that are smaller than x. Then the prime number theorem asserts that

$$\pi(x) \sim \frac{x}{\log x}.$$

We write

$$f(x) = 0(x^\alpha) \text{ as } x \to \infty \qquad (\alpha \text{ fixed real number}),$$

if there is a constant C such that $|f(x)| \leq Cx^\alpha$ for all x sufficiently large. This is equivalent to the statement:

$$\varlimsup_{x \to \infty} \frac{|f(x)|}{x^\alpha} < \infty.$$

If

$$\lim_{x \to \infty} \frac{|f(x)|}{x^\alpha} = 0,$$

then we write

$$f(x) = o(x^\alpha) \qquad \text{as } x \to \infty.$$

Thus, (1) is equivalent to

$$\frac{F(x)}{G(x)} = 1 + o(1) \qquad \text{as } x \to \infty.$$

We shall be interested in the asymptotic behavior of integrals of the form

$$\int_0^\alpha \phi(x,t) \, dt$$

as $x \to \infty$. Such integrals occur in many applications. There are several different methods of treating such integrals. We shall give here the *Laplace method*. The main result is stated in the following theorem.

THEOREM 1. *Let $f(t)$ be a function having two continuous derivatives in an interval $0 \leq t < \alpha$ where α is either a real number or $\alpha = \infty$. Assume:*

(i) *$f(t)$ is monotone increasing in some interval $0 \leq t < \delta_0$, and $f(t) \geq f(\delta_0)$ if $\delta_0 \leq t < \alpha$.*

(ii) *$f(0) = f'(0) = 0, f''(0) > 0$.*

Let $g(t)$ be a continuous function in $0 \leq t < \alpha$. If the integral

$$I(x) = \int_0^\alpha g(t) e^{-x f(t)} \, dt \tag{3}$$

is absolutely convergent for some point $x = x_0 \geq 0$, then it is also absolutely convergent for any $x > x_0$, and

$$\sqrt{x} \, I(x) \to g(0) \sqrt{\frac{\pi}{2 f''(0)}} \qquad \text{as } x \to \infty. \tag{4}$$

From (4) it follows that if $g(0) \neq 0$, then

$$I(x) \sim g(0) \sqrt{\frac{\pi}{2 x f''(0)}} \qquad \text{as } x \to \infty. \tag{5}$$

Proof. We begin with a very special case:

$$J(x) = \int_0^\beta e^{-x b t^2} \, dt \qquad (\beta > 0, b > 0).$$

If we substitute $s = \sqrt{xb} \, t$, then we get

$$\sqrt{bx} \, J(x) = \int_0^{\sqrt{bx} \, \beta} e^{-s^2} \, ds \to \frac{\sqrt{\pi}}{2} \qquad \text{if } x \to \infty.$$

Thus

$$\sqrt{x} \int_0^\beta e^{-x b t^2} \, dt \to \frac{1}{2} \sqrt{\frac{\pi}{b}} \qquad \text{as } x \to \infty. \tag{6}$$

Consider now the general case. Since

$$|g(t) e^{-x f(t)}| \leq |g(t)| \, e^{-x_0 f(t)} \qquad (x > x_0)$$

and since the integral

$$\int_0^\beta |g(t)| \, e^{-x_0 f(t)} \, dt$$

is convergent, we deduce, by the comparison test, that the integral (3) is absolutely convergent if $x > x_0$.

Let ε be any number satisfying $0 < \varepsilon < f''(0)$. By continuity of $f''(t)$ and of $g(t)$, there is a $\delta > 0$ such that $\delta < \alpha$,

$$f''(0) - \varepsilon < f''(t) < f''(0) + \varepsilon \qquad \text{if } 0 \le t \le \delta, \tag{7}$$

and

$$g(0) - \varepsilon < g(t) < g(0) + \varepsilon \qquad \text{if } 0 \le t \le \delta. \tag{8}$$

By Taylor's formula, if $0 < t \le \delta$, then

$$f(t) = f(0) + f'(0)t + f''(\theta t)\frac{t^2}{2} \qquad (0 < \theta < 1).$$

Recalling that $f(0) = f'(0) = 0$ and using (7), we deduce

$$[f''(0) - \varepsilon]\frac{t^2}{2} \le f(t) \le [f''(0) + \varepsilon]\frac{t^2}{2} \qquad (0 \le t \le \delta). \tag{9}$$

We may assume that δ has been chosen so small that $\delta < \delta_0$, where δ_0 is the number occurring in the assumption (i). Since $f(t)$ is monotone for $0 \le t \le \delta_0$ and $f(t) \ge f(\delta_0)$ if $\delta_0 < t < \alpha$, it follows that

$$f(t) \ge f(\delta_0) \ge f(\delta) \qquad \text{if } \delta_0 < t < \alpha,$$
$$f(t) \ge f(\delta) \qquad \text{if } \delta \le t \le \delta_0.$$

Since, by (9), $f(\delta) > 0$, we conclude that

$$f(t) \ge f(\delta) > 0 \qquad \text{if } \delta \le t < \alpha. \tag{10}$$

We now break up the integral $I(x)$ into two integrals:

$$I(x) = I_1(x) + I_2(x),$$

$$I_1(x) = \int_0^\delta g(t)e^{-xf(t)}\,dt, \qquad I_2(x) = \int_\delta^\alpha g(t)e^{-xf(t)}\,dt. \tag{11}$$

Using (10) it follows that, for $\delta \le t < \alpha$,

$$e^{-xf(t)} = e^{-xf(t)/2}e^{-xf(t)/2} \le e^{-x_0f(t)}e^{-xf(\delta)/2}$$

if $x \ge 2x_0$. Consequently,

$$|I_2(x)| \le e^{-xf(\delta)/2}\int_\delta^\alpha |g(t)|\,e^{-x_0f(t)}\,dt \le Ke^{-xf(\delta)/2} \tag{12}$$

where

$$K = \int_0^\alpha |g(t)|\,e^{-x_0f(t)}\,dt.$$

To evaluate $I_1(x)$ we multiply both sides of (8) by $e^{-xf(t)}$ and integrate over t, $0 \le t \le \delta$. We get

$$[g(0) - \varepsilon]\int_0^\delta e^{-xf(t)}\,dt \le I_1(x) \le [g(0) + \varepsilon]\int_0^\delta e^{-xf(t)}\,dt.$$

Estimating $f(t)$ by (9), we find

$$[g(0) - \varepsilon] \int_0^\delta e^{-x[f''(0)+\varepsilon]t^2/2} \, dt \leq I_1(x) \leq [g(0) + \varepsilon] \int_0^\delta e^{-x[f''(0)-\varepsilon]t^2/2} \, dt.$$

Combining this with (12) and (11), we obtain

$$[g(0) - \varepsilon] \int_0^\delta e^{-x[f''(0)+\varepsilon]t^2/2} \, dt - Ke^{-xf(\delta)/2}$$

$$\leq I(x) \leq [g(0) + \varepsilon] \int_0^\delta e^{-x[f''(0)-\varepsilon]t^2/2} \, dt + Ke^{-xf(\delta)/2}. \tag{13}$$

Using the special case (6) and the fact that

$$\sqrt{x}\, e^{-xf(\delta)/2} \to 0 \qquad \text{if } x \to \infty,$$

we find, after multiplying both sides of (13) by $\sqrt{x}$ and letting $x \to \infty$,

$$[g(0) - \varepsilon]\sqrt{\frac{\pi}{2[f''(0) + \varepsilon]}} \leq \varliminf_{x \to \infty} \sqrt{x}\, I(x)$$

$$\leq \varlimsup_{x \to \infty} I(x) \leq [g(0) + \varepsilon]\sqrt{\frac{\pi}{2[f''(0) - \varepsilon]}}.$$

This is true for any $0 < \varepsilon < f''(0)$. Letting $\varepsilon \to 0$, we get

$$g(0)\sqrt{\frac{\pi}{2f''(0)}} \leq \varliminf_{x \to \infty} \sqrt{x}\, I(x) \leq \varlimsup_{x \to \infty} \sqrt{x}\, I(x) \leq g(0)\sqrt{\frac{\pi}{2f''(0)}}.$$

This gives (4).

As an application of Theorem 1 we shall prove the *generalized Stirling's formula*:

THEOREM 2.   *As* $x \to \infty$,

$$\Gamma(x + 1) \sim \sqrt{2\pi}\, x^{x + 1/2} e^{-x}. \tag{14}$$

If we take, in (14), $x = n$, n a positive integer, then we obtain Stirling's formula.

Proof.   Recall that

$$\Gamma(x + 1) = \int_0^\infty e^{-u} u^x \, du.$$

Substituting $u = x(1 + t)$, we get

$$\Gamma(x + 1) = \int_{-1}^\infty e^{-x(1+t)} [x(1 + t)]^x x \, dt = x^{x+1} e^{-x} \int_{-1}^\infty e^{-xt}(1 + t)^x \, dt.$$

Hence

$$x^{-x-1}e^x\Gamma(x+1) = \int_{-1}^{\infty} e^{-xt}(1+t)^x \, dt$$

$$= \int_{-1}^{0} e^{-xt}(1+t)^x \, dt + \int_{0}^{\infty} e^{-xt}(1+t)^x \, dt \qquad (15)$$

$$\equiv I_1(x) + I_2(x).$$

As for the integral $I_1(x)$, we substitute $t = -s$ and obtain

$$I_1(x) = \int_0^1 e^{xs}(1-s)^x \, ds = \int_0^1 e^{-x[-s-\log(1-s)]} \, ds.$$

We now can apply Theorem 1 with $f(t) = -t - \log(1-t)$, $g(t) = 1$. Indeed, $f(0) = f'(0) = 0$, $f'(t) = t/(1-t) > 0$ if $0 < t < 1$ so that $f(t)$ is monotone increasing, and $f''(t) = 1/(1-t)^2$ so that $f''(0) = 1$. We get

$$I_1(x) \sim \sqrt{\frac{\pi}{2x}}. \qquad (16)$$

Next we write $I_2(x)$ in the form

$$I_2(x) = \int_0^{\infty} e^{-x[t-\log(1+t)]} \, dt.$$

It is easily verified that all the assumptions of Theorem 1 are satisfied with $f(t) = t - \log(1+t)$, $g(t) = 1$, $\alpha = \infty$. Applying this theorem, we get

$$I_2(x) \sim \sqrt{\frac{\pi}{2x}}.$$

Combining this with (16) and (15), the assertion (14) follows.

PROBLEMS

1. Find the values of α and β for which the following relations hold:

   (a) $x \sin(1/x) = O(x^\alpha)$ as $x \to \infty$;

   (b) $x^2 \sin(1/\sqrt{x}) = o(x^\beta)$ as $x \to \infty$.

2. Prove the following asymptotic relations as $x \to \infty$:

   (a) $\displaystyle\int_0^1 (\cos t)^x \, dt \sim \sqrt{\pi/2x}$;   (b) $\displaystyle\int_0^1 (1-t^2)^x \cos t \, dt \sim \sqrt{\pi/4x}$;

   (c) $\displaystyle\int_0^1 e^{x\cos t} \, dt \sim e^x\sqrt{\pi/2x}$;   (d) $[\Gamma(x+1)]^{1/x} \sim x/e$.

3. The *Laplace transform* of a function $g(t)$ is the function $\hat{g}(x) = \int_0^\infty e^{-xt}g(t)\, dt$. Prove that if $g(t) = h(t)/\sqrt{t}$ where $h(t)$ is continuous and bounded in $0 \le t < \infty$, and if $h(0) \ne 0$, then

$$\hat{g}(x) \sim \sqrt{\frac{\pi}{x}}\, h(0) \qquad \text{as } x \to \infty.$$

4. Prove

   (a) $\Gamma(x + \alpha) \sim \sqrt{2\pi}\, x^{x+\alpha-1/2} e^{-x} \qquad \text{as } x \to \infty;$

   (b) $\dfrac{\Gamma(x + \alpha)}{\Gamma(x + \beta)} \sim x^{\alpha-\beta} \qquad \text{as } x \to \infty;$

   (c) $\Gamma(x) = \lim\limits_{n \to \infty} \dfrac{n!\, n^x}{x(x + 1)\cdots(x + n)} \qquad \text{if } x \ne -m,\, m = 0,1,2,\dots.$

5. Let $f(t)$ be a function whose first $2n$ derivatives are continuous for $0 \le t < \alpha$, where α is either a positive number or $\alpha = \infty$. Assume that $f(t)$ is monotone increasing in some interval $0 \le t < \delta_0$ and that $f(t) \ge f(\delta_0)$ if $\delta_0 \le t < \alpha$. Assume also that

$$f^{(j)}(0) = 0 \qquad \text{if } j = 0,1,\dots,2n - 1; \qquad f^{(2n)}(0) > 0.$$

Let $g(t)$ be a continuous function in $0 \le t < \alpha$. Prove: If the integral

$$\int_0^\alpha g(t) e^{-xf(t)}\, dt$$

is absolutely convergent for some $x = x_0$, then it is also absolutely convergent for any $x > x_0$, and

$$x^{1/2n} \int_0^\alpha g(t) e^{-xf(t)}\, dt \to \frac{1}{2n} \Gamma\!\left(\frac{1}{n}\right) \left[\frac{(2n)!}{f^{(2n)}(0)} \right]^{1/2n} g(0).$$

[*Hint:* Extend step by step the proof of Theorem 1; make use of Problem 3 in Section 8.9].

4. FOURIER SERIES

A series of the form

$$\frac{1}{2} a_0 + \sum_{n=1}^{\infty} (a_n \cos nx + b_n \sin nx) \tag{1}$$

is called a *trigonometric series*. Suppose the series is convergent for each x, and denote its sum by $f(x)$. Then $f(x)$ is a *periodic function* with period 2π, that is, $f(x + 2\pi) = f(x)$.

Using the formulas

$$2 \cos \alpha \sin \beta = \sin(\alpha + \beta) - \sin(\alpha - \beta),$$
$$2 \sin \alpha \sin \beta = \cos(\alpha - \beta) - \cos(\alpha + \beta),$$
$$2 \cos \alpha \cos \beta = \cos(\alpha + \beta) + \cos(\alpha - \beta),$$

we find that, for any nonnegative integers m and n,

$$\int_0^{2\pi} \sin mx \cos nx \, dx = 0,$$

$$\frac{1}{\pi} \int_0^{2\pi} \sin mx \sin nx \, dx = \begin{cases} 0 & \text{if } m \neq n, \\ 1 & \text{if } m = n \geq 1, \end{cases}$$

$$\frac{1}{\pi} \int_0^{2\pi} \cos mx \cos nx \, dx = \begin{cases} 0 & \text{if } m \neq n, \\ 1 & \text{if } m = n \geq 1, \\ 2 & \text{if } m = n = 0. \end{cases} \tag{2}$$

If the trigonometric series (1) is uniformly convergent to $f(x)$ then, by multiplying it by $\cos nx$ and integrating term by term over the interval $0 \leq x \leq 2\pi$, we get, after making use of the formulas (2),

$$a_n = \frac{1}{\pi} \int_0^{2\pi} f(x) \cos nx \, dx \qquad (n = 0,1,2,...). \tag{3}$$

Similarly,

$$b_n = \frac{1}{\pi} \int_0^{2\pi} f(x) \sin nx \, dx \qquad (n = 1,2,...). \tag{4}$$

DEFINITION.   Let $f(x)$ be an integrable function defined in the interval $0 < x < 2\pi$. The numbers a_n, b_n defined by (3) and (4) are called the *Fourier coefficients* of f, and the trigonometric series (1) [with a_n, b_n defined by (3) and (4)] is called the *Fourier series* of f. We write

$$f(x) \sim \frac{1}{2} a_0 + \sum_{n=1}^{\infty} (a_n \cos nx + b_n \sin nx). \tag{5}$$

For each x, the Fourier series may either converge or diverge.

From the discussion given above it follows that if a trigonometric series (1) is uniformly convergent to a function $f(x)$, then the trigonometric series coincides with the Fourier series of $f(x)$.

EXAMPLE 1.   Let $f(x) = x$ for $0 < x < 2\pi$. Then $(1/\pi) \int_0^{2\pi} x \, dx = 2\pi$;

$$\frac{1}{\pi} \int_0^{2\pi} x \cos nx \, dx = \left[\frac{1}{\pi} \cdot \frac{x \sin nx}{n} \right]_0^{2\pi} - \frac{1}{\pi} \int_0^{2\pi} \frac{\sin nx}{n} \, dx = \left[\frac{1}{\pi} \cdot \frac{\cos nx}{n^2} \right]_0^{2\pi} = 0$$

$$\text{if } n \geq 1;$$

$$\frac{1}{\pi} \int_0^{2\pi} x \sin nx \, dx = \left[-\frac{1}{\pi} \cdot \frac{x \cos nx}{n} \right]_0^{2\pi} + \frac{1}{\pi} \int_0^{2\pi} \frac{\cos nx}{n} \, dx$$

$$= -\frac{1}{\pi} \cdot \frac{2\pi}{n} + \left[\frac{1}{\pi} \cdot \frac{\sin nx}{n^2} \right]_0^{2\pi} = -\frac{2}{n}.$$

Thus

$$x \sim \pi - 2 \sum_{n=1}^{\infty} \frac{\sin nx}{n}. \tag{6}$$

EXAMPLE 2.   Let $f(x)$ be a function defined on the real line. f is said to be an *even* function if $f(-x) = f(x)$, and an *odd* function if $f(-x) = -f(x)$. Now, for any periodic function $g(x)$ with period 2π, and for any number α,

$$\int_0^{\alpha} g(x) \, dx = \int_0^{\alpha} g(x + 2\pi) \, dx = \int_{2\pi}^{2\pi + \alpha} g(y) \, dy.$$

Therefore

$$\int_0^{2\pi} g(x) \, dx = \int_0^{\alpha} g(x) \, dx + \int_{\alpha}^{2\pi} g(x) \, dx = \int_{2\pi}^{2\pi + \alpha} g(x) \, dx + \int_{\alpha}^{2\pi} g(x) \, dx$$

$$= \int_{\alpha}^{2\pi + \alpha} g(x) \, dx. \tag{7}$$

If, in particular, $g(x)$ is an odd function, then

$$\int_0^{2\pi} g(x) \, dx = \int_{-\pi}^{\pi} g(x) \, dx = \int_0^{\pi} g(x) \, dx + \int_{-\pi}^{0} g(x) \, dx.$$

Substituting in the last integral $x = -y$ and using the fact that $g(-y) = -g(y)$, we find that the last integral reduces to

$$-\int_0^{\pi} g(x) \, dx.$$

Consequently,

$$\int_0^{2\pi} g(x) \, dx = 0 \qquad \text{if } g(x) \text{ is periodic (period } 2\pi) \text{ and odd.} \tag{8}$$

Suppose now that $f(x)$ is periodic with period 2π. If $f(x)$ is odd, then $f(x)\cos nx$ is also odd and, by (8),

$$a_n = \frac{1}{\pi} \int_0^{2\pi} f(x) \cos nx \, dx = 0 \qquad \text{for } n = 0,1,2,\dots .$$

It follows that

$$f(x) \sim \sum_{n=1}^{\infty} b_n \sin nx.$$

This Fourier series for f is called a *sine series*. Similarly, if $f(x)$ is an even function, then its Fourier series is a *cosine series*:

$$f(x) \sim \frac{1}{2} a_0 + \sum_{n=1}^{\infty} a_n \cos nx.$$

Let

$$T_n(x) = \frac{c_0}{2} + \sum_{k=1}^{n} (c_k \cos kx + d_k \sin kx)$$

where the c_k, d_k are any numbers.

PROBLEM

Given an integrable function $f(x)$ $(0 < x < 2\pi)$ and a positive integer n, find c_k, d_k such that T_n gives the " best approximation " to f in the sense that it minimizes the integral

$$\int_0^{2\pi} [f(x) - T_n(x)]^2 \, dx.$$

The answer is given in the following theorem:

THEOREM 1. *Let*

$$S_n(x) = \frac{1}{2} a_0 + \sum_{k=1}^{n} (a_k \cos kx + b_k \sin kx) \tag{9}$$

where the a_k, b_k are the Fourier coefficients of f. Then, for any numbers c_k, d_k,

$$\int_0^{2\pi} [f(x) - T_n(x)]^2 \, dx \geq \int_0^{2\pi} [f(x) - S_n(x)]^2 \, dx. \tag{10}$$

Equality holds if, and only if, $c_k = a_k$ and $d_k = b_k$ for all k.

Proof.   Using (2) and the definition of a_k, b_k, we have

$$\int_0^{2\pi} [f(x) - T_n(x)]^2 \, dx = \int_0^{2\pi} \left[f(x) - \left(\frac{c_0}{2} + \sum_{k=1}^n (c_k \cos kx + d_k \sin kx) \right) \right]^2 dx$$

$$= \int_0^{2\pi} f^2(x) \, dx - 2 \int_0^{2\pi} f(x) \left[\frac{c_0}{2} + \sum_{k=1}^n (c_k \cos kx + d_k \sin kx) \right] dx$$

$$+ \int_0^{2\pi} \left[\frac{c_0}{2} + \sum_{k=1}^n (c_k \cos kx + d_k \sin kx) \right]$$

$$\times \left[\frac{c_0}{2} + \sum_{m=1}^n (c_m \cos mx + d_m \sin mx) \right] dx$$

$$= \int_0^{2\pi} f^2(x) \, dx - \pi \left[a_0 c_0 + 2 \sum_{k=1}^n (a_k c_k + b_k d_k) \right]$$

$$+ \pi \left[\frac{c_0^2}{2} + \sum_{k=1}^n (c_k^2 + d_k^2) \right]$$

$$= \int_0^{2\pi} f^2(x) \, dx - \pi \left[\frac{a_0^2}{2} + \sum_{k=1}^n (a_k^2 + b_k^2) \right]$$

$$+ \pi \left[\frac{(c_0 - a_0)^2}{2} + \sum_{k=1}^n ((c_k - a_k)^2 + (d_k - b_k)^2) \right].$$

Specializing the above computations to $T_n = S_n$, we get

$$\int_0^{2\pi} [f(x) - S_n(x)]^2 \, dx = \int_0^{2\pi} f^2(x) \, dx - \pi \left[\frac{a_0^2}{2} + \sum_{k=1}^n (a_k^2 + b_k^2) \right]. \qquad (11)$$

Consequently,

$$\int_0^{2\pi} [f(x) - T_n(x)]^2 \, dx = \int_0^{2\pi} [f(x) - S_n(x)]^2 \, dx$$

$$+ \pi \left\{ \frac{(c_0 - a_0)^2}{2} + \sum_{k=1}^n [(c_k - a_k)^2 + (d_k - b_k)^2] \right\}.$$

This gives the assertion of the theorem.

From (11) it follows that

$$\frac{a_0^2}{2} + \sum_{k=1}^n (a_k^2 + b_k^2) \leq \frac{1}{\pi} \int_0^{2\pi} f^2(x) \, dx.$$

Taking $n \to \infty$, we obtain *Bessel's inequality*:

$$\frac{a_0^2}{2} + \sum_{k=1}^\infty (a_k^2 + b_k^2) \leq \frac{1}{\pi} \int_0^{2\pi} f^2(x) \, dx. \qquad (12)$$

We now come to the main object of this section, namely, establishing the convergence of a Fourier series of f, under some conditions on f.

DEFINITION.   Suppose there is a partition

$$\tau_0 = \alpha < \tau_1 < \tau_2 < \cdots < \tau_{r-1} < \tau_r = \beta \tag{13}$$

of an interval $[\alpha,\beta]$, such that $f(x)$ and $f'(x)$ are uniformly continuous in each interval (τ_{i-1},τ_i). Then we say that f and f' are *piecewise continuous* in (α,β). We shall need the following:

RIEMANN–LEBESGUE LEMMA. *Let g and g' be piecewise continuous functions in an interval (α,β). Then*

$$\int_\alpha^\beta g(x)\cos nx\,dx \to 0 \qquad if\ n \to \infty, \tag{14}$$

$$\int_\alpha^\beta g(x)\sin nx\,dx \to 0 \qquad if\ n \to \infty. \tag{15}$$

The lemma is true for any integrable function g, but we shall need it only in the special case where g and g' are piecewise continuous.

Proof.   We introduce a partition (13) of $[\alpha,\beta]$ such that g and g' are uniformly continuous in each interval (τ_{i-1},τ_i). To prove (14) it suffices to show that

$$\int_{\tau_{i-1}}^{\tau_i} g(x)\cos nx\,dx \to 0 \qquad if\ n \to \infty. \tag{16}$$

By integration by parts,

$$\int_{\tau_{i-1}}^{\tau_i} g(x)\cos nx\,dx = \left[\frac{g(x)\sin nx}{n}\right]_{\tau_{i-1}+0}^{\tau_i-0} - \frac{1}{n}\int_{\tau_{i-1}}^{\tau_i} g'(x)\sin nx\,dx.$$

Since $g(x)$ and $g'(x)$ are uniformly bounded in the interval (τ_{i-1},τ_i), it follows that

$$\left|\int_{\tau_{i-1}}^{\tau_i} g(x)\cos nx\,dx\right| \le \frac{C}{n} \qquad (C\ \text{constant}).$$

This yields (16) and, therefore, also (14).

The proof of (15) is similar.

We can now state the main result on the convergence of a Fourier series:

THEOREM 2. *Let f be a periodic function with period 2π. Assume that f and f' are piecewise continuous in $0 < x < 2\pi$. Denote by $S_n(x)$ the partial sum of the Fourier series of $f(x)$. Then, for any x,*

$$\lim_{n\to\infty} S_n(x) = \frac{f(x+0) + f(x-0)}{2}. \tag{17}$$

COROLLARY. *At each point x where f is continuous,*

$$f(x) = \frac{1}{2} a_0 + \sum_{m=1}^{\infty} (a_m \cos mx + b_m \sin mx). \tag{18}$$

Proof. Set

$$D_n(\theta) = \frac{1}{2} + \cos \theta + \cos 2\theta + \cdots + \cos n\theta. \tag{19}$$

It can be verified by induction on n that

$$D_n(\theta) = \frac{\sin(n + 1/2)\theta}{2 \sin \frac{1}{2}\theta}. \tag{20}$$

We write $S_n(x)$ in a different form:

$$S_n(x) = \frac{1}{2\pi} \int_0^{2\pi} f(t)\, dt$$

$$+ \frac{1}{\pi} \sum_{k=1}^{n} \left[\cos kx \int_0^{2\pi} f(t) \cos kt\, dt + \sin kx \int_0^{2\pi} f(t) \sin kt\, dt \right]$$

$$= \frac{1}{\pi} \int_0^{2\pi} f(t) \left[\frac{1}{2} + \sum_{k=1}^{n} (\cos kx \cos kt + \sin kx \sin kt) \right] dt$$

$$= \frac{1}{\pi} \int_0^{2\pi} f(t) \left[\frac{1}{2} + \sum_{k=1}^{n} \cos k(t - x) \right] dt$$

$$= \frac{1}{\pi} \int_0^{2\pi} f(t) D_n(t - x)\, dt.$$

Substituting $t - x = u$, the last integral becomes

$$\int_{-x}^{2\pi - x} f(x + u) D_n(u)\, du.$$

Since $f(x + u)$ and $D_n(u)$ are both periodic in u with period 2π, we can apply (7) with $g(u) = f(x + u)D_n(u)$ and $\alpha = -x$. We conclude that the last integral is equal to

$$\int_0^{2\pi} f(x + u)D_n(u)\, du.$$

We have thus proved that

$$S_n(x) = \frac{1}{\pi} \int_0^{2\pi} f(x + u)D_n(u)\, du. \tag{21}$$

Consider the integral

$$\int_\pi^{2\pi} f(x + u)D_n(u)\, du.$$

If we substitute $u = 2\pi - v$ and make use of the relations $f(x + 2\pi - v) = f(x - v)$, $D_n(2\pi - v) = D_n(v)$, we get

$$\int_\pi^{2\pi} f(x + u)D_n(u)\, du = \int_0^\pi f(x - v)D_n(v)\, dv. \tag{22}$$

Writing (21) in the form

$$S_n(x) = \frac{1}{\pi} \int_0^\pi f(x + u)D_n(u)\, du + \frac{1}{\pi} \int_\pi^{2\pi} f(x + u)D_n(u)\, du$$

and using (22), we find that

$$S_n(x) = \frac{2}{\pi} \int_0^\pi \frac{f(x + u) + f(x - u)}{2}\, D_n(u)\, du. \tag{23}$$

From the definition of $D_n(\theta)$ it immediately follows that

$$\frac{2}{\pi} \int_0^\pi D_n(\theta)\, d\theta = 1. \tag{24}$$

Therefore

$$S_n(x) - \frac{f(x + 0) + f(x - 0)}{2} = \frac{2}{\pi} \int_0^\pi g(x,u)D_n(u)\, du \tag{25}$$

where

$$g(x,u) = \frac{f(x + u) + f(x - u)}{2} - \frac{f(x + 0) + f(x - 0)}{2}. \tag{26}$$

By the mean value theorem, for any $x + u$ in some interval $(x, x + \gamma_1)$ where both f and f' are uniformly continuous,

$$|f(x + u) - f(x + 0)| = |f'(x + \theta u)|u \le M|u|$$

where M is a constant independent of x,u. Similarly, for any $x - u$ in some interval $(x - \gamma_2, x)$,

$$|f(x - u) - f(x - 0)| \le Mu.$$

It follows that for any u in $(0,\gamma)$, where $\gamma = \min(\gamma_1, \gamma_2)$,

$$|g(x,u)| \le Mu. \tag{27}$$

We shall establish the assertion (17) on the basis of (25). For any $0 < \delta < \pi$, $\delta < \gamma$, we break the integral in (25) into two parts:

$$\int_0^\delta g(x,u)D_n(u)\, du + \int_\delta^\pi g(x,u)D_n(u)\, du \equiv I_1(x) + I_2(x). \tag{28}$$

Using (20), we can write

$$|I_1(x)| \le \int_0^\delta \frac{|g(x,u)|}{u} \frac{n}{2\sin\frac{1}{2}u}\left|\sin\left(n + \frac{1}{2}\right)u\right|\, du.$$

The function $u/(2\sin\frac{1}{2}u)$ is monotone increasing in the interval $0 \le u \le \pi$, and its value at $u = \pi$ is $\pi/2$. Hence

$$\frac{u}{2\sin\frac{1}{2}u} \le \frac{\pi}{2}.$$

We also have $|\sin(n + \frac{1}{2})u| \le 1$. Using these inequalities and (27), we get

$$|I_1(x)| \le \int_0^\delta M\frac{\pi}{2}\, du = \frac{M\pi}{2}\delta.$$

Let ε be any positive number. We fix δ such that $M\delta < \varepsilon/2$. Then

$$|I_1(x)| \le \frac{\pi}{2}\cdot\frac{\varepsilon}{2}. \tag{29}$$

In the interval $\delta \le u \le \pi$ the function

$$h(u) = \frac{g(x,u)}{2\sin\frac{1}{2}u}$$

and its derivative are piecewise continuous. Applying the Riemann–Lebesgue lemma, we conclude that

$$I_2(x) = \int_\delta^\pi \frac{g(x,u)}{2\sin\frac{1}{2}u}\sin\left(n + \frac{1}{2}\right)u\, du = 2\int_{\delta/2}^{\pi/2}\frac{g(x,2v)}{2\sin v}\sin(2n + 1)v\, dv \to 0$$

if $n \to \infty$. Hence, for some positive number N (depending on ε),

$$|I_2(x)| < \frac{\pi}{2}\cdot\frac{\varepsilon}{2} \quad \text{if } n \ge N.$$

Combining this with (29) and with (28) and (25), we see that

$$\left| S_n(x) - \frac{f(x+0) + f(x-0)}{2} \right| < \varepsilon \qquad \text{if } n \geq N.$$

This completes the proof of (17).

EXAMPLE. Applying Theorem 2 to the function in Example 1, we obtain the interesting formula

$$\sum_{n=1}^{\infty} \frac{\sin nx}{n} = \frac{\pi - x}{2} \qquad \text{if } 0 < x < 2\pi. \tag{30}$$

Substituting $x = \pi/2$, we get

$$\frac{1}{1} - \frac{1}{3} + \frac{1}{5} - \frac{1}{7} + - \cdots = \frac{\pi}{4}.$$

Under some conditions on f, one can show that its Fourier series can be integrated and differentiated term by term.

The expansion of a function into a Fourier series has many applications. For instance, it enables one to solve problems in partial differential equations. We shall give one example.

The problem is to find a function $u(x,t)$ satisfying:

$$u_{xx} - u_t = 0 \qquad \text{if } 0 < x < \pi, \quad 0 < t < T, \tag{31}$$

$$u(x,0) = f(x) \qquad \text{if } 0 < x < \pi, \tag{32}$$

$$u(0,t) = u(\pi,t) = 0 \quad \text{if } 0 < t < T, \tag{33}$$

such that $u(x,t)$ is continuous in the rectangle $0 \leq x \leq \pi$, $0 \leq t \leq T$. The function f is given.

To solve this problem, we assume that $f(0) = f(\pi) = 0$ and extend f to $(-\pi,0)$ as an odd function. We next extend it to the whole real line as a periodic function of period 2π. Since f is odd, its Fourier series is a sine series. Suppose f satisfies the conditions of Theorem 2 and suppose it is also continuous. Then

$$f(x) = \sum_{n=1}^{\infty} b_n \sin nx. \tag{34}$$

A formal solution of (31)–(33) now is given by

$$u(x,t) = \sum_{n=1}^{\infty} b_n e^{-n^2 t} \sin nx. \tag{35}$$

It can be proved that this is indeed a solution of (31)–(33).

PROBLEMS

1. Verify the formula (20).

2. Find the Fourier series of the function
$$f(x) = \begin{cases} 1 & \text{if } 0 < x < \pi, \\ -1 & \text{if } \pi < x < 2\pi. \end{cases}$$

Use Theorem 2 to deduce that

$$\sum_{n=1}^{\infty} \frac{\sin(2n+1)x}{2n+1} = \begin{cases} \pi/4 & \text{if } 0 < x < \pi, \\ 0 & \text{if } x = 0, x = \pi, \\ -\pi/4 & \text{if } \pi < x < 2\pi. \end{cases}$$

Substitute $x = \pi/2$, $x = \pi/3$, $x = \pi/6$ to obtain

$$1 - \frac{1}{3} + \frac{1}{5} - \frac{1}{7} + - \cdots = \frac{\pi}{4},$$

$$1 - \frac{1}{5} + \frac{1}{7} - \frac{1}{11} + \frac{1}{13} - + \cdots = \frac{\pi}{2\sqrt{3}},$$

$$1 + \frac{1}{5} - \frac{1}{7} - \frac{1}{11} + \frac{1}{13} + \frac{1}{17} - - + \quad + = \frac{\pi}{3}.$$

3. Find the Fourier series of the function
$$f(x) = \begin{cases} x & \text{if } 0 < x < \pi, \\ x - 2\pi & \text{if } \pi < x < 2\pi. \end{cases}$$

Use Theorem 2 to deduce that

$$\sum_{n=1}^{\infty} (-1)^{n+1} \frac{\sin nx}{n} = \begin{cases} \dfrac{x}{2} & \text{if } 0 \le x < \pi, \\ 0 & \text{if } x = \pi, \\ \dfrac{x}{2} - \pi & \text{if } \pi < x \le 2\pi. \end{cases}$$

4. Find the Fourier series of the function
$$f(x) = \begin{cases} x & \text{if } 0 < x < \pi, \\ 2\pi - x & \text{if } \pi < x < 2\pi, \end{cases}$$

and deduce that

$$\sum_{n=1}^{\infty} \frac{\cos(2n+1)x}{(2n+1)^2} - \begin{cases} \dfrac{\pi^2}{8} - \dfrac{\pi x}{4} & \text{if } 0 \le x \le \pi, \\ \dfrac{\pi x}{4} - \dfrac{3\pi^2}{8} & \text{if } \pi \le x \le 2\pi. \end{cases}$$

Taking $x = 0$, it follows that

$$1 + \frac{1}{3^2} + \frac{1}{5^2} + \frac{1}{7^2} + \cdots = \frac{\pi^2}{8}.$$

Use this relation to show that

$$1 + \frac{1}{2^2} + \frac{1}{3^2} + \frac{1}{4^2} + \frac{1}{5^2} + \cdots = \frac{\pi^2}{6}. \tag{36}$$

5. Let $f(x) = x^2/4$ for $-\pi < x < \pi$ and extend $f(x)$ to $(\pi, 2\pi)$ by $f(2\pi - x) = f(-x)$ $(0 < x < \pi)$. Prove that

$$f(x) = \frac{\pi^2}{12} - \left(\frac{\cos x}{1^2} - \frac{\cos 2x}{2^2} + \frac{\cos 3x}{3^2} - + \cdots \right).$$

Taking $x = \pi$, (36) follows. Taking $x = 0$, one gets

$$\frac{1}{1^2} - \frac{1}{2^2} + \frac{1}{3^2} - \frac{1}{4^2} + - \cdots = \frac{\pi^2}{12}.$$

6. Let $f(x) = |x|$ if $-\pi < x < \pi$, and extend $f(x)$ to $(\pi, 2\pi)$ by $f(2\pi - x) = f(-x)$ $(0 < x < \pi)$. Find the Fourier series of f and deduce the formula

$$x = \frac{\pi}{2} - \frac{4}{\pi} \sum_{n=1}^{\infty} \frac{\cos(2n+1)x}{(2n+1)^2}.$$

7. Prove that for any noninteger α,

$$\cos \alpha x = \frac{2\alpha \sin \pi\alpha}{\pi} \left[\frac{1}{2\alpha^2} + \sum_{n=1}^{\infty} (-1)^n \frac{\cos nx}{\alpha^2 - n^2} \right] \qquad (-\pi < x < \pi).$$

8. Prove that

$$e^x = 2 \frac{\sinh \pi}{\pi} \left[\frac{1}{2} + \sum_{n=1}^{\infty} (-1)^n \frac{\cos nx - n \sin nx}{1 + n^2} \right] \qquad (-\pi < x < \pi).$$

9. Let $f(x) = (\pi^2 - x^2)^2$ if $-\pi < x < \pi$, $f(2\pi - x) = f(-x)$ if $0 < x < \pi$. Show that

$$f(x) \sim \frac{8}{15} \pi^4 + 48 \sum_{n=1}^{\infty} \frac{(-1)^{n+1}}{n^4} \cos nx.$$

Deduce that

$$\frac{1}{1^4} - \frac{1}{2^4} + \frac{1}{3^4} - \frac{1}{4^4} + - \cdots = \frac{7\pi^4}{720}.$$

10. Let $f(t)$ be a periodic function of period 2π and assume that all the derivatives of f of orders $\leq m$ are continuous functions. Prove that the Fourier coefficients a_n, b_n of f satisfy

$$|a_n| \leq \frac{C}{n^m}, \qquad |b_n| \leq \frac{C}{n^m} \qquad (n = 1,2,...)$$

where C is a constant. Deduce that the Fourier series of f can be differentiated term by term $m - 2$ times.

11. Let f be given by (34) and assume that $\sum |a_n| < \infty$. Define $u(x,t)$ by (35). Prove that:

   (i) $u(x,t)$ is continuous in the rectangle $0 \leq x \leq \pi, 0 \leq t \leq T$;

   (ii) u_x, u_{xx}, u_t exist if $t > 0$ and (31) holds;

   (iii) u satisfies (32) and (33).

5. LAGRANGE MULTIPLIERS

Let $f(x)$ and $g_1(x),...,g_m(x)$ be continuous functions defined in a domain D of R^n. Let x^0 be a point in D having the following property: There is a neighborhood D_0 of x^0 contained in D such that $f(x) \geq f(x^0)$ for any x in D_0 for which

$$g_1(x) = 0,...,g_m(x) = 0. \qquad (1)$$

Then we call x^0 a point of *relative minimum* of f *subject to* (or *under*) *the conditions* (1). Similarly, we define the concepts of a point of relative maximum and of a point of relative extremum of f subject to the conditions (1).

The conditions (1) are also called *constraints*.

EXAMPLE. Find the box of largest volume that is contained inside the ellipsoid

$$\frac{x^2}{a^2} + \frac{y^2}{b^2} + \frac{z^2}{c^2} = 1, \qquad (2)$$

assuming that each edge of the box is parallel to one of the coordinate axes. It is clear that each corner of the box must lie on the ellipsoid. Denote by (x,y,z) the corner with positive coordinates. Then the volume of the box is $V = (2x)(2y)(2z) = 8xyz$. Thus the problem is to maximize $V = 8xyz$ subject to the constraint (2).

The method of *Lagrange multipliers* provides necessary conditions for a point to be a point of relative extremum. We give it first in the case $m = 1$.

THEOREM 1. *Let f and g be continuously differentiable functions in a domain D of R^n. Assume that* grad $g(x) \neq 0$ *for any x in D. If f attains a relative* *extremum under the constraint* $g = 0$ *at a point* x^0 *of* D, *then there exists a number* λ *such that*

$$\sum_{i=1}^{n} \frac{\partial f(x^0)}{\partial x_i} + \lambda \frac{\partial g(x^0)}{\partial x_i} = 0 \qquad (1 \leq i \leq n). \tag{3}$$

If we add to the system (3) the equation

$$g(x^0) = 0, \tag{4}$$

then we obtain $n + 1$ equations for the $n + 1$ unknowns $x_1{}^0,\ldots,\ x_n{}^0,\ \lambda$, where $(x_1{}^0,\ldots,x_n{}^0) = x^0$.

Proof. By assumption, grad $g(x^0) \neq 0$. Assume, for definiteness, that $\partial g(x^0)/\partial x_n \neq 0$. By the implicit function theorem, we can solve $g(x) = 0$ in a neighborhood of x^0 by $x_n = \phi(x_1,\ldots,x_{n-1})$, where ϕ is continuously differentiable. Set

$$x^0 = (x_1{}^0,\ldots,x_{n-1}^0,x_n{}^0) = (\bar{x}^0,x_n{}^0).$$

The function $f(x_1,\ldots,x_{n-1},\phi(x_1,\ldots,x_{n-1}))$ attains a relative extremum at the point $\bar{x}^0$. Therefore

$$\frac{\partial f(x^0)}{\partial x_i} + \frac{\partial f(x^0)}{\partial x_n} \frac{\partial \phi(\bar{x}^0)}{\partial x_i} = 0 \qquad (1 \leq i \leq n - 1). \tag{5}$$

Differentiating the identity $g(x_1,\ldots,x_{n-1},\phi(x_1,\ldots,x_{n-1})) = 0$ with respect to x_i and substituting $x_j = x_j{}^0\ (1 \leq j \leq n - 1)$, we get the equations

$$\frac{\partial g(x^0)}{\partial x_i} + \frac{\partial g(x^0)}{\partial x_n} \frac{\partial \phi(\bar{x}^0)}{\partial x_i} = 0 \qquad (1 \leq i \leq n - 1). \tag{6}$$

If we substitute $\partial \phi(\bar{x}^0)/\partial x_i$ from (6) into (5) and define

$$\lambda = - \frac{(\partial f(x^0))/(\partial x_n)}{(\partial g(x^0))/(\partial x_n)}$$

we obtain Equations (3).

Consider now the case of m constraints (1). We assume that $m \leq n$.

THEOREM 2. *Let f and $g_1,\ldots,g_m$ be continuously differentiable functions in a domain D of R^n. Assume that for any point x of D there are m numbers $i_1,\ldots,i_m$ among the numbers $1,2,\ldots,n$ such that*

$$\frac{\partial(g_1,\ldots,g_m)}{\partial(x_{i_1},\ldots,x_{i_m})} \neq 0 \text{ at } x.$$

If f attains a relative extremum under the constraints (1) at a point x^0 of D, then there exist numbers $\lambda_1, ..., \lambda_m$ such that

$$\frac{\partial f(x^0)}{\partial x_i} + \sum_{j=1}^{m} \lambda_j \frac{\partial g_j(x^0)}{\partial x_i} = 0 \qquad (1 \leq i \leq n). \tag{7}$$

If we add to the n equations (7) the m equations (1) at x^0, we obtain $n + m$ equations for the $n + m$ unknowns $x_1{}^0, ..., x_n{}^0, \lambda_1, ..., \lambda_m$. The numbers $\lambda_1, ..., \lambda_m$ are called the *Lagrange multipliers*.

One can prove Theorem 2 by the method of proof of Theorem 1. The details are omitted.

EXAMPLE. Consider the problem posed above of maximizing $V = 8xyz$ under the constraint (2), assuming $x > 0$, $y > 0$, $z > 0$. Equations (3) become

$$8yz + 2\lambda \frac{x}{a^2} = 0,$$

$$8zx + 2\lambda \frac{y}{b^2} = 0, \tag{8}$$

$$8xy + 2\lambda \frac{z}{c^2} = 0.$$

Multiplying the first equation by x, the second by y, and the third by z, and adding, we obtain, after making use of (2):

$$12xyz + \lambda = 0, \qquad \text{or} \quad \lambda = -12xyz.$$

If we substitute this into (8), we get

$$yz(a^2 - 3x^2) = 0,$$
$$zx(b^2 - 3y^2) = 0,$$
$$xy(c^2 - 3z^2) = 0.$$

This gives $x = a/\sqrt{3}$, $y = b/\sqrt{3}$, $z = c/\sqrt{3}$. Now V, as a continuous function on the compact set given by $x \geq 0$, $y \geq 0$, $z \geq 0$ and by (2), must attain a maximum. The maximum cannot be attained at points (x, y, z) where one of the coordinates is zero, for $V = 0$ at such points. It follows that the maximum of V must be attained at $x = a/\sqrt{3}$, $y = b/\sqrt{3}$, $z = c/\sqrt{3}$. It is equal to $8abc/3^{3/2}$.

PROBLEMS

1. Find the maximum of $x^2 y^2 z^2$ subject to the constraint $x^2 + y^2 + z^2 = 3$. Conclude that

$$x^2 y^2 z^2 \leq \frac{x^2 + y^2 + z^2}{3}.$$

2. Generalize the result of the previous problem, and deduce the *geometric-arithmetic inequality*:

$$(a_1 a_2 \cdots a_n)^{1/n} \le \frac{a_1 + a_2 + \cdots + a_n}{n},$$

where $a_1, \ldots, a_n$ are nonnegative numbers.

3. Find the maximum of $x + y + z$ subject to the conditions:

$$\frac{a}{x} + \frac{b}{y} + \frac{c}{z} = 1, \qquad x > 0, y > 0, z > 0 \quad (a, b, c \text{ positive numbers}).$$

4. Find the maximum of $xyz/(a^3 x + b^3 y + c^3 z)$ subject to the condition $xyz = d$; a, b, c, d are positive numbers.

5. Find the minimum of $x^3 + y^3 + z^3$ subject to the constraints $ax + by + cz = 1$, $x > 0$; a, b, c are positive numbers.

6. Let p, q be positive numbers. Show that the minimum of $x^p/p + y^q/q$ under the constraint $xy = 1$ is $1/p + 1/q$. Deduce that if $1/p + 1/q = 1$ then $ab \le a^p/p + b^q/q$ for all $a \ge 0$, $b \ge 0$.

7. Find the minimum of $x_1{}^2 + \cdots + x_n{}^2$ subject to the condition

$$a_1 x_1 + \cdots + a_n x_n = 1 \qquad (a_1{}^2 + \cdots + a_n{}^2 > 0).$$

8. Find the maximum of $\left(\sum_{i=1}^{n} a_i x_i \right)^2$ under the condition

$$\sum_{i=1}^{n} x_i{}^2 = 1 \qquad \left(\sum_{i=1}^{n} a_i{}^2 > 0 \right).$$

6. THE DIVERGENCE THEOREM IN R^n

An x_n-*projectable hypersurface* S in R^n has the form

$$x_n = \phi(x_1, \ldots, x_{n-1}), \qquad (x_1, \ldots, x_{n-1}) \in D,$$

where D is a closed domain in R^{n-1}. We shall assume throughout this section that D is bounded and its boundary has content zero in R^{n-1}. If ϕ is continuous in D and if its first derivatives are piecewise continuous in D, then we say that S is *piecewise continuously differentiable*. Given a continuous function $g(x)$ on S, we define the *surface integral* of g over S by

$$\iint_S \cdots \int g \, dS = \iint_D \cdots \int g(x_1, \ldots, x_{n-1}, \phi(x_1, \ldots, x_{n-1})) \tag{1}$$

$$\times (1 + \phi_{x_1}{}^2 + \cdots + \phi_{x_{n-1}}^2)^{1/2} \, dx_1 \cdots dx_{n-1};$$

the number of integrals on each side is $n - 1$. We also write

$$dS = (1 + \phi_{x_1}^2 + \cdots + \phi_{x_{n-1}}^2)^{1/2} \, dx_1 \cdots dx_{n-1} \tag{2}$$

and call dS the *element of surface area*. One considers surface integrals (1) also in cases where the first derivatives $\partial\phi/\partial x_i$ are not piecewise continuous in D but are piecewise continuous in every closed domain lying in the interior of D. The integral on the right-hand side of (1) is then taken as an improper integral.

If $g \equiv 1$ in (1), then the surface integral is called the *surface area* of S. Let G be a domain in R^n, given by

$$\{x = (x_1,\ldots,x_{n-1},x_n);\ \phi_1(x_1,\ldots,x_{n-1}) < x_n$$
$$< \phi_2(x_1,\ldots,x_{n-1}), (x_1,\ldots,x_{n-1}) \in D\}$$

where D is a bounded domain in R^{n-1} and $\phi_1 < \phi_2$ in D. Denote by $\bar{D}$ the closure of D. If the hypersurfaces

$$S_i : x_n = \phi_i(x_1,\ldots,x_n), (x_1,\ldots,x_{n-1}) \in \bar{D} \qquad (i = 1,2)$$

are piecewise continuously differentiable, then we say that the domain G is x_n-*projectable*. We then define

$$\cos(x_n,v) = \left\{1 + \left(\frac{\partial\phi_i}{\partial x_1}\right)^2 + \cdots + \left(\frac{\partial\phi_i}{\partial x_{n-1}}\right)^2\right\}^{-1/2} \text{ on } S_i,$$
$$\cos(x_n,v) = 0 \text{ on } S - (S_1 \cup S_2)$$

where S is the boundary of G.

Similarly, we define the concept of x_i-projectable domain and the function $\cos(x_i,v)$ on the boundary.

If a domain G in R^n is x_i-projectable for $i = 1,2,\ldots,n$, then we say that G is a *simple domain*.

We now can state the divergence theorem in R^n:

THEOREM 1.   *Let G be a simple domain in R^n with boundary S. Let $P = (P_1,\ldots,P_n)$ be a continuous vector valued function in $G \cup S$, and assume that $\partial P_i/\partial x_i$ are uniformly continuous in G. Then*

$$\iint\cdots\int_G \left[\sum_{i=1}^n \frac{\partial P_i}{\partial x_i}\right] dV = \iint\cdots\int_S \left[\sum_{i=1}^n P_i \cos(x_i,v)\right] dS. \tag{3}$$

The last formula can also be written in the form

$$\iint\cdots\int_G \text{div } \mathbf{P} \, dV = \iint\cdots\int_S \mathbf{P} \cdot v \, dS \tag{4}$$

where $v = (\cos(x_i,v),\ldots,\cos(x_n,v))$, div $P = \sum_{i=1}^n \partial P_i/\partial x_i$.

The proof of Theorem 1 is similar to the proof of the divergence theorem in R^3.

Theorem 1 can be used to prove the change of variables formula in multiple integrals. We have done this in Sections 9.6 and 9.10 for $n = 2$ and $n = 3$.

PROBLEMS

1. Let $f(x)$ be a continuous function on the unit sphere S: $\|x\| = 1$, in R^n. If $f(-x) = -f(x)$ for every x in S, then $\iint_S \cdots \int f \, dS = 0$.

2. Prove that the divergence theorem holds in a ball $\|x\| < R$, the surface integral being taken as an improper integral. [*Hint:* Compare Example 3 in Section 9.9.]

3. Denote by $S_m(r)$ the surface area of the sphere $\|x\| = r$ in R^m. Prove

$$S_{2n}(r) = \frac{2\pi^n}{(n-1)!} r^{2n-1},$$

$$S_{2n-1}(r) = \frac{4^n \pi^{n-1}(n-1)!}{2(2n-2)!} r^{2n-2}.$$

[*Hint:* Use the result of Problem 11, Section 8.5.]

4. Denote by $V_n(r)$ the volume of the ball $\|x\| < r$ in R^n and denote by $S_n(r)$ the surface area of the sphere $\|x\| = r$ in R^n. Prove that

$$V_n(r) = \frac{r}{n} S_n(r).$$

[*Hint:* Compare Example 1 in Section 9.9.]

7. HARMONIC FUNCTIONS

Let D be a bounded domain in R^n, $n \geq 2$. A function $u(x)$ is said to be *harmonic* in D if its first two derivatives are continuous in D, and

$$\Delta u \equiv \frac{\partial^2 u}{\partial x_1^2} + \frac{\partial^2 u}{\partial x_2^2} + \cdots + \frac{\partial^2 u}{\partial x_n^2} = 0 \text{ in } D.$$

There are "many" harmonic functions in D. In fact, let S be a sphere $\|x\| = R$ containing D, and let $\sigma(y)$ be any continuous function defined on S. Then, if $n \neq 2$, the function

$$u(x) = \iint \cdots \int\limits_{S} \frac{\sigma(y)}{\|x - y\|^{2-n}} \, dS$$

is harmonic in D. If $n = 2$, the function

$$u(x) = \int_{S} \sigma(y) \log \frac{1}{\|x - y\|} \, dS$$

is harmonic in D (here S is the circle $y_1{}^2 + y_2{}^2 = R^2$).

Let $f(x)$ be a continuous function defined on the boundary ∂D of D. Consider the problem of finding a harmonic function $u(x)$ in D, such that $u(x)$ is continuous in $D \cup \partial D$ and satisfies the *boundary condition*

$$u = f \quad \text{on } \partial D.$$

This problem is called the *Dirichlet problem*. It is one of the most important problems in mathematics. There are many methods of solving it, and there are far-reaching generalizations of it. What we shall do here is prove that the Dirichlet problem has at most one solution. This would follow from the following *maximum principle*:

THEOREM 1.   *Let $u(x)$ be harmonic in a bounded domain D, and continuous in $D \cup \partial D$. Then the maximum of u in $D \cup \partial D$ is attained at a point of ∂D, that is*

$$\max_{x \in D \cup \partial D} u(x) = \max_{x \in \partial D} u(x). \tag{1}$$

Suppose the theorem is true. If u_1 and u_2 are harmonic functions in D, continuous in $D \cup \partial D$, and satisfying

$$u_1 = u_2 = f \quad \text{on } \partial D,$$

then $u = u_1 - u_2$ is harmonic in D, continuous in $D \cup \partial D$, and $u = 0$ on ∂D. Applying (1) to u and to $-u$, we see that $u \equiv 0$ in D. Hence $u_1 \equiv u_2$ in D. This proves the uniqueness of solution of the Dirichlet problem.

Proof of Theorem 1.   Suppose the assertion of the theorem is false. Then there is a point x^0 in D such that

$$u(x^0) > \max_{x \in \partial D} u(x).$$

Thus

$$u(x^0) > \max_{x \in \partial D} u(x) + \alpha$$

for some $\alpha > 0$. Let $\varepsilon > 0$ be so small that

$$\varepsilon \|x - x^0\|^2 \leq \alpha \qquad \text{for all } x \in D \cup \partial D.$$

Consider the function

$$v(x) = u(x) + \varepsilon \|x - x^0\|^2.$$

It satisfies

$$v(x^0) = u(x^0) > \max_{x \in \partial D} u(x) + \alpha$$

$$> \max_{x \in \partial D} u(x) + \varepsilon \max_{x \in \partial D} \|x - x^0\|^2 \geq \max_{x \in \partial D} v(x).$$

Consequently, the maximum of $v(x)$ in $D \cup \partial D$ is not attained at any point of ∂D. Let y be a point of D where this maximum is attained. Then

$$\frac{\partial v(y)}{\partial x_i} = 0 \qquad (i = 1,...,n),$$

$$\frac{\partial^2 v(y)}{\partial x_i{}^2} \leq 0 \qquad (i = 1,...,n).$$

In particular,

$$\Delta v(y) \leq 0. \tag{2}$$

But, for any x in D,

$$\Delta v(x) = \Delta u(x) + \varepsilon \Delta \|x - x^0\|^2 = \Delta u(x) + 2n\varepsilon = 2n\varepsilon > 0.$$

This contradicts (2).

PROBLEMS

1. Let u be a continuous function in $D \cup \partial D$, D being a bounded domain in R^n, and assume that the first two derivatives of u are continuous in D. Prove:

(a) If $\Delta u \geq 0$ in D, then u satisfies (1);

(b) If $\Delta u + c(x)u = 0$ in D where $c(x) \leq 0$ in D, and if $u = 0$ on ∂D, then $u \equiv 0$ in D;

(c) If $\Delta u = f$ in D, $u = 0$ on ∂D, then there is a constant C, independent of u, such that

$$\max_{x \in D \cup \partial D} |u(x)| \leq C \max_{x \in D \cup \partial D} |f(x)|.$$

[*Hint:* Apply (a) to $v = \pm u - c(1 - e^{\varepsilon x_1})(\max_{D \cup \partial D} |f|).$]

2. Let $f(\theta)$ be a twice continuously differentiable function on the boundary
    of the disc $D : x^2 + y^2 < 1$. Let

    $$f(\theta) \sim \frac{a_0}{2} + \sum_{n=1}^{\infty} (a_n \cos n\theta + b_n \sin n\theta).$$

    (a) Prove that the function

    $$u(r,\theta) = \frac{a_0}{2} + \sum_{n=1}^{\infty} (a_n \cos n\theta + b_n \sin n\theta)r^n$$

    is harmonic in D, continuous in $D \cup \partial D$, and $u = f$ on ∂D.

    (b) Show that

    $$u(r,\theta) = \frac{1}{2\pi} \int_0^{2\pi} f(\alpha) \frac{1 - r^2}{1 - 2r \cos(\alpha - \theta) + r^2} \, d\alpha.$$

8. ORDINARY DIFFERENTIAL EQUATIONS

We shall denote an indefinite integral $h(x)$ of a function $g(x)$ by $\int g(x) \, dx$.

Let $f(x,y,z)$ be a function defined in some open set Ω of R^3. Let I be a
fixed interval on the real line and consider functions $y(x)$ continuously
differentiable on I, for which $(x,y(x),y'(x))$ belongs to Ω if $x \in I$. The function
$f(x,y(x),y'(x))$ is then well defined on I. We shall be concerned with such
functions $y(x)$ for which

$$f(x,y(x),y'(x)) = 0 \tag{1}$$

for all $x \in I$. Equation (1) is called an *ordinary differential equation*, and a
function $y(x)$ satisfying (1) is called a *solution* of the differential equation (1).

EXAMPLE 1.   The solutions of the differential equation

$$y' - f(x) = 0$$

are given by

$$y = \int f(x) \, dx + C \qquad (C \text{ constant}).$$

EXAMPLE 2.   The differential equation

$$\phi(y)y' + \psi(x) = 0 \tag{2}$$

is said to have *separated variables*. If $y = y(x)$ is a solution of (2), then

$$\phi(y(x))\frac{dy}{dx} + \psi(x) = 0.$$

Integrating with respect to x, we get

$$\int \phi(y(x))\frac{dy}{dx} dx + \int \psi(x) dx = C,$$

or

$$\int \phi(y) dy + \int \psi(x) dx = C. \tag{3}$$

Thus the solutions of (2) satisfy the implicit equation (3). Conversely, if $y = y(x)$ satisfies (3) then, by differentiating (3) with respect to x, we obtain Equation (2) for $y = y(x)$.

EXAMPLE 3.   Consider the equation

$$M(x,y) + N(x,y)y' = 0 \tag{4}$$

where (x,y) vary in a simply connected domain D. If

$$\frac{\partial M}{\partial y} = \frac{\partial N}{\partial x},$$

then Equation (4) is called an *exact equation*. In that case the line integral

$$u(x,y) = \int_{(a,b)}^{(x,y)} M\, dx + N\, dy$$

$[(a,b)$ being a fixed point in $D]$ is independent of the path and it is thus a function of (x,y) only. Furthermore,

$$\frac{\partial u}{\partial x} = M, \qquad \frac{\partial u}{\partial y} = N.$$

It follows that if $y = y(x)$ is a solution of (4), then

$$\frac{d}{dx} u(x,y(x)) = u_x(x,y(x)) + u_y(x,y(x))y'(x) = 0. \tag{5}$$

Thus

$$u(x,y(x)) = C. \tag{6}$$

Conversely, if $y(x)$ is a function satisfying (6), then it satisfies (5). Consequently, it is a solution of (4). We thus have shown that the solutions $y(x)$ of (4) are given implicitly by (6).

EXAMPLE 4.   The equation

$$\frac{dy}{dx} + P(x)y = Q(x) \tag{7}$$

is called a *linear differential equation*. If we multiply both sides of (7) by $e^{\int P(x)\,dx}$, we obtain

$$\frac{d}{dx}\left[e^{\int P(x)\,dx}y\right] = e^{\int P(x)\,dx}Q(x). \tag{8}$$

Suppose $y = y(x)$ is a solution of (7). Then it satisfies (8). Integrating both sides of (8), we get

$$e^{\int P(x)\,dx}y = \int \left[e^{\int P(x)\,dx}Q(x)\right]dx + C,$$

or

$$y = Ce^{-\int P(x)\,dx} + e^{-\int P(x)\,dx}\int \left[e^{\int P(x)\,dx}Q(x)\right]dx. \tag{9}$$

Conversely, if y is given by (9) then, as is easily verified, it satisfies (7).

In all the examples given above there are infinitely many solutions of each differential equation. In fact, the solutions form a family depending on a real parameter C. If we impose a condition that $y(x_0) = y_0$, the solution will usually be unique.

In what follows we shall consider differential equations that can be written in the form

$$y' = f(x,y). \tag{10}$$

We shall prove, under some conditions, that there exists a unique solution of (10) satisfying a condition

$$y(x_0) = y_0. \tag{11}$$

A function $f(x,y)$ defined in a domain D of R^2 is said to satisfy the *Lipschitz condition* with respect to y, if there is a constant K such that

$$|f(x,y) - f(x,y')| \le K|y - y'| \qquad \text{for all } (x,y) \in D, (x,y') \in D. \tag{12}$$

THEOREM 1.   *Let $f(x,y)$ be a continuous function in a domain D of R^2. Assume that f is bounded by a constant M and that it satisfies the Lipschitz condition* (12). *Let R be a rectangle*

$$R : |x - x_0| \le \alpha, \qquad |y - y_0| \le \beta$$

contained in D, such that $M\alpha \le \beta$. Then there exists a unique solution of (10) *for $|x - x_0| \le \alpha$, satisfying* (11).

Proof. If $y(x)$ is a solution of (10) and (11) for $|x - x_0| \leq \alpha$, then, by integrating (10) and using (11), we get

$$y(x) = y_0 + \int_{x_0}^{x} f(t,y(t)) \, dt \qquad (|x - x_0| \leq \alpha). \tag{13}$$

Conversely, if $y(x)$ is a continuous solution of (13), then it satisfies (10) (for $|x - x_0| \leq \alpha$) and (11). Thus it remains to prove that the "integral equation" (13) has a unique solution.

Define a sequence $\{y_n(x)\}$ successively by

$$\begin{aligned} y_0(x) &\equiv y_0 & (|x - x_0| \leq \alpha), \\ y_{n+1}(x) &= y_0 + \int_{x_0}^{x} f(t,y_n(t)) \, dt & (|x - x_0| \leq \alpha). \end{aligned} \tag{14}$$

We shall prove that this sequence is well defined, and that $\lim y_n(x)$ exists and is the desired solution of (13). This method of constructing a solution is called the *method of successive approximations*.

It is clear that $y_1(x)$ is well defined and

$$|y_1(x) - y_0| = \left| \int_{x_0}^{x} f(t,y_0) \, dt \right| \leq M |x - x_0| \leq M\alpha \leq \beta.$$

Proceeding by induction, we assume that $y_m(x)$ is well defined, that

$$|y_m(x) - y_0| \leq \beta \qquad \text{if } |x - x_0| \leq \alpha, \tag{15}$$

and that

$$|y_m(x) - y_{m-1}(x)| \leq MK^{m-1} \frac{|x - x_0|^{m-1}}{(m-1)!} \qquad (|x - x_0| \leq \alpha), \tag{16}$$

for $m = 1, 2, \dots, n$. We shall prove that $y_{n+1}(x)$ is well defined and that (15) and (16) hold for $m = n + 1$.

From (15) with $m = n$ we see that the points $(x, y_n(x))$ belong to R if $|x - x_0| \leq \alpha$. Hence $y_{n+1}(x)$ is well defined by (14), and

$$|y_{n+1}(x) - y_0| \leq \left| \int_{x_0}^{x} f(t,y_n(t)) \, dt \right| \leq M |x - x_0| \leq M\alpha \leq \beta.$$

It remains to prove (16) for $m = n + 1$. Using (12) and (16) for $m = n$, we can write

$$\begin{aligned} |y_{n+1}(x) - y_n(x)| &= \left| \int_{x_0}^{x} [f(t,y_n(t)) - f(t,y_{n-1}(t))] \, dt \right| \\ &\leq K \left| \int_{x_0}^{x} |y_n(t) - y_{n-1}(t)| \, dt \right| \\ &\leq MK^n \left| \int_{x_0}^{x} \frac{|t - x_0|^{n-1}}{(n-1)!} \, dt \right| = MK^n \frac{|x - x_0|^n}{n!}. \end{aligned}$$

Thus (16) holds for $m = n + 1$.

From (16) we get

$$|y_{m+1}(x) - y_m(x)| \le MK^m \frac{\alpha^m}{m!} \quad \text{for } m = 0,1,2,\dots .$$

Employing the Weierstrass M-test, we conclude that the series

$$y_0(x) + \sum_{m=0}^{\infty} (y_{m+1}(x) - y_m(x))$$

is uniformly convergent to some continuous function $y(x)$. Since the partial sums of this series are the functions $y_n(x)$, it follows that

$$y_n(x) \to y(x) \text{ uniformly} \quad \text{for } |x - x_0| \le \alpha. \tag{17}$$

If we now let $n \to \infty$ in (14) and use (17), we obtain Equation (13) for $y(x)$.

It remains to prove uniqueness. Let $\bar{y}(x)$ be another solution of (13). Then

$$|y(x) - \bar{y}(x)| = \left| \int_{x_0}^{x} [f(t,y(t)) - f(t,\bar{y}(t))] \, dt \right|$$

$$\le K \left| \int_{x_0}^{x} |y(t) - \bar{y}(t)| \, dt \right|. \tag{18}$$

Using this inequality we can prove by induction on n that

$$|y(x) - \bar{y}(x)| \le LK^n \frac{|x - x_0|^n}{n!} \quad (|x - x_0| \le \alpha) \tag{19}$$

where $L = \max\limits_{|x-x_0| \le \alpha} |y(x) - \bar{y}(x)|$. Taking $n \to \infty$, we deduce that $y(x) - \bar{y}(x) \equiv 0$.

An ordinary differential equation of *order n* has the form

$$f(x,y,y',y'',\dots,y^{(n)}) = 0. \tag{20}$$

A function $y(x)$ (x varies in some interval I) is called a *solution* of (20), if

$$f(x,y(x),y'(x),y''(x),\dots,y^{(n)}(x)) = 0 \text{ for all } x \text{ in } I.$$

EXAMPLE.   Consider the differential equation

$$a_0 y^{(n)} + a_1 y^{(n-1)} + \cdots + a_{n-1}y' + a_n y = 0 \tag{21}$$

where $a_0, a_1, \dots, a_n$ are constants and $a_0 \ne 0$. A function $y = e^{\lambda x}$ is a solution of (21) if, and only if, λ satisfies the polynomial equation

$$a_0 \lambda^n + a_1 \lambda^{n-1} + \cdots + a_{n-1}\lambda + a_n = 0.$$

If this equation has n distinct roots, $\lambda_1, \lambda_2, \ldots, \lambda_n$, then

$$y = C_1 e^{\lambda_1 x} + C_2 e^{\lambda_2 x} + \cdots + C_n e^{\lambda_n x}$$

is a solution of (21), for any constants $C_1, C_2, \ldots, C_n$.

PROBLEMS

Solve the equations:

(a) $y' = \dfrac{xy}{x-1}$;   (b) $e^{x^2+y} + \dfrac{y}{x} y' = 0$;   (c) $y' = \dfrac{x}{y\sqrt{1-x^2}}$.

2. Equation (4) is called a *homogeneous equation* if $M(x,y)$ and $N(x,y)$ are homogeneous functions of (x,y) of the same degree. Show that the substitution $y = ux$ reduces it to an equation in u and x with separated variables. Use this substitution to solve the following equations:

(a) $y' = \dfrac{x^2 + y^2}{2x^2}$;   (b) $(x^2 + y^2) - 2xyy' = 0$.

3. Solve the linear equations:

(a) $(x^2 - x)y' + (1 - 2x)y + x^2 = 0$;

(b) $xy' + 2y = (3x + 2)e^{3x}$;

(c) $y' + y \tan x = \sin 2x$.

4. *Bernoulli's equation* is the differential equation

$$y' + P(x)y = Q(x)y^n.$$

If $n = 0$ or $n = 1$, it becomes a linear equation. If $n \neq 0,1$, define $u = y^{1-n}$ and show that the equation is reduced to the linear equation

$$\frac{1}{1-n} u' + P(x)u = Q(x).$$

Solve the equations:

(a) $y' + y = xy^2$;   (b) $y' - y + y^2(x^2 + x + 1) = 0$.

5. Solve the exact equations:

(a) $(x + y^2)y' + (y - x^2) = 0$;   (b) $\dfrac{xy + 1}{y} + \dfrac{2y - x}{y^2} y' = 0$;

(c) $(2xy^4 + \sin y) + (4x^2y^3 + x \cos y)y' = 0$.

6. Prove the inequality (19).

7. Solve the second-order equations:

(a) $y'' - 3y' = 6$;   (b) $xy'' - y' = x^4$.

9. THE RIEMANN–STIELTJES INTEGRAL

Let $f(x)$ be a bounded function defined in a bounded, closed interval $a \leq x \leq b$. Let $g(x)$ be a function of bounded variation in $a \leq x \leq b$. Take any partition

$$x_0 = a < x_1 < x_2 < \cdots < x_{n-1} < x_n = b \tag{1}$$

of $[a,b]$ and choose any points ξ_i in $[x_{i-1}, x_i]$. Form the sum

$$T = \sum_{i=1}^{n} f(\xi_i)[g(x_i) - g(x_{i-1})].$$

Suppose that there exists a number K having the following property: For any $\varepsilon > 0$ there is a $\delta > 0$ such that

$$|T - K| < \varepsilon \qquad \text{if max } (x_i - x_{i-1}) < \delta.$$

Then we say that the *Riemann–Stieltjes integral* of f with respect to g over $a \leq x \leq b$ exists. The number K is called the Riemann–Stieltjes integral of f with respect to g over $a \leq x \leq b$, and is denoted by

$$(RS) \int_a^b f(x) \, dg(x). \tag{2}$$

If $g(x) = x$, the above definition coincides with the definition of the Riemann integral $\int_a^b f(x) \, dx$.

Similarly, one can define the Darboux–Stieltjes integral. We begin by forming sums

$$S = \sum_{i=1}^{n} M_i[g(x_i) - g(x_{i-1})], \tag{3}$$

$$S = \sum_{i=1}^{n} m_i[g(x_i) - g(x_{i-1})], \tag{4}$$

where

$$M_i = \sup_{x_{i-1} \leq x \leq x_i} f(x), \qquad m_i = \inf_{x_{i-1} \leq x \leq x_i} f(x).$$

Denote by J the infimum of the set of all the numbers S obtained by taking all possible partitions of $[a,b]$. Denote by I the supremum of the set of all the numbers s obtained by taking all possible partitions of $[a,b]$. One can show (by exactly the same arguments as in Section 4.1) that $J \geq I$. If $J = I$, then we say that the *Darboux–Stieltjes integral* of f with respect to g over $a \leq x \leq b$ exists. The number J is then called the Darboux–Stieltjes integral of f with respect to g over $a \leq x \leq b$. We denote it by

$$(DS) \int_a^b f(x)\, dg(x). \tag{5}$$

The results of Sections 4.1 and 4.2 can be extended with minor changes to the present more general case. In particular, we have:

THEOREM 1.   *The Darboux–Stieltjes integral* (5) *exists if, and only if, for any $\varepsilon > 0$ there is a partition* (1) *of* $[a,b]$ *such that the corresponding sums* S, s *(defined by* (3) *and* (4)*) satisfy* $S - s < \varepsilon$.

THEOREM 2.   *Let g be a monotone increasing function. Then the Riemann–Stieltjes integral* (2) *exists if, and only if, the Darboux–Stieltjes integral* (5) *exists, and, in that case, the two integrals are equal.*

When the integral (2) and (5) are equal, we denote their common value by

$$\int_a^b f(x)\, dg(x).$$

The next theorem establishes the existence of the Riemann–Stieltjes integral.

THEOREM 3.   *Let $f(x)$ be a continuous function in $a \leq x \leq b$ and let $g(x)$ be a function of bounded variation in $a \leq x \leq b$. Then the Riemann–Stieltjes integral* (2) *and the Darboux–Stieltjes integral* (5) *exist and they are equal.*

The proof is left to the student.

THEOREM 4.   *If $f(x)$ is a continuous function in $a \leq x \leq b$, and if $g(x)$ is of bounded variation in $a \leq x \leq b$, then*

$$\left| \int_a^b f(x)\, dg(x) \right| \leq \max_{a \leq x \leq b} |f(x)| \cdot \bigvee_a^b (g).$$

The proof is left to the student.

THEOREM 5.   *If $f(x)$ is continuous and $g(x)$ is continuously differentiable in $a \leq x \leq b$, then*

$$\int_a^b f(x) \, dg(x) = \int_a^b f(x)g'(x) \, dx. \tag{6}$$

The proof is left to the student.

THEOREM 6.   *If $f(x)$ and $g(x)$ are continuous and of bounded variation in $a \leq x \leq b$, then*

$$\int_a^b f(x) \, dg(x) + \int_a^b g(x) \, df(x) = \left[f(x)g(x) \right]_a^b. \tag{7}$$

This formula is called the formula of *integration by parts*. If f and g are continuously differentiable then, by Theorem 5, (7) reduces to the integration by parts formula

$$\int_a^b f(x)g'(x) \, dx + \int_a^b f'(x)g(x) \, dx = \left[f(x)g(x) \right]_a^b.$$

Proof of Theorem 6.   Take any partition (1) of $[a,b]$ and any points ξ_i in (x_i, x_{i+1}). Writing the sum

$$T = \sum_{i=0}^{n-1} f(\xi_i)[g(x_{i+1}) - g(x_i)]$$

in the form

$$\sum_{i=0}^{n-1} f(\xi_i)g(x_{i+1}) - \sum_{i=0}^{n-1} f(\xi_i)g(x_i)$$

we get

$$T = - \sum_{i=1}^{n-1} g(x_i)[f(\xi_i) - f(\xi_{i-1})] + f(\xi_{n-1})g(x_n) - f(\xi_0)g(x_0).$$

Taking now a sequence of partitions with mesh tending to zero, (7) follows.

PROBLEMS

1. Prove Theorem 3.

2. Prove Theorem 4.

3. Prove Theorem 5.

4. The function

$$f(x) = \begin{cases} 0 & \text{if } -1 \le x \le 0, \\ 1 & \text{if } 0 < x \le 1 \end{cases}$$

is piecewise continuous in $-1 \le x \le 1$. The function

$$g(x) = \begin{cases} 0 & \text{if } -1 \le x < 0, \\ 1 & \text{if } 0 \le x \le 1 \end{cases}$$

is monotone increasing in $-1 \le x \le 1$. Prove that the Riemann–Stieltjes integral $\int_{-1}^{1} f(x) dg(x)$ does not exist.

5. Prove that if $a < b < c$ and if each of the integrals

$$\int_a^b f \, dg, \qquad \int_b^c f \, dg, \qquad \int_a^c f \, dg$$

exists, then

$$\int_a^b f \, dg + \int_b^c f \, dg = \int_a^c f \, dg.$$

Note that the example of the preceding problem shows that the existence of $\int_a^b f \, dg$ and of $\int_b^c f \, dg$ does not imply the existence of $\int_a^c f \, dg$.

6. If $\int_a^b f_1 \, dg$ and $\int_a^b f_2 \, dg$ exist, then $\int_a^b (f_1 + f_2) \, dg$ exists and

$$\int_a^b f_1 \, dg + \int_a^b f_2 \, dg = \int_a^b (f_1 + f_2) \, dg.$$

ANSWERS TO PROBLEMS

Chapter 1, Section 4

8(a).	$-1,1.$	8(b).	$0,1.$	8(c).	$\frac{1}{2},1.$
8(d).	$-\frac{1}{2},1.$	8(e).	$0,\frac{1}{2}$	8(f).	$2,3.$

Chapter 1, Section 5

5(a).	$\frac{1}{2}.$	5(b).	$0.$	5(c).	$0.$
5(d).	$0.$	5(e).	$\frac{3}{5}.$	5(f).	$\frac{1}{2}.$
6(a).	$1,3.$	6(b).	$1.$	6(c).	$-1,1.$
6(d).	$\infty.$	10.	$-1,0,1.$		

Chapter 1, Section 6

11(a).	$3.$	11(b).	$0.$	11(c).	$1.$

Chapter 1, Section 7

5. $\lim b_n = 2.$

6. $\lim b_n = \frac{1}{2}(1 + \sqrt{1 + 4c}).$

7. $\frac{1}{2}.$

11. $2.$

Chapter 1, Section 8

1. $[a,b]$.   2. $\left\{-1, \dfrac{(-1)^n n}{n+1} \text{ where } n = 1,2,...\right\}$.   4. Yes.

Chapter 2, Section 1

6. $x = -2$.

10. $\infty, -\infty, -\infty, 0, 0$.

11(a). 0.

11(b). $\frac{1}{2}$.

11(c). $\frac{1}{2}$.

11(d). 1.

Chapter 2, Section 2

3. 0.

4. $\left\{(2n + 1)\dfrac{\pi}{2}\right\}$; $\{\infty, -\infty\}$.

5(a). $\left\{n\pi + \dfrac{\pi}{4}\right\}$; $\{-\infty, \infty\}$.

5(b). $\left\{n\pi - \dfrac{\pi}{4}\right\}$; $\{-\infty, \infty\}$.

5(c). $\{0\}$; $[-1,1]$.

5(d). $\left\{n\pi + \dfrac{\pi}{3}\right\}$; $\{-\infty, \infty\}$.

5(e). $\{0\}$; $[-1,1]$.

5(f). No points of discontinuity.

Chapter 2, Section 3

2(b). 1,2.

2(c). $-7, -1$.

5. Not necessarily. For instance, $f(x) = 1 - 1/x$ in $[1,\infty)$.

Chapter 2, Section 6

8. $1; 0$.

Chapter 2, Section 7

2. $f(x) = \sin x$.

3(a). $\delta = \varepsilon$.

Chapter 3, Section 1

4(a). $y = 4x - 4$.

4(b). $y = 1$.

4(c). $y = -x + \dfrac{\pi}{2}$.

4(d). $y = -\dfrac{x}{2.5^{3/2}} + \dfrac{1}{\sqrt{2}}$.

4(e). $y = \left(\sqrt{2} + \dfrac{1}{2\sqrt{2}}\right)x - \dfrac{1}{2\sqrt{2}}$.

4(f). $y = 0$.

Chapter 3, Section 2

1(c). $-\dfrac{x \sin\sqrt{1 + x^2}}{\sqrt{1 + x^2}}$.

1(d). $m \sin^{m-1} x \cdot \cos x$.

1(e). $-\dfrac{1}{\cos^2 x}$.

1(f). $-\dfrac{2\cos x}{(\sin x - 1)^2}$.

3(a). $\dfrac{x}{(1 - x^2)^{3/2}}$.

3(b). $-\dfrac{x}{(1 + x^2)^{3/2}}$.

3(c). $-\dfrac{x}{\sqrt{1 + x^2}} - \dfrac{x\sqrt{1 - x^2}}{(1 + x^2)^{3/2}}$.

Chapter 3, Section 3

3(a). $x^x(1 + \log x)$.

3(b). $x^{\sin x}\left(\cos x \cdot \log x + \dfrac{\sin x}{x}\right)$.

3(c). $\dfrac{x}{\sqrt{1 + x^2}}\, e^{\sqrt{1 + x^2}}$.

3(d). $-\dfrac{1}{\sqrt{1 - x^2}}$.

Chapter 3, Section 4

1(a). $n!$

1(b). For n even, $(-1)^{n/2}3^n \sin 3x$; for n odd, $(-1)^{(n-1)/2}3^n \cos 3x$.

1(c). $\dfrac{(-1)^n n!}{x^{n+1}}$.

1(e). e^x.

1(f). $\dfrac{(2n)!x^n}{n!}$.

1(g). $\dfrac{(-1)^{n-1}(n-1)!}{x^n}$.

4(a). $e^x(n + x)$.

4(b). For n even, $n \sin x + (-1)^{n/2}x \sin x$; for n odd, $n \sin x + (-1)^{(n-1)/2}x \cos x$

Chapter 3, Section 7

1(a). 0.

1(b). $\dfrac{1}{\sqrt{e}}$.

1(c). 0.

1(d). $\frac{1}{2}$.

1(e). $\dfrac{\pi}{2}$.

1(f). $\dfrac{\pi}{4}$.

1(g). $\dfrac{e}{2}$.

1(h). $\log\dfrac{a}{b}$.

1(i). 2.

1(j). 2.

2(a). 1.

2(b). $e^{1/6}$.

2(c). 0.

2(d). ∞.

2(e). 1.

2(f). $1/e$.

2(g). 0.

2(h). e.

3(a). $\dfrac{1}{2\log(a/b)}$.

3(b). $\dfrac{\log a}{\log b}$.

3(c). -1.

3(d). 2.

3(e). 2.

3(f). ∞.

Chapter 3, Section 8

3(a). $(-1)^n(1 + x)^{-n-1}$.

3(b). $(-1)^n \dfrac{1\cdot 3\cdots(2n + 1)}{2\cdot 4\cdots(2n + 2)}(1 + x)^{-(2n+3)/2}$.

Chapter 3, Section 9

1(a). Min. at 0,1; max. at $\frac{1}{2}$.

1(b). Max. at -2; min. at $\frac{1}{2}$.

1(c). Monotone increasing.

1(d). Min. at 0; min. at $\pm \sqrt[4]{(n-\frac{1}{2})\pi}$ for n even; max. at $\pm \sqrt[4]{(n-\frac{1}{2})\pi}$ for n odd.

1(g). n even: min at 0, max. at $\pm \sqrt{n/2}$;
n odd: inflection at 0, max. at $\sqrt{n/2}$, min. at $-\sqrt{n/2}$.

1(h). Min. at 0. 3. $\sqrt{a^2 + b^2}$.

4. m, n even: min. at a,c, max. at $b = \dfrac{na + mc}{m + n}$; m, n odd: min. at b. m even, n odd: max. at a, min. at b. m odd, n even: max. at b, min. at c.

Chapter 4, Section 4

6(a). $\dfrac{b^2 - a^2}{2}$. 6(b). $\frac{3}{2}$.

6(c). 1. 6(d). $e - 1$.

Chapter 4, Section 5

1(d). $\frac{1}{3}$. 1(e). $\dfrac{e - 1}{2}$.

Chapter 4, Section 6

1(a). $x \tan^{-1} x - \log \sqrt{1 + x^2}$. 1(b). $e^x(x^2 - 2x + 2)$.

1(c). $x^{n+1}\left(\log x - \dfrac{1}{n + 1}\right)$. 1(d). $\dfrac{\tan^{-1} x}{2} (1 + x^2) - \dfrac{x}{2}$.

Chapter 4, Section 6

1(e). $-\dfrac{x^2 \cos ax}{a} + \dfrac{2x \sin ax}{a^2} + \dfrac{2 \cos ax}{a^3}$.

1(f). $x \sin^{-1} x + \sqrt{1 - x^2}$.

1(h). $\dfrac{x}{2} [\sin(\log x) - \cos(\log x)]$.

1(j). $\dfrac{\sqrt{x} - \sqrt{x + 1}}{\sqrt{x(x + 1)}}$.

Chapter 4, Section 8

1(a). Integral does not exist. 1(b). π.

1(d). $\frac{1}{2}$. 1(e). Integral does not exist.

1(f). $5/e$.   1(h). $\frac{1}{2}$.

1(i).   Integral does not exist.   1(j). $\dfrac{2\pi}{\sqrt{3}}$.

Chapter 5, Section 5

5.   Set $f(x) = \lim f_n(x),\ I = \lim \int_a^b f_n(x)\ dx$.

    (a) $f(x) = 0,\ I = \dfrac{\pi}{4}$.   (b) $f(x) = 0,\ I = b$.

    (c) $f(x) = 0$ if $x \neq 0,\ I = \frac{1}{2}$.

Chapter 5, Section 6

1(a).   1.   1(b). $\dfrac{1}{|\alpha|}$.   1(c).   0.

1(d). ∞.   1(e).   1.   1(f). ∞.

4(c). $\dfrac{x(x+1)}{(1-x)^3}$.   4(e). $-\int_0^x \{[t + \log(1 - t)]/t^2\}dt$.

Chapter 5, Section 6

4(g). $\dfrac{2}{x}\int_0^x \left\{\dfrac{[t + \log(1 - t)]}{t^2}\right\} dt - \dfrac{1}{x}\log(1 - x)$.

4(h). $\dfrac{x}{1 - x^4}$.

Chapter 5, Section 7

10. $\dfrac{1}{x}\int_0^x \dfrac{e^t - 1}{t}\ dt$.

Chapter 5, Section 8

1(a).   Convergence if $\alpha > 2$.   1(b).   Convergence.
1(c).   Convergence.   1(e).   Divergence.
1(h).   Convergence.   1(k).   Divergence.
3(a).   Convergence if $\alpha > 0$.   3(b).   Convergence if $\alpha < 1$.
3(c).   Convergence if $1 < \alpha < 2$.   3(d).   Convergence if $0 < \alpha < 2$.

Chapter 6, Section 2

1(a). $\{(-1,0), (1,0)\}$.

1(b). $\left\{\left(\dfrac{1}{m},0\right),\ \left(0,\dfrac{1}{n}\right),\ (0,0),\ m,n = 2,4,6...\right\}$.

1(c). $\left\{\left(\dfrac{(-1)^n}{n},0\right) ; \left(0,\dfrac{(-1)^m}{m}\right),\ (0,0);\ m,n = 1,2,...\right\}$.

Chapter 6, Section 4

1. Yes.   2. No.   3. No.   8. Yes.

Chapter 6, Section 5

1(a). Limit does not exist.   1(b). 0.
1(c). Limit does not exist.   1(d). Limit does not exist.

Chapter 7, Section 4

3(a). Min. at its only critical point.
3(b). There are five critical points.
 4. $\sqrt{2}$.
 6. There are two critical points of relative maximum, two of relative minimum, and four of neither.

Chapter 7, Section 7

1(a). $v = 2 \cosh u$.   1(b). $w = u^2 - 2v$.
1(c). $v = 1 + e^{2u}$.   1(d). $v = (1 + u^2)^{-1}$.
2(a). $uv = u + v$.   2(c). Independent.
2(d). $v = \tan u$.

Chapter 8, Section 3

2(a). $\left(\dfrac{2a}{3}, \dfrac{b}{3}\right)$.   2(b). $\left(\dfrac{4a}{3\pi}, 0\right)$.   2(c). $\left(\dfrac{a}{2}, \dfrac{2b}{5}\right)$.

 3. $\frac{1}{3}$.   5(a). $\frac{1}{2}(e - 1)$.   5(b). $\dfrac{1}{2}\left(1 - \dfrac{1}{e}\right)$.

Chapter 8, Section 4

 6. $\dfrac{\pi b a^2}{2}$.   7. $\dfrac{\pi}{48}$.

10. $(\frac{3}{8}a, \frac{3}{8}b, \frac{3}{8}c)$.   12. $M\dfrac{b^2 + c^2}{5}$, M = volume of the ellipsoid.

Chapter 8, Section 5

 2. $\dfrac{4\pi}{15}(a^2 + b^2)\, a\, b\, c$.   3. 6.   4. $2\frac{1}{8}$.

 7. 1.   8. $\frac{1}{2}$.

Chapter 8, Section 8

1. $\dfrac{1}{\sqrt{1-a^2}}.$

Chapter 8, Section 9

11. $\dfrac{\Gamma(1-\alpha)}{\alpha}.$

Chapter 9, Section 1

6. $8a.$

7. $r\sqrt{a^2+b^2}.$

8. $\sqrt{2}\,\sinh c.$

9. $2a\pi^2.$

11. $8a.$

Chapter 9, Section 2

3(a). $\frac{8}{3}.$

3(b). $\frac{4}{3}.$

3(c). $8.$

3(d). $-\dfrac{\pi}{2}.$

3(e). $\frac{4}{3}.$

3(f). $\frac{8}{15}.$

4(b). $\dfrac{207}{20}.$

4(c). $3.$

4(d). $\dfrac{\pi}{2}(a^2+1).$

5. $\dfrac{647}{210}.$

Chapter 9, Section 3

4. $2w+\frac{8}{3}.$

Chapter 9, Section 4

1(a). $\dfrac{1}{2}\left(1-y^2-\dfrac{y^2+1}{x^2}\right).$

1(c). $\frac{1}{2}x^2+\dfrac{x}{y}+2\log y.$

1(d). $x^2+2xy-yz^2.$

Chapter 9, Section 5

4(a). $-56.$

4(b). $0.$

4(c). $0.$

4(e). $0.$

Chapter 9, Section 6

1. $\sqrt{3}.$

2. $\dfrac{8}{3\pi}.$

Chapter 9, Section 7

1(a). $\dfrac{x^2}{a^2} + \dfrac{y^2}{b^2} = z^2$.

1(b). $\dfrac{x^2}{a^2} + \dfrac{y^2}{b^2} - \dfrac{z^2}{c^2} = 1$.

1(c). $xy = z$.

2(a). Tangent: $\dfrac{x\,x_o}{a^2} + \dfrac{y\,y_o}{b^2} + \dfrac{z\,z_o}{c^2} = 1$.

2(b). Normal: $(-2b\,u_o \cos v_o,\, 2a\,u_o \sin v_o,\, ab)$.

5(a). $\dfrac{\pi}{6}(5^{3/2} - 1)$.

5(b). $\sqrt{2}\pi$.

5(c). $\tfrac{2}{3}\pi(2\sqrt{2} - 1)$.

7. $4\pi^2 ab$.

8. $4\sqrt{2}\,\pi a^2$.

9. $\pi(\sqrt{2} + \log(1 + \sqrt{2}))$.

11. $4\pi c\,\sqrt{ab}$.

Chapter 9, Section 8

3(a). $\dfrac{\pi}{2}$.

3(b). π.

3(c). $\dfrac{4\pi}{3}$.

3(e). $\sqrt{2}$.

5. $2\pi\sigma\,a(2 - \sqrt{2})$.

Chapter 9, Section 9

4. $\dfrac{8\pi a^4}{3}$.

5. 3.

6. $\dfrac{8\pi}{3}$.

14. $\dfrac{r^m}{m + 3}$.

Chapter 9, Section 10

3. $\dfrac{315}{32}\,\pi$.

4. $\dfrac{21}{8}\,\pi$.

Chapter 9, Section 11

1(a). 6π.

1(b). $\dfrac{\pi\alpha^2}{\sqrt{2}}$.

1(c). 0.

6(a). $\dfrac{x^2 + y^2 + z^2}{z} - 3$.

6(b). $(xy - z)e^{-xy}$.

6(c). $x^2 y + y \log z$.

7. 0.

Chapter 10, Section 5

1. 27.

3. $(\sqrt{a} + \sqrt{b} + \sqrt{c})^2$.

4. $\dfrac{d^2}{3abc}$.

5. $(a^{3/2} + b^{3/2} + c^{3/2})^{-2}$.

7. $(\sum a_i^2)^{-1}$.

8. $\sum a_i^2$.

Chapter 10, Section 8

1(a). $y = C(x - 1)e^x$.

1(b). $e^{x^2} - 2(y + 1)e^{-y} = C$.

1(c). $y^2 + 2\sqrt{1 - x^2} = C$.

2(a). $\dfrac{2x}{y - x} + \log |x| = C$.

2(b). $x^2 - y^2 = Cx$.

3(a). $y = C(x^2 - x) + x$.

3(b). $y = e^{3x} + \dfrac{C}{x^2}$.

3(c). $y = C \cos x - 2 \cos^2 x$.

4(a). $y = C e^x + x + 1$.

4(b). $\dfrac{1}{y} = C e^{-x} + x^2 - x + 2$.

5(a). $y^3 + 3xy - x^3 = C$.

5(b). $\dfrac{x^2}{2} + \dfrac{x}{y} + 2 \log |y| = C$.

5(c). $x^2 y^4 + x \sin y = C$.

7(a). $y = C_1 + C_2 e^{3x} - 2x$.

7(b). $y = C_1 + C_2 x^2 + \dfrac{x^5}{15}$.

INDEX